Interactions of Pacific tuna fisheries

Volume 1 — Summary report and papers on interaction

Proceedings of the First FAO Expert Consultation
on Interactions of Pacific Tuna Fisheries
3-11 December 1991
Nouméa, New Caledonia

Edited by
Richard S. Shomura
Hawaii Institute of Marine Biology
University of Hawaii, Manoa
Honolulu, Hawaii, USA
Jacek Majkowski
FAO Fisheries Department
Sarah Langi
Neiafu, Vava'u
Tonga

FAO
FISHERIES
TECHNICAL
PAPER

336/1

Food
and
Agriculture
Organization
of
the
United
Nations

Rome, 1994

M-43

ISBN 92-5-103453-2

iii

PREPARATION OF THIS DOCUMENT

This publication results from the First FAO Expert Consultation on Interactions of Pacific Tuna Fisheries hosted in Noumea, New Caledonia by the South Pacific Commission in cooperation with the Institut Français de Recherche Scientifique pour le Développement en Coopération from 3 to 11 December 1991. The Consultation was organized by the FAO Trust Fund project: "Cooperative Research on Interactions of Pacific Tuna Fisheries" in close collaboration with regional and national institutions involved in tuna fisheries research in the Pacific (see Acknowledgements).

The information presented at the Consultation was compiled by TUNET, a network of ten Working Groups organized by the FAO project. That information was contributed by scientists of the regional and national institutions studying tuna stocks and fisheries mainly in the Pacific, but also outside of the region.

<u>Distribution</u>

FAO Fisheries Department
FAO Regional Fisheries Officers
FAO fisheries projects and programmes
International fisheries organizations
National fisheries departments
National fisheries research laboratories
Members of TUNET (FAO's network of working groups
        studying tuna fisheries interactions in the Pacific)

Shomura, R.S.; Majkowski, J.;Langi, S. (eds.)
Interactions of Pacific tuna fisheries. Proceedings of the first FAO Expert
Consultation on Interactions of Pacific Tuna Fisheries. 3-11 December 1991.
Noumea, New Caledonia. Volume 1: summary report and papers on
interaction.
FAO Fisheries Technical Paper. No. 336, Vol.1. Rome, FAO. 1993. 326p.

ABSTRACT

This publication presents papers and discussions of the First FAO Expert
Consultation on Interactions of Pacific Tuna Fisheries held in Noumea, New
Caledonia from 3 to 11 December 1991. The objectives of the Consultation
included:
- the identification and documentation of concern related to interactions
  among fisheries directed at tuna and tuna-like species in the Pacific,
- the classification of these interactions,
- the review of information on them and methods for their study, and
- the formulation of recommendations for future research.

Volume 1 contains:
- the Summary Report of the Consultation,
- a review paper on methods for studying interactions in tuna fisheries,
- thirteen papers presenting new methods and case studies on such
  interactions, and
- seven reviews on fisheries interactions related to individual stocks of
  Pacific tuna and tuna-like species.

Volume 2 includes:
- eleven review papers on the biology, population dynamics and fisheries
  associated with the stocks of Pacific tuna, which are supplemented by
- four additional papers on specific fisheries.

The information contained herein demonstrates the potential for interactions
occurring between and among the fisheries directed at tunas and tuna-like
species. Empirical evidence for such interactions, however, has been
available for only few fisheries, and these interactions have been quantified
for even fewer fisheries. It is unclear whether interactions are insignificant
among fisheries directed at tuna and tuna-like species or whether scientists
are unable to detect these interactions possibly due to various changes to
fisheries and resources, resulting in a too-variable background which conceals
the effects of interactions.

ACKNOWLEDGEMENTS

The editors of this document would like to thank Convenors of TUNET's
Working Groups, Session Chairmen, Rapporteurs and participants of the
Consultation for their effort and collaboration in preparing the Summary Report
of the Consultation (see Volume 1 of this document). Thanks are also due to
authors and referees of the papers for their valuable contributions. Valuable
assistance and advice were received from Dr William Bayliff, Ms Mary Lynne
Godfrey, and Mr Robert Harman in the editing and from Mr Harman in the final
formatting and lay-out of the document.

The Consultation and its Proceedings were made possible through the close
cooperation of tuna scientists in the Pacific region. Funds for the organization
of the Consultation were provided by the Government of Japan and by FAO.
Technical expertise, data, and computer facilities for the preparatory work and
the Consultation were contributed by many institutions, especially:

- Commonwealth Scientific and Industrial Research Organization (Hobart, Australia),
- FAO/UNDP Regional Fisheries Support Programme (Suva, Fiji),
- Indo-Pacific Tuna Programme (Colombo, Sri Lanka),
- Inter-American Tropical Tuna Commission (La Jolla, USA),
- International Commission for the Conservation of Atlantic Tunas (Madrid, Spain),
- Institut Français de Recherche Scientifique pour le Développement en Cooperation (Noumea, New Caledonia),
- Ministry of Agriculture and Fisheries (Wellington, New Zealand),
- National Marine Fisheries Service (La Jolla and Honolulu, USA),
- National Research Institute of Far Seas Fisheries (Shimizu, Japan),
- South Pacific Commission (Noumea, New Caledonia), and
- South Pacific Forum Fisheries Agency (Honiara, Solomon Islands).

National research laboratories of many countries of Latin America, Southeast
Asia, and the South Pacific also contributed significantly to the work before and
during the Consultation.

Particular thanks are extended to local organizers from the institutions hosting
the Consultation, Dr Anthony Lewis of the South Pacific Commission in
Noumea, New Caledonia and Mr Renaud Pianet of the Institut Francais de
Recherche Scientifique pour le Développement en Coopération also in Noumea.

The editors would like to acknowledge the assistance and encouragement of the
staff of the Fishery Resources and Environment Division and the Operations
Service of the FAO Fisheries Department (Rome, Italy), the FAO/UNDP Regional
Fisheries Support Programme (Suva, Fiji), and the Indo-Pacific Tuna Programme
(Colombo, Sri Lanka) and particularly to Dr John Caddy, Dr Serge Garcia, Mr
Robert Gillett, Mr Andhi Isarankura, Dr Yasuhisa Kato, Mrs Christiane Lagrange-
Hall, Mr Toshifumi Sakurai, Mr Mitsuo Yesaki, and Mr Hugh Walton.

INTERACTIONS OF PACIFIC TUNA FISHERIES

Proceedings of the First FAO Expert Consultation on
Interactions of Pacific Tuna Fisheries

PREFACE

Tunas and tuna-like species are extremely valuable commercially, especially albacore, bigeye, northern and southern bluefin, skipjack, and yellowfin tuna. Collectively, these species are referred to as principal market tuna species, and are prized for canning, sashimi (raw fish dishes), and other products. The lesser known tuna species, however, should not be discounted because they provide considerable in-country commerce and are important sources of protein in some parts of the world. In recent years, the Pacific Ocean has become the dominant ocean for tuna landings. Between 1980 and 1991, the annual catch of tuna and tuna-like species in the Pacific increased by 68% to about 3 million metric tons (mt). The 1991 Pacific total catch represented about 68% of the world's catch of these species. While these very high catches in the Pacific are impressive, some recent studies suggest that there is potential for still higher sustainable catches of some species.

The increases in the catches of Pacific tunas and tuna-like species have resulted from both intensification and expansion of existing fisheries, and the development of new fisheries. These changes have led to overlap of areas of operations of large and small-scale fisheries, as well as competition for the same tuna resources by large-scale fisheries using different gear. Detecting or predicting even this most direct type of interaction, however, is difficult and presents a serious research challenge. Presently, two or more tuna fisheries may be operating simultaneously on the same stock in overlapping geographical areas, targeting fish of similar sizes. In such a situation, changes in the fishing intensity or pattern of one fishery may affect the catches of the other fisheries. A further factor in fisheries interaction among tunas is the ability of many tuna species to undertake rapid, long distance movements or migrations across or even between oceans. Under these circumstances, fisheries operating in different exclusive economic zones and on the high seas may significantly affect each other.

The knowledge of fisheries interactions is essential for rational management of fisheries. The principal market tuna species and many tuna-like species are recognized by the United Nations Convention for the Law of the Sea (UNCLOS) as highly migratory. Recently, considerable attention has been directed to the need for rational management of fisheries for highly migratory species and resources that straddle adjacent exclusive economic zones (EEZs). Such management would enhance economic and social benefits to the countries involved in fishing, processing, and trade of these resources. Presently, small-scale tuna fisheries exist in many developing countries in the Pacific, and many of these fisheries operate in the same areas as the large industrial tuna fisheries (purse seine, pole and line, and longline).

From 6 to 8 May 1992, the International Conference on Responsible Fishing was held in Cancun, Mexico, leading to the Cancun Declaration. From 7 to 15 September 1992, FAO organized the Technical Consultation on High Seas Fishing held in Rome, Italy, to consider technical issues related to such fishing. As a consequence of these meetings, FAO is involved in addressing the issue of flag of convenience. This issue is of major relevance to tuna fisheries and their management because many tuna vessels use such flags of convenience to avoid restrictive measures imposed by certain countries. Also, FAO actively participates in the development of a Code of Conduct of Responsible Fishing, which will apply to both the high seas and economic exclusive zones.

In a broader context, fisheries issues were considered at the United Nations Conference on Environment and Development (UNCED) held in Rio de Janeiro, Brazil, from 3 to 14 June 1993. The outcome of this Conference is relevant to fisheries directed at tuna and tuna-like species. The existing programme of action on environment and development (referred to as Agenda 21 or the Rio Declaration) and the two Conventions on Biodiversity and Climate Change are now open for ratification. These initiatives represent an important commitment at the highest national political level to resolve a wide range of problems associated with rational use of marine resources.

According to UNCLOS, fisheries management needs to be based on the best available scientific information. The recent attention directed to fisheries management of highly migratory and high seas resources has pointed the need for scientific information on interactions of fisheries directed at tuna and tuna-like species. This need has also become evident at recent regional and international meetings of fisheries scientists and administrators.

In the Pacific, where most catch of tuna and tuna-like species is taken, there is an additional urgent need to integrate available information and to coordinate fisheries research. The Pacific is the only ocean where there is neither a single fisheries body nor a technical programme directed to tuna and tuna-like species that encompass the entire ocean. Some of the Pacific stocks of tuna and tuna-like species are only partly covered by existing fisheries bodies and programmes in terms of their areas of distribution. This situation promoted FAO to initiate a project: "Cooperative Research on Interactions of Pacific Tuna Fisheries" and to create a network of ten working groups of scientists (TUNET) to provide direction and to facilitate the implementation of the project.

To provide an information base for the execution of the project, FAO organized the First FAO Expert Consultation on Interactions of Pacific Tuna Fisheries hosted in Noumea, New Caledonia, by the South Pacific Commission with collaboration of the Institut Francais de Recherche Scientifique pour le Developpement en Cooperation (ORSTOM) from 3 to 11 December 1991. The Consultation was preceded by a preparatory meeting held in Noumea in late 1989. The success of the Consultation was due to the close collaboration and contribution of many other institutions; these institutions and the host organizations are duly acknowledged in the Summary Report presented in Volume I.

The objectives of the First FAO Expert Consultation on Interactions of Pacific Tuna Fisheries included describing the concerns related to interactions of Pacific fisheries directed at tuna and tuna-like species, classifying these interactions, reviewing all available information on them and the methods applied to their study, and making recommendations for future research.

The information presented in the proceedings demonstrates that there is a potential for interactions occurring between and among the fisheries directed at tunas and tuna-like species. Empirical evidence for such interactions, however, has been available for only few fisheries, and these interactions have been quantified for even fewer fisheries. It is unclear whether interactions are insignificant among fisheries directed at tuna and tuna-like species or whether scientists are unable to detect these interactions possibly due to various changes to fisheries and resources, resulting in a too-variable background which conceals the effects of interactions.

These "Proceedings of the First FAO Expert Consultation on Interactions of Pacific Tuna Fisheries" are provided in two volumes:

- Volume 1 contains the Summary Report of the Consultation. Volume 1 also includes a review paper on methods for studying interactions in tuna fisheries, thirteen papers presenting new methods and case studies on such interactions, and seven reviews on fisheries interactions related to individual stocks of Pacific tunas and tuna-like species.

- Volume 2 includes eleven review papers on the biology, population dynamics, and fisheries associated with the Pacific tuna resources. These reviews are supplemented by four additional papers on specific fisheries.

INTERACTIONS OF PACIFIC TUNA FISHERIES

Proceedings of the First FAO Expert Consultation on
Interactions of Pacific Tuna Fisheries

VOLUME 1

CONTENTS

page

**FIRST FAO EXPERT CONSULTATION ON
INTERACTIONS OF PACIFIC TUNA FISHERIES**

Noumea, New Caledonia
3-11 December 1991

SUMMARY REPORT

1.   INTRODUCTION

Since the early 1970s, there has been a marked expansion in tuna fishing throughout the world's oceans. This has been especially true in the central and western Pacific Ocean where purse seining became the dominant method of tuna fishing. In addition to the increase in catch, the number of countries actively engaged, directly or indirectly, in tuna fishing increased dramatically in the Pacific. The increase in tuna fishing activities has increased the potential for fishery interactions.

The problem of tuna interactions prompted the Food and Agriculture Organization (FAO) to organize a preliminary meeting of the "Expert Consultation on Interactions of Pacific Ocean Tuna Fisheries". The meeting was held in Noumea, New Caledonia, from 30 October to 3 November 1989. The status of the major tuna resources and fisheries in the Pacific was reviewed and it became apparent that fishery interaction was a major problem that needed to be addressed. Several small-scale tuna fisheries operated in the same region as large-scale, distant water tuna fleets, and in some regions, large-scale tuna fishing using different gears operated in the same areas.

On the basis of the preliminary meeting, FAO organized a follow-up meeting which was held in Noumea, New Caledonia, from 3-11 December 1991. This report represents the results of this meeting.

2.   OBJECTIVES

The general objective of the Consultation was to enhance the capacity of countries involved in tuna fishing in the Pacific, especially developing countries, to address in a scientific manner the problems of interactions of tuna fisheries in the region, and to optimize the benefits from these fisheries.

The specific objectives noted for the Consultation were:

a)   Identify the major types of interactions among tuna fisheries.

b)   Identify the scientific problems related to these interactions that are relevant to tuna fisheries management in the Pacific.

c)   Review methods and experiences used to address interaction problems, especially the applicability and effectiveness of these methods.

d)   Review existing information on tuna fisheries interaction and determine deficiencies in the available information and scientific understanding of interactions.

e)   Make recommendations for:

    1)   improvements to the existing methods and for development of new methods,

    2)   other future research (*e.g.*, collection and processing of data and biological samples, and application of methods), and

    3)   future activities of the FAO-executed Japan Trust Fund Project entitled "Cooperative Research on Interactions of Pacific Tuna Fisheries".

3.   ACKNOWLEDGEMENTS

The FAO organized the Consultation and formulated the Provisional Programme. In preparation for the Consultation, FAO organized an informational network (TUNET) of tuna scientists and ten working groups, the latter to carry out preparatory work for the Consultation. Major funding support for the Consultation and for specific preparatory work was provided by Japan under an FAO Trust Fund project entitled "Cooperative Research on Interactions of Pacific Tuna Fisheries". FAO provided some additional funds to support participation of some scientists at the Consultation. The South Pacific Commission (SPC) with cooperation from the Office de la Recherche Scientifique et Technique Outre-Mer (ORSTOM) hosted the Consultation and provided secretarial and technical support. Several major institutions contributed technical expertise, data and computer facilities for the preparatory work for the Consultation. These institutions included:

CSIRO	Commonwealth Scientific and Industrial Research Organization (Australia)
FAO/UNDP RFSP	FAO/UNDP Regional Fisheries Support Programme
FFA	Forum Fisheries Agency
IATTC	Inter-American Tropical Tuna Commission
ICCAT	International Commission for the Conservation of Atlantic Tunas
IPTP	Indo-Pacific Tuna Development and Management Programme
MAF	Ministry of Agriculture and Fisheries (New Zealand)
NMFS	National Marine Fisheries Service (USA)
NRIFSF	National Research Institute of Far Seas Fisheries (Japan)
ORSTOM	Office de la Recherche Scientifique et Technique Outre-Mer (France)
SPC	South Pacific Commission

Many tuna experts participated in the Consultation with the full support of their organisations; these experts provided the technical expertise necessary for the success of the Consultation. National research laboratories of many countries of Latin America, Southeast Asia, and the South Pacific contributed significantly to the preparatory work of the Consultation; many scientists of these countries also participated actively in the

Consultation. Finally, FAO/UNDP RFSP was responsible for all administrative and financial matters relating to the Consultation, and provided local coordination and logistic arrangements for the Consultation.

4.   OFFICIAL OPENING

The Consultation was officially opened on 3 December 1991 in the auditorium of the ORSTOM by the Chair, Mr. Richard Shomura, who welcomed participants and honoured guests. Mr. Shomura introduced Dr. Jacek Majkowski of FAO, who thanked ORSTOM and the South Pacific Commission (SPC) for their generosity in co-hosting the meeting. Dr. Majkowski provided the background information on the Consultation and the history of efforts leading to this current meeting. He noted that a preliminary meeting, kindly hosted by SPC, in 1989 established the framework for the present Consultation. Then he acknowledged the contributions of various institutions in organizing the Consultation (see section 3.). In particular, Dr. Majkowski expressed thanks to the government of Japan for financing an FAO Trust Fund project sponsoring the Consultation. Also, he outlined the objectives of the Consultation and emphasized their relevance to developing coastal countries of the Pacific.

The Chair then introduced Mr. Jean Fages of ORSTOM. Mr. Fages welcomed participants and stressed the importance of the topic of tuna fisheries interactions to the Pacific, and the importance of the success of the Consultation to the Pacific islands.

Mr. Shomura then introduced Dr. Antony Lewis, who welcomed the participants on behalf of SPC. Dr. Lewis noted that the production of tunas from the South Pacific has increased markedly in recent years and that the tuna resources have become very important to the economy of the Pacific island communities.

The Chair then introduced Mr. Jacques Iekawe of the government of New Caledonia. Mr. Iekawe stressed the increasing importance of the ocean and adjacent EEZs to the people of the Pacific. He mentioned that the open-ocean resources, such as tuna, are becoming increasingly important economically to island nations. He also described how New Caledonia is now looking beyond the traditional reef and lagoon fisheries as the economic expectations of young people increase. Mr. Iekawe pointed out the importance of research programme such as the SPC fisheries programme and consultations such as this one to help develop the regional fisheries of the Pacific. He recommended the continued financial support of such programme.

At an earlier preliminary meeting, organizational details were discussed. The provisional agenda was presented and accepted (Appendix A). The Chair, Mr. Shomura, introduced the Vice-Chair of the Consultation, Dr. Ziro Suzuki, and the Convenor, Dr. Majkowski from FAO, Rome. Dr. Majkowski welcomed all participants on behalf of FAO and reviewed the objectives for the Consultation. He noted that FAO desires a significant output from this meeting, and suggested that during the meeting several groups of participants may wish to formulate proposals for projects to address tuna fisheries interactions in the Pacific. He also discussed report formats for each session, noting that they should be "output-oriented". The Chair reported that Dr. George Boehlert would serve as rapporteur for the plenary sessions.

A list of participants, list of organizers of the Consultation, and a list of documents for the Consultation are given in Appendices B, C, and D, respectively.

5.   METHODS FOR STUDYING TUNA FISHERIES INTERACTIONS

The technical sessions of the Consultation was opened with a session on methodology. The session was chaired by Dr. Pierre Kleiber; vice-chair was Dr. John Sibert and rapporteur Dr. Chris Boggs.

5.1   <u>Types of Tuna Fisheries Interactions</u>

Fishery interactions can be classified by their mechanisms (Kleiber, this document). This classification can be useful in conceptualizing fishery interactions and in determining appropriate methods to quantify them. The following mechanisms of interaction were suggested:

a)   Direct resource-mediated interaction, such as when the catch of one fishery directly influences the fish resource available to another fishery.

b)   Non-resource-mediated interaction, such as when fishing gear of one fishery physically interferes with gear from another fishery, or when production by one fishery depresses the profitability of another.

Resource-mediated interaction may also be indirect, such as when the catch of a prey species affects the abundance or availability of a predator species which is a target of another fishery. Such interactions have been little studied in relation to tunas. The research discussed at this Consultation focused on direct, resource-mediated interaction. Other research has been conducted on modifications of fisheries operations and on economic aspects, in order to identify and resolve conflicts of gear and of economic interests. These are two types of non-resource-mediated interaction.

Mechanisms of direct interaction between resources may be complex, and depend on the biological characteristics (growth, movement, depth distribution, *etc.*) of the exploited fish. These mechanisms may further be complicated by age structure, for example catches of young fish by one fishery may affect the catches of older fish in another fishery. In many situations where fisheries do not overlap geographically, or do so partially, the intensity of interaction depends on movement of the fish from one fishery to another. Where fisheries overlap, interaction may be most intense when there is little net movement of the fish.

Components of population turnover, mortality, recruitment, immigration, and emigration, all affect the rate at which fish become available to fisheries exploiting the population. For fisheries covering a wide area, immigration and emigration tend to be relatively less important components of turnover than in limited-range fisheries. A "rule of thumb" is that when fishing mortality is low relative to turnover (low exploitation rate), the magnitude of interaction between competing fisheries tends to be small. The actual degree of interaction depends on characteristics of the two fisheries, including their relative levels of exploitation, and the relative geographical scales and geographical arrangement (distance apart, degree of overlap, and conformation) of their areas of operation.

5.2   Methods of assessing interactions

Approaches to assessing fishery interactions can be classified into the following types:

a)   An empirical approach of regulating the activity of one fishery to observe the effect on the performance of another.

b)   Tagging experiments to describe the effect one fishery has on other fisheries.

c)   Analyses of tagging data to estimate the movement parameters that govern the potential for fishery interaction.

d)   Statistical analyses of fishery data to look for relationships between the activity of one fishery and the performance of another.

e)   Simulation models using a broad range of information about population dynamics to observe the effects of one fishery on another under various conditions.

Catch was explored as a measure of fishery activity (instead of effort) in an attempt to find a negative correlation between activity and the performance of certain limited-range fisheries in Hawaii (Boggs, this document). The method showed some utility in theory, but in practice, fishery performance as measured by catch per unit of effort (CPUE) was dominated by exogenous influences. In attempting to measure fishery interaction in a limited-range fishery, it may be feasible to account for exogenous influences on performance through indexing local performance in relation to performance in more wide-ranging fisheries.

Two papers (Hearn and Mazanov, this document; Majkowski, Hearn and Sandland, this document) presented methods of tagging a representative portion of the catch of one fishery. In this method, the number or weight of recaptured fish in the second fishery represents the magnitude of interaction. The advantage of this method is that it requires no fishery statistics or complex modelling, and the logic for the estimation of interaction is so clear that it can be understood by the fishermen whose cooperation is needed to achieve the results. The ideal case requires tagged fish recaptured by the first fishery to be re-released, or replacement fish tagged. Alternatively, some mathematical adjustments need to be introduced to account for the recapture of tagged fish retained by the first fishery.

Information on tuna movements has not been widely used in population-dynamics models and stock-assessment techniques. The importance of considering fish movements in assessing interaction was a theme of the working group on methods, which considered several methods of integrating the estimation of movement parameters with the estimation of population dynamics parameters (integrated models) to provide improved methods of extracting such information from a combination of tagging data and catch and effort statistics. These new methods provide powerful new tools for the quantification of fishery interaction.

Dr. Pierre Kleiber presented a case study of interaction among skipjack fisheries along the coast of west Africa (Kleiber and Fonteneau, this document). A population dynamics and movement model was used in one (north-south) dimension. Movement included components of diffusion and seasonally-varying advection. They first fit the model to tagging data, and then modified it to deal with untagged fish. The model was then used to predict catch under a variety of effort regimes to investigate the effect of changing effort in one fishery on the catch of other fisheries.

Dr. John Sibert presented a method for fitting tag returns predicted from a two-dimensional advection-diffusion model to observed tag returns (Sibert and Fournier, this document). The estimation procedure uses an alternating direction-implicit method to solve a partial differential equation. The method appears to give good estimates of a regionally- and seasonally-varying movement pattern. It is sensitive to differences in movement patterns, so that the significance of these differences can be statistically tested.

Mr. Carlos Salvadó presented a method by which tagged-fish data are used to determine the transition probability densities of the fish through space and time, independent of the particular effort regime extant during the tagging experiment (Salvadó, this document). The resulting probability density function can be used to construct expressions of population processes such as catch rate density for any assumed effort regime, which allows the estimation of fishery interaction. The moments of the probability density function can be used to estimate parameters that are varying in space and time, such as diffusivity, advective velocity, and mortality.

Dr. Richard Deriso discussed recent research on quantifying movement of yellowfin tuna in the eastern Pacific Ocean (EPO). One presentation was a summary of the paper "A Markov movement model of yellowfin tuna in the eastern Pacific Ocean and some analyses for international management" (Deriso, Punsly and Bayliff, 1991, in *Fisheries Research* 11: 375-395). That paper describes a maximum-likelihood approach applied to tagging data, and then shows some mean residence-time calculations relevant to effects of changes in the minimum retention size of yellowfin by the fishery. The second presentation was a summary of research at the Inter-American Tropical Tuna Commission (IATTC) on the development of a general simulation model of the purse-seine fishery for tunas of the EPO. Sub-models are included for yellowfin tuna, skipjack tuna, dolphin mortality, and fleet dynamics. A novel feature of this research effort is the use of age-specific yellowfin abundance estimates stratified by time and area to estimate movement rates.

5.3   Problems in Quantifying Tuna Fishery Interactions

In analyzing fishery statistics to look for relationships between one fishery's performance and the activity of another fishery, the choice of method and indices for fishery activity and performance will alter the results obtained. Modern computer software has made correlation analysis easy, but interpreting the results is still difficult. Ultimately, most correlations are subject to several interpretations because more than one factor, or unknown factors, may be hypothesized to account for the observed relationship.

Confusion and misinterpretation are common in attempts to identify or quantify interaction through correlation analyses. Positive correlation between the performance of two sectors has been interpreted as indicating a potential for interaction and negative correlation as indicating the existence of interaction, but neither may be correct. A positive correlation indicates only the similar availability of the fish exploited by the two sectors, suggesting (but not proving) that they are exploiting the same resource. A negative correlation indicates high availability in one fishery when availability is low in another, which could result from fish distribution changes. However, convincing evidence of interaction can come from negative correlations between an index of relative abundance in a fishery, and the activities of other fisheries or all fisheries combined.

Another problem with correlation analyses is that environmental influences on fish availability and abundance are often great enough to obscure correlations between fishery performance and activity, or to give the impression of fishery interaction where none exists. Undocumented changes in fishing gear efficiency can have the same effect. Acknowledging or accounting for such effects is important when considering evidence of fishery interaction derived from correlation analyses.

Some drawbacks of tagging experiments are:

a)   They are expensive.

b)   It is difficult to tag the larger-sized tunas important in many fisheries (large tuna may behave differently than small tuna).

c)   There are persistent questions regarding the interpretation of low recovery rates (which may be due to tagging mortality, tag shedding, non-reporting, or a low rate of exploitation).

d)   Estimates of movement and population dynamics parameters from tagging studies are specific to the conditions existing at the time of the study. Such estimates may be useful for simple linear extrapolation, but to investigate conditions that are different from those existing at the time of the study requires the development of models that capture the effects of the different conditions.

Even with integrated procedures for estimating movement and population dynamics parameters from tagging data and complete fisheries statistics, questions remain regarding the statistical properties of the estimates. Furthermore, the ability of simulation models to use these parameters to predict real-world situations will to a great extent depend on the development of models to explain fish movement, behaviour, and population-dynamics variations due to ecological changes. To the extent that fish behaviour and population dynamics during the tagging studies are typical, the current generation of simulation models should be useful for estimating fisheries interactions under current conditions. However, the usefulness of these models in simulating fisheries under widely different future conditions needs to be explored.

5.4 <u>Recommendations</u>

The working group did not prioritize the following recommendations:

a) Complete and accurate statistics on species composition, size composition, catch and effort, by area, time, and gear should be collected for use in assessing the level of exploitation. [The exploitation rate is an indicator of potential for fishery interaction, and its determination is an important preliminary to more definitive assessment of fishery interaction (Figure 1).]

b) Methods of stock assessment should be improved to include considerations of spatial structure and dynamics of the resource and the fishing fleets. [Although measurements of tuna movements exist, there has been little progress in using such data in stock assessment.]

c) Further development, testing, and improvement of integrated approaches to modelling fisheries using tagging data, fisheries statistics, and biological and economic data should be carried out. [These models have wide applicability in modelling other fisheries scenarios and in extrapolating beyond the conditions under which the model parameters were estimated.]

d) Three key elements in addressing interaction problems (and other stock assessment questions) are data gathering, data analysis, and modelling. It is recommended that these activities be conducted simultaneously, because they all inform each other. [Data are necessary for analyses to estimate parameters to put into simulation models, but simulation models should be used in designing experiments and sampling regimes for gathering data.]

e) Tagging experiments should be conducted because they are a powerful source of information on interactions unavailable from normal fishery statistics. In designing and conducting tagging experiments, consideration should be given to using existing experienced personnel and equipment. [Tagging studies are needed to estimate exploitation rates, residence times, possible vertical segregation of exploited stocks, movement patterns, and the environmental variables that produce variation in movements.]

f) The statistical properties of movement and population-dynamics estimates from different methods of tagging-data analysis should be further explored, but great care should be taken in making extrapolations from such estimates.

g) The different approaches to the analysis and synthesis of tagging data should be compared by analysing the same simulated data and the ability to extrapolate from such estimates should also be tested through simulation.

h) Generally-applicable software for generating simulated, tag-return and fishery data should be written for the development and evaluation of various simulation models.

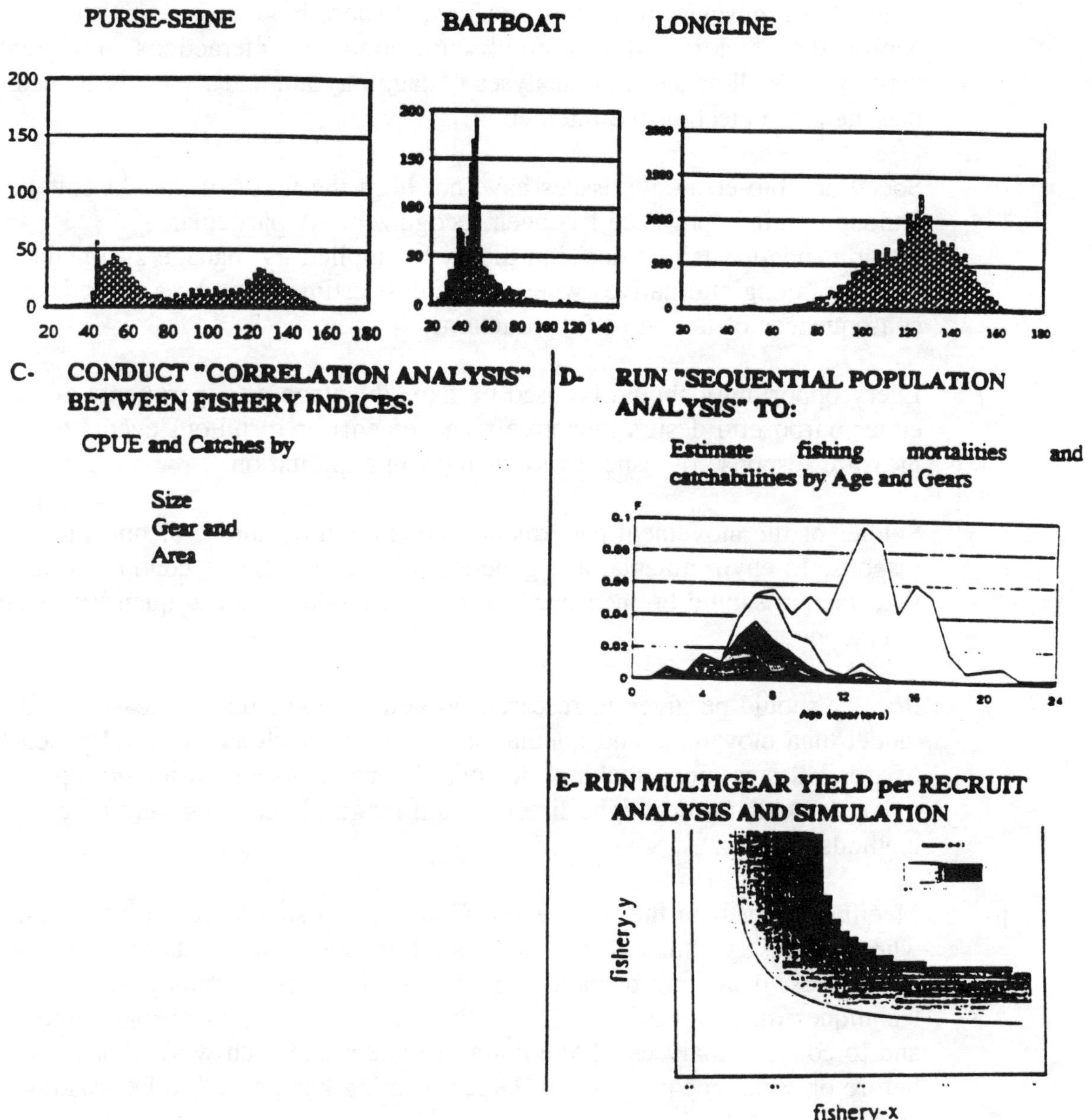

Figure 1. Assessing the stock-wide level of exploitation.

i) Work should continue on methods that can quantify fishery interactions, even when the interactions are small. [Such estimates may provide the basis for extrapolating or forecasting the magnitude of fishery interaction under different exploitation regimes, if due regard is given to potential stochastic and non-linear effects.]

j) The confounding effects of tag shedding, tag-induced mortality, and non-reporting of tags should be the subject of active research, and these factors should be included in all analyses of tagged-fish dynamics.

k) Several alternative methods should be applied to any analysis of interaction to test the robustness of the conclusions. [There is great potential for confounding factors and noise to obscure important interactions. Integrated models, as well as separate analyses of tagging data and fishery statistics, may help in detecting the interaction.]

l) Social and bio-economic issues have not been the focus of the Consultation, although their importance has been recognized. A particular recommendation is that decision theory be applied by managers to make choices among alternatives where some cost estimate can be assigned to the consequences of taking different actions.

m) Every opportunity should be used to take advantage of experiments of either purposeful design (adaptive management) or fortuitous events to measure response of fisheries to changes in exploitation.

n) Studies of the movement patterns and population dynamics of tunas in response to environmental and genetic variables should be conducted, and such factors should be incorporated into the models used to quantify fishery interaction.

o) Priority should be given to research on new technologies to measure and model tuna movement and population dynamics, such as the development of archival tags, new tracking methods, hydro-acoustics, synthetic- aperture radar, LIDAR (laser - light directing and ranging) scanning, and other methods.

p) Meetings other than the present full Consultation should be promoted and, where necessary, funded, to enable small groups of scientists to assemble and process data from disparate sources, to develop new analytical techniques (or adapt existing ones), to implement computer programmes, and to conduct analyses. [Attempting to undertake such work shortly before or simultaneously with a larger meeting has proved to be frustrating and inefficient.]

q) Because of the similarity of the problems to be solved in assessing tuna fishery interactions in all oceans, expertise on the subject should be sought worldwide.

6. PACIFIC SKIPJACK TUNA

The chair for the session on Pacific skipjack tuna was Dr. Richard Deriso; Dr. John Hampton served as co-chair, and Dr. Talbot Murray as rapporteur.

6.1 Fishery Components

It is currently believed that skipjack tuna in the Pacific Ocean belong to a single population. Larvae occur in tropical and subtropical waters of the western Pacific Ocean (WPO), central Pacific Ocean (CPO) and, to a much lesser extent, the eastern Pacific Ocean (EPO). Changes in allele frequencies suggest either different subpopulations in the WPO and EPO or a cline across the Pacific, but the data are insufficient to distinguish between the two cases.

Fisheries for skipjack are widely distributed, with EPO catches occurring in nearshore waters and well offshore in the region of 10° N. The EPO catches are taken primarily by purse seine, with about 36,280 mt to 163,260 mt caught per year. It is thought that the EPO fishery is sustained largely by recruitment from the CPO and temporary residence in the EPO. In the WPO, fisheries are more diverse, and include industrial, artisanal, and subsistence fisheries, using purse seine (group and single seine), pole-and-line, troll, and other gear. The WPO fisheries are found mostly in equatorial waters west of 180°; the skipjack catch has increased steadily and is currently about 800,000 mt per year. In addition, a seasonal fishery by Japanese pole-and- line vessels occurs in northern subtropical waters of the WPO, generating annual catches of up to 150,000 mt.

6.2 Scope of Fishery Interactions

In the EPO, the extent of fisheries interactions is considered small. Yield-per-recruit (Y/R) analyses for the EPO indicate that increasing effort and catching skipjack as soon as they are available maximises the Y/R. The short residence time of skipjack in the EPO suggests that interactions are likely to be low. Some concern exists that recruitment to the EPO is from fish in the CPO which, in turn, may receive recruits from the WPO. This recruitment pattern raises questions as to whether recent increasing yields in the WPO could reduce recruitment to the CPO and subsequently to the EPO.

The diverse range of industrial, artisanal, and subsistence skipjack fisheries operating in the same or adjacent areas of the WPO and the large harvests suggests that the scope for fisheries interactions is high in this area. Potential interactions may exist in the WPO between (a) industrial purse-seine and pole-and-line fisheries, (b) distant water fishing nations (DWFNs) and locally-based industrial fisheries, (c) industrial and the diverse artisanal and subsistence fisheries, and (d) fisheries operating in adjacent Exclusive Economic Zones (EEZs).

Industrial pole-and-line and purse-seine fisheries have operated in overlapping WPO areas since the early 1980s. Although not adjusted for changes in fishing power, the variable pole-and-line catch per unit effort (CPUE) has had an upward trend throughout the 1980s. Several DWFN purse-seine and pole-and-line fleets operate broadly throughout the WPO, and overlap with locally-based Pacific island and

Associated Southeast Asian Nations' (ASEAN) fisheries. Locally-based fisheries are mostly small pole-and-line fisheries, with potential for direct competition for fish schools. In the WPO, a diverse range of mostly small artisanal and subsistence fisheries operate, in some cases, in proximity to industrial fisheries, suggesting a potential for interactions. To date, the low tag-recovery rate from artisanal and subsistence fisheries suggests that interactions due to large-scale DWFN activity are probably small, and these interactions would likely be overwhelmed by other factors influencing local skipjack abundance. The exception to this would be in cases of direct competition for schools between artisanal and industrial fishing vessels. The potential for interactions between the municipal Philippines skipjack fishery and industrial fisheries may be greater, given the larger contribution of municipal fisheries to total catches, relative to the artisanal fisheries in Pacific island states. Analysis of data from early tagging experiments indicate that in some areas a substantial percentage of tagged fish released in the EEZ of one country can be recovered in the EEZ of an adjacent country. While movement between most countries appears to be small, the potential for interaction among fisheries in some EEZs exists.

6.3   Importance for Fisheries Management

Previous management regimes have operated on a stock-wide basis in the EPO with the goal of operating at maximum sustainable yield (MSY), without individual fishery- allocation considerations. Given the short residence time of skipjack in the EPO and Y/R considerations, allocation among fisheries sectors has not been a problem.

In the WPO the diversity of artisanal, subsistence, local industrial, and DWFN industrial fisheries has led to the perception among several sectors that catches by one group affect those of another. Estimating the extent of interaction between different fishery components is therefore of greater importance in the WPO because of possible resource-allocation implications. The generally limited operating range and fishing power of artisanal and subsistence fisheries makes even occasional interaction with industrial fisheries a concern for management.

6.4   Methods Applied for Studying Interactions

Skipjack CPUE trends in industrial fisheries operating primarily within EEZ areas have generally not provided evidence of significant interactions between purse-seine and pole-and-line fisheries. The CPUE data are not available for most artisanal and subsistence fisheries nor for most industrial fishing by DWFNs outside of EEZs. Another difficulty with the use of CPUE data is that only nominal catch and effort data are available, so CPUE cannot be standardised for changes in fishing power and other factors affecting performance.

The extensive tagged-fish release and recapture data from the South Pacific Commission's (SPC) Skipjack Survey and Assessment Programme (SSAP) in the early 1980s are appropriate for the investigation of some interaction issues. The SSAP data have already been used to identify the potential for interactions among some adjacent EEZ fisheries. The SPC's Regional Tuna Tagging Project (RTTP), which is currently underway, and its national components (*e.g.*, Solomon Islands' In-Country Tagging Study) have specifically focussed on questions of interaction between fisheries. The RTTP will provide data to allow estimation of the interaction between purse-seine and pole-and-line fisheries in the WPO. Development and testing of models incorporating movement based on tag recoveries and related analytical approaches are continuing, and it is hoped that such models can be applied to interaction questions in WPO skipjack fisheries.

6.5   Information on Interactions

During the late 1950s and early 1960s, baitboats and purse seiners frequently competed for tuna resources in nearshore waters in the EPO. However, purse seiners now make up over 97% of the current fleet capacity and gear competition, for skipjack in particular, is virtually nonexistent in the EPO. Similarly, purse seiners now account for most of the catching capacity in the WPO; in 1990 purse seiners caught 66 percent of the estimated total skipjack catch of 785,000 mt. However, the possibility of interaction between purse-seine and baitboat fisheries is of concern in countries like the Solomon Islands and Fiji, where domestic fisheries have existed for some time. In the case of the Solomon Islands, purse-seine fishing has increased in recent years, and the SPC has been collaborating with the Solomon Islands government in a tagging experiment designed to estimate the magnitude of the interaction between purse seiners and baitboats.

Although gear interactions are largely minor, interactions among geographical areas (particularly EEZs) are of concern in some areas. In the WPO, the main area of operation of the purse-seine and baitboat fisheries is composed primarily of the largely-contiguous EEZs of Philippines, Indonesia, Palau, Federated States of Micronesia, Papua New Guinea, the Solomon Islands, Nauru, Kiribati, and the Marshall Islands. The degree of interaction among areas such as these will be determined by controlling factors such as the size of the areas, the distances between them, skipjack movement rates, the natural mortality rate, and the intensity of the fisheries. There has been some controversy regarding movement rates of skipjack and their possible effects on spatially-separated fisheries. However, some specific analyses have been carried out. Using SSAP tagging data, Sibert (pers. commun.) calculated a series of interaction coefficients based on the proportions of total throughput in receiver EEZs derived from immigration from donor EEZs. Most of the coefficients are low, indicating that with movement patterns prevailing when the SSAP data were gathered, there was generally little potential for fishery interaction. Not surprisingly, most cases of significant exchange occurred between adjacent EEZs. In particular, the results suggested that 37% of throughput in the Marshall Islands EEZ at the time of tagging resulted from immigration from Federated States of Micronesia. Relatively high interaction coefficients were also observed for Northern Mariana Islands-Federated States of Micronesia and to a lesser extent Palau-Federated States of Micronesia, and Papua New Guinea-Solomon Islands, indicating some potential for fishery interaction between those countries. The only case of a relatively high interaction coefficient for widely-separated areas was New Zealand-Fiji.

This relatively simple representation of interaction does not explicitly specify the controlling factors noted above. A more rigorous method to estimate interaction between two countries was derived at SPC and applied to Papua New Guinea and the Solomon Islands, both of which had substantial baitboat fisheries for skipjack at the time of the tagging project. Estimations were made of exchange rates between the two EEZs, losses from natural mortality and movement to other areas, the proportions that remained resident and lived, and the proportions that were caught locally on a monthly basis. The

Solomon Islands stock was found to be relatively stable, with a low rate of natural mortality and emigration (resulting in high survival) and low rate of movement to Papua New Guinea. The Papua New Guinea stock was found to be more dynamic, with a higher rate of natural mortality and emigration (lower survival), but with a low rate of movement to the Solomon Islands. It was estimated from these results that an increase in the catch of 1,000 mt in either EEZ would result in a decrease of only 1-3 mt in the steady-state catch of the other.

Incomplete availability of catch and effort data has hindered a more thorough analysis of skipjack movement and its interpretation with respect to interaction in the WPO.

6.6   Recommendations

Further work is required that is relevant to both EPO and WPO skipjack tuna fisheries in the following areas:

a)   Further development, testing, and implementation of tag-recovery models and, more generally, population-dynamics models which incorporate movements.

b)   Fishery data compilation in the WPO, particularly with regard to industrial fisheries catch and effort data from high-seas areas.

c)   Development of abundance indices. [Adequate data are needed to standardize CPUE trends for changes in fishing power and other factors in pole-and-line and purse-seine fisheries to improve interpretations of changes in CPUE and increase the use of these data.]

d)   Information on patterns of recruitment and origins. [Information on the source of fish being exploited by a fishery would assist in the interpretation of interactions.]

The first two items are considered to be of the highest priority by those working in the WPO, whereas the latter two items are of higher priority to those working in the EPO.

6.7   Institutional Arrangements and Future Research

Research in the EPO is conducted primarily by the Inter-American Tropical Tuna Commission (IATTC) and scientists of Mexico and other countries. In the WPO tuna research is carried out by the SPC and the National Research Institute of Far Seas Fisheries (NRIFSF) of Japan. The WPO skipjack tuna research is to some extent coordinated with research underway in ASEAN countries through the Western Pacific Fisheries Consultative Committee (WPFCC).

7.   EASTERN PACIFIC YELLOWFIN TUNA

The chair for the session on eastern Pacific yellowfin tuna was Dr. Richard Deriso; the vice-chair was Dr. Alex Wild, and the rapporteur Dr. Norman Bartoo.

7.1   Fishery Components

Yellowfin tuna in the eastern Pacific Ocean (EPO) are considered to be a single stock. In the EPO, the stock is currently producing near its estimated maximum sustainable production of 297,500 mt. The fish do not exhibit pronounced tendencies for predictable movements in east-west or north-south directions, although the movements do not appear to be random. The overall implication of tagging information is that yellowfin do not usually undertake migrations in excess of several hundred miles, although the tagging experiments do not cover the entire size range or geographical range of the population.

Yellowfin tuna in the EPO are fished primarily by purse seines, although a limited number of baitboats is in operation, and a small longline catch occurs. Total catches of yellowfin in the EPO since 1985 have been between 226,000 mt and 302,000 mt annually. Longline catches during the same period are small, limited to a few thousand tons. Baitboat catches are small and coastal in distribution.

Purse seiners fish on schools of yellowfin associated with dolphins or with floating objects and on free-swimming schools, taking fish about 40 to 150 cm long. Longlines catch yellowfin of about 90 to 150 cm in length. Details on the size of yellowfin caught in specific types of association are given by Wild (this document).

7.2   Scope of Interactions

In the EPO there exists a potential for interaction between longline and purse-seine gears. This interaction may have been greatest in the 1960s, when the distribution of effort by the two gears overlapped to some degree. Currently, fishermen are targetting bigeye tuna by fishing deeper and in more southerly latitudes. The CPUE effort for longlines and purse seines show nearly-parallel declining trends through the 1970s and early 1980s. The declines in the longline catch and CPUE are thought to be caused by the purse seines (resource-mediated interaction) and by changes in recruitment to the population. In recent years, increased abundance of the yellowfin resource and decreased effort in the purse-seine fishery has effected an increase in CPUE. The longline CPUE, however, has recovered only two-thirds as much as the purse-seine CPUE, but it should be noted that yellowfin is only an incidental catch for the longline fishery.

The potential exists for interaction between purse-seine and baitboat gears. Currently, the baitboat fleet is small and stable in numbers and catch, and operates near shore. Purse seiners also operate near shore and on similar-size fish, so there is a potential for localized interaction.

7.3   Importance for Fisheries Management

The yellowfin resource in the EPO is producing near its maximum sustainable production with the current yield per recruit.

7.4   Methods for Studying Interactions

Various methods for studying interactions have been undertaken with differing degrees of success. Statistical measures, such as CPUE time series for the purse-seine and longline fleets, have suggested that interactions exist. The CPUE data for these gears span a period of more than 30 years. A movement model incorporating predicted movement through advection and diffusion, driven by existing tagging data, has been used to address the effects of changes in minimum size of capture and the effects on local fisheries through resource redistribution. A general simulation model for tuna in the EPO is also being developed. This age-structured model, based on area-time strata, addresses interactions between yellowfin and skipjack tunas and dolphins. This model incorporates fleet dynamics and fish movements, and also allows different scenarios to be tested.

7.5   Recommendations

The following topics were considered to be equally important:

a)   Further development of simulation methodology and applications is warranted. Movement models and age-structured models can be used to address interaction questions in a cost-effective manner.

b)   Additional data on timing and movement of fish between fisheries are needed to provide parameters for models. This is best accomplished by tagging experiments. Modification of previous experimental designs and methodology is needed to improve tag-recovery rates throughout the fisheries. In addition, technical advances are needed for tagging methods which would permit the successful tagging of purse-seine-caught yellowfin, particularly in offshore areas.

c)   Further study of the spawning and maturity of yellowfin is needed to better understand the distribution in space and time of reproduction of the stock. This can be examined for possible spawner, environmental, and recruit relationships.

d)   Collection of basic fishery data needs to be continued on a regular basis for all segments of the fishery to provide a basis for analysis of interactions, as well as other topics.

e)   Ecological studies of the species components in the purse-seine and baitboat fisheries should be examined for species interactions.

7.6   Institutional Arrangements and Future Research

Future research is likely to be conducted by a number of institutions. These include, but are not limited to, the Inter-American Tropical Tuna Commission, several institutions in Mexico (*e.g.*, Instituto Nacional de Pesca, Centro de Investigaciones Biologicas, and Centro Interdisciplinario de Ciencias Marinas), the US National Marine Fisheries Service, and the National Research Institute of Far Seas Fisheries of Japan.

8.   WESTERN PACIFIC YELLOWFIN TUNA

The chair of the session on western Pacific yellowfin tuna was Dr. Ziro Suzuki; Mr. Atilio Coan was the vice-chair, and Mr. Peter Ward the rapporteur.

8.1   Fishery Components

Yellowfin tuna are distributed widely in the tropical Pacific and, during the summer, also occur in higher latitudes. The western Pacific yellowfin tuna fishery is characterised by a significant expansion in fishing effort and catches during the 1980s. The term "western Pacific" is used here to refer to all waters of the Pacific west of 150°W, including the Philippines and eastern Indonesia. Diverse fishing methods (purse seine, longline, baitboat, and handline) are used by fishermen of many nations to catch yellowfin tuna in the region. Components of the western Pacific yellowfin tuna fishery are described by Suzuki (this document).

Purse seining, predominantly by distant-water fishing nations, accounted for over 50% of the 338,868 mt of yellowfin tuna caught in the region in 1990. Significant catches were also taken by various methods in the Philippines (62,146 mt, mainly with ringnet) and Indonesia (57,995 mt). The rapid expansion of purse-seining has raised concern over its possible effects on other components of the tuna fishery.

8.2   Scope of Interactions

There are many possibilities for interactions between tuna fisheries in the western Pacific. Several potential interactions for yellowfin tuna in the western Pacific were identified by the Consultation and are discussed below.

8.2.1 Purse-seine affecting longline

Concern has been expressed by groups using longlines (Japan, Korea, Taiwan, and several coastal states) that purse seining may affect longline catch rates. According to the system of classification developed by Hampton (this document), this would be a "Type B Interaction"; the effect one component (purse seine) fishing yellowfin tuna at an early stage in its life cycle has upon a component using different gear (longline) at a later stage. There may also be instances of more direct competition, where the two groups operate in the same area and the size composition of the catch overlaps. For example, purse seining on free-swimming schools of a wide size range sometimes overlaps with longlining in the same area.

8.2.2 Offshore activities affecting coastal activities

Many coastal states are concerned over the possible adverse effects of offshore activities on their commercial and subsistence fisheries for yellowfin tuna. There are several offshore activities, specifically purse seining, that are of particular concern. There is a range of coastal operations that might be affected, from surface activities such as subsistence trolling and handline, to small-scale commercial longlining. Interaction occurs where these local, surface fisheries compete with purse seining for yellowfin tuna at the same stage of their life cycle in adjacent or surrounding areas.

8.2.3 Purse seine interacting with purse seine

Direct competition occurs between vessels of different nationalities using the same fishing gear, *e.g.*, purse seines, when they are active in the same areas. There may also be more diffuse interactions, with subtle variations in targetting between operations. For example, USA seiners, which sometimes target free-swimming schools over a wide area, might be affected by the Japanese who concentrate only on log-associated schools in more discrete areas.

8.2.4 Commercial activities affecting artisanal or recreational components

Coastal states are often concerned over the effects of commercial fishing within their economic zones on artisanal or recreational components. Various commercial operations may be responsible for interactions; these interactions would be direct and mainly affect surface activities of artisanal and recreational components. Concern for these interactions, for example, have led to anglers successfully lobbying for area closures to prevent competition from longliners.

8.2.5 Coastal fishing affecting longline

A unique situation of a local, small-scale operation affecting commercial fishing exists with the ringnet fishery in the Philippines. Large amounts of very small (20-40 cm) yellowfin tuna are taken by ringnets, and declines in longline catch rates in nearby areas two or three years later might be evidence of the adverse affects of ringnets. Rigorous experimental design and special techniques are required to investigate the effect of this component on offshore fishing activities.

8.3   Importance of Interactions for Fisheries Management

Despite growing fishing effort, there is still no evidence of a decline in the abundance of yellowfin tuna in the region. The situation of rapidly expanding effort on a spatially-heterogeneous resource has led many fisheries administrators to be more concerned with possible interactions than overall stock condition. This concern has been expressed by some coastal nations and international bodies.

8.4   Methods Applied for Studying Interactions

The Consultation noted general uncertainty as to whether there was evidence of interaction in components of the western Pacific fishery. Analyses of interaction have been limited to the issue of purse seines interacting with longlines through examination of abundance indices for negative correlations. More than 10 years ago, Lenarz and Zweifel (1979) suggested that the total yield of yellowfin tuna from coexisting surface and longline fishing was greater than that by either gear operating alone. More recently, Hilborn (1989) showed that the total yield would be maximised if there was little mixing between components of the stock exploited by longline and surface gears. If the stock was available to both fishing methods however, a longline fishery would maximise yield.

A preliminary study by Sibert (pers. commun.) found negative cross-correlations between longline catch rates and purse-seine catches by the Japanese in the Federated States of Micronesia (FSM). Results, however, were generally inconclusive, with inverse correlations in some analyses. In an analysis of catch and effort data, Hampton (1988) found no evidence of fishing activities affecting the stock available to surface gears during 1978-88.

More recently, Suzuki (this document) noted that the Japanese longline component had been relatively stable during the 1980s. The effects of large catches of yellowfin tuna by purse seines during 1987 and 1988 might be apparent in a decline in longline catch rates in 1989. Decline in longline catch rates tended to occur in areas where the two fisheries overlap, supporting the hypothesis that purse-seining was having an adverse affect on longlining. No decline of catch rates however, was apparent in the Korean component at that time (Park *et al.*, this document). Medley (this document) assumed that purse-seine catches affect longline catch rates in the same area, and developed a model to estimate the effect on catches. Medley also provided a decision framework for assessing the costs and benefits of longlining and purse-seining.

The South Pacific Commission (SPC) tagging projects will add significantly to knowledge of movement patterns and provide data for modelling population dynamics. Over 30,000 yellowfin tuna have been tagged and released throughout the western Pacific; tags from over 2,500 recaptures of these have so far been received. The Consultation noted the release of significant numbers of large (longer than 60 cm) yellowfin tuna in the Coral Sea, which may aid the study of recruitment to longline components.

8.5   Scientific Problems and Research Priorities

8.5.1 Fishery statistics and abundance indices

The coverage of data-collection programmes and the quality of catch, effort, and size data were inadequate for many of the analyses required to answer interaction questions. The Consultation noted the importance of accurate species identification in Indonesian and Philippines data and in purse-seine reports, particularly for distinguishing yellowfin tuna from bigeye tuna.

Accurate abundance indices were required for correlation analyses of time-series data for the longline and purse-seine components. Detailed description of developments in fishing methods and investigation of their influence on catch rates were necessary for refining abundance indices.

Trends in abundance depicted by indices may be masked by environmental noise. The large-scale effects of extraneous influences (such as El Niño events) on apparent abundance must be taken into account.

8.5.2 Movement patterns

Movement patterns are complex and influenced by various biological and environmental factors, *e.g.*, age, reproductive behaviour, water temperature, fish aggregating devices (FADs), and distribution of forage organisms. Archival and sonic tags show promise for studying movement patterns, especially for large yellowfin tuna.

Investigation of interaction between surface fisheries and longline operations highlighted specific problems faced in studying movement patterns. Reports of tagged yellowfin recaptured by longline are rare. For example, not one of the 2,500 yellowfin recaptures reported to the SPC has come from a longliner, whereas at least 80 of these recaptured fish were 90 cm or larger, the size at which yellowfin tuna are recruited to the longline fishery. This is perplexing because recaptures of other tuna species by longline are not uncommon, *e.g.*, albacore, northern bluefin, and southern bluefin. It would be unlikely that a conspiracy existed among fishermen to not report tags, which extended across countries, and for yellowfin tuna, but not other species. There was concern over the implications of the lack of returns of tags from the longline fishery. The Consultation, therefore, recommended further investigation of reporting of tag recaptures in the longline fishery. [If non-reporting is significant, then remedial action might be considered, such as observer programmes to gauge non-reporting, increasing tag rewards, and promotion of tagging programmes directly with longline crews.]

8.5.3 Stock structure and recruitment

Detailed knowledge of stock structure is a useful starting point for studying interaction. Stock structure of yellowfin tuna in the region however, is poorly understood. One hypothesis is that Pacific yellowfin tuna comprise three stocks, roughly corresponding to the western Pacific, central Pacific, and eastern Pacific, but their exact boundaries and the level of mixing between them is not known. The relationship of yellowfin tuna taken in Indonesian and Philippines waters to the rest of the western Pacific also requires investigation. Analyses of morphometric relationships indicate local heterogeneity, yet tag-recapture studies show the potential for yellowfin tuna to move great distances. An understanding of the heterogeneity of local groups or sub-populations within the stock is essential for assessing interaction between components of the fishery.

8.5.4 Biological parameters

In addition to movement and intensity of exploitation, the degree of interaction will be dependent on growth and natural mortality. In the absence of comprehensive studies in the western Pacific, assessments of interaction have used estimates of age and growth of yellowfin tuna from other regions, such as the eastern Pacific. Age and growth in the western Pacific, however, may be different to that in the eastern Pacific, and assessments of interaction based on these parameters may thus be misleading. Similarly, schedules for age-dependent mortality are required. Data collected by SPC tagging projects will be useful in this regard.

8.5.5 Analytical tools

In the long term, an age-structured model for yellowfin tuna in the western Pacific is essential for effectively addressing interaction questions. Such a model should incorporate vulnerability schedules for the various gears used, and movement, particularly exchange between surface and deep components of the stock. Yield-per-recruit analyses that evaluate multiple gears would also provide a guide to administrators seeking advice on optimum fishing regimes.

8.6 <u>Recommendations</u>

The following problems and recommendations are not necessarily listed in order of priority:

a) Inadequate fishery statistics.

1) SPC and IPTP should continue to improve collection systems and quality of data, particularly for high-seas operations[1].

b) Uncertainty over movement patterns and mixing.

1) analyse data from SPC tagging projects,
2) investigate non-reporting in longline fishery,
3) conduct studies using sonic and/or archival tags, particularly for large fish, and
4) investigate fine-scale genetic structure of the population(s).

c) Poor indices of abundance.

1) quantify changes in fishing efficiency by collecting information on developments in targets of fisheries and in fishing gear and practices, and
2) investigate the relationship between environmental conditions and abundance.

d) Uncertainty over biological parameters.

1) conduct comprehensive studies of age and growth, and
2) study natural mortality

e) *Ad hoc* approach to interaction questions.

1) develop comprehensive age-structured models.

---

[1] *The current US Multilateral Treaty on Fisheries is already providing high-seas data for all USA purse-seine vessels operating within the treaty area. It was noted that a set of Minimum Terms and Conditions (MTCs) is currently being applied by the 16 member states of the Forum Fisheries Agency (FFA). The 16 member states of FFA are currently attempting to apply a revised set of MTCs for access agreements. If accepted, these would include provision of high seas data by foreign fishing vessels operating within the FFA region.*

8.7 Institutional Arrangements for Future Research

Identification of problems and investigation of interaction depends to a large extent on comprehensive catch, effort, and size data. Most interaction problems identified by the Consultation involve DWFNs. The SPC and IPTP, their member nations, and others taking yellowfin tuna in the region, must continue to establish data collection and validation systems for these activities. In the absence of any formal arrangement for the coordination of tuna research in the western Pacific, these organisations should encourage participation by all parties involved in western Pacific yellowfin tuna fisheries to cooperate in relevant research and exchange of data.

The first meeting of the Western Pacific Yellowfin Tuna Research Group was held in June 1991. The group will provide an important international forum for analysing data and providing advice on yellowfin tuna, including advice on interaction. Results of SPC tagging programmes will form an important data base for many of the analyses required. Scientists may need to develop collaborative projects through the stock assessment group and SPC.

8.8 References

The following papers were cited in this session:

Hampton, J. 1988. Status of tuna fisheries in the western and central Pacific Ocean. IPTP Report of the 2nd Southeast Asian Tuna Conference and 3rd Meeting of Tuna Research Groups in the Southeast Asian Region, pp 187-200.

Hilborn, R. 1989. Yield estimation for spatially connected populations: an example of surface and longline fisheries for yellowfin tuna. *North Amer.J.Fish.Mgt.* 9(4):402-10.

Lenarz, W., and J.R. Zweifel. 1979. A theoretical examination of some aspects of the interactions between longline and surface fisheries for yellowfin tuna, *Thunnus albacares*. *Fish.Bull.NOAA-NMFS*, 76(4):807-25.

9. NORTH PACIFIC ALBACORE

The session on North Pacific albacore was chaired by Dr. Norman Bartoo; the rapporteurs were Dr. William Bayliff and Mr. Atilio Coan.

9.1 Fishery Components

There are four principal fisheries for North Pacific albacore: the USA/Canadian troll fishery in the central to eastern North Pacific, the Japanese pole-and-line fishery that operates in waters off the Japanese homeland and further offshore in the Kuroshio Extension waters, the large- and small-mesh drift gillnet fishery in the central and western North Pacific, and the longline fishery of several nations which takes place over a wide area of the North Pacific. Pole-and-line and troll fisheries began in the early 1900s, longline fisheries have operated since before World War II, and drift gillnet fisheries started in 1981. The highest catches, in excess of 100,000 mt per year, occurred in the early 1970s when the surface fisheries took over 80% of this catch. However, since the mid-1970s, CPUE, catch, and effort of the pole-and-line and troll fisheries have decreased. Currently, the catch from all gear is approximately 50,000 mt.

Albacore in the North Pacific are considered to be one stock. The drift gillnet and troll fisheries take 3- and 4-year-old fish, the pole-and-line fishery takes mainly 5-year-olds, and the longline fishery takes mostly fish older than 5 years old. There are overlaps in age composition among fisheries.

9.2   Scope of Interactions

Albacore in the North Pacific have been extensively tagged in the commercial USA troll and Japanese pole-and-line fisheries. Tagged fish have been recovered by the pole-and-line, troll, and longline fisheries. These results show that albacore migrate freely back and forth across the North Pacific through areas commonly exploited by all gears. Further, since the pole-and-line, troll, and drift gillnet fisheries take fish of about the same age, there is considerable opportunity for interaction among these fisheries. Since the longline fishery takes almost entirely older fish, it can also be affected by the other fisheries, and may in turn affect recruitment to the surface fisheries. After the start of the drift gillnet fishery, drift gillnet-marked albacore began to be caught in the troll fishery, indicating direct interaction between these fisheries.

9.3   Importance of Interactions for Fisheries Management

Since albacore migrate through areas exploited by the pole-and-line, drift gillnet, and troll fisheries, and then through areas exploited by longlines, concerns have always been present about the effects of fisheries on each other. Catches and CPUE of albacore fisheries in the North Pacific have decreased considerably since the mid-1970s. While the cause for this decrease is not known, it has caused interest in the role played by environmental factors on abundance/availability.

9.4   Methods for Studying Interactions

Tagging has yielded more information about interactions than has any other technique. Two types of tagging are utilized. The first is conventional tagging, which is employed in the troll and pole-and-line fisheries. The second is accidental tagging, which results when fish make contact with drift gillnets but manage to escape. The net markings on the fish constitute the "tag". Accidental tagging is not as useful as conventional tagging because the number of fish tagged is not known, and the locations and dates of tagging cannot be determined, so the methods used in conventional tagging analyses and models are not useful for analysing accidental tagging data.

In addition, statistical data have been used to study interactions. The drift gillnet fishery began during the 1980s, so the catches by the other fisheries before and after the drift gillnet fishery began were compared to attempt to determine the reduction in the catches of the other fisheries caused by the drift gillnet fishery. The drift gillnet fishery was recently terminated, so the catches by other fisheries before and after termination can be compared. However, the effects of the drift gillnet fishery may be masked by the effects of other factors, such as environmental conditions. Simulation models have also been used to estimate interactions by incorporating age-structured movement and mortality.

9.5   Scientific Problems to be Resolved and their Priorities

There is considerable opposition, particularly from environmental groups, to the drift gillnet fishery. The extent of the catches of non-target species should be determined, as should the effects of the catches on the abundance of those species. Also, the amounts of albacore which drop out of the nets should be estimated.

9.6   Recommendations

Further work is required in the following areas:

a)   A variety of research should be undertaken to determine the interactions between the various fisheries. Catches of the albacore fishery before, during, and after the drift gillnet fishery should be compared in order to determine the effects of the drift gillnet fishery on troll, pole-and-line, and longline fisheries. Loss rates from drift gillnets should be estimated and used in the interaction study.

b)   Continued analyses of the tagging data should be conducted to quantify interactions and movements.

c)   Simulation modelling of tagging data should be conducted to further quantify interactions among gear types.

d)   Interactions between gears should be analysed by age-structured methods leading to yield-per-recruit analyses by fishery and time period.

9.7   Institutional Arrangements for Future Research

Research is being conducted principally by the fisheries agencies of the USA, Canada, Japan, Taiwan, and Korea, and continuation is encouraged. Data and results of analyses on the status of albacore stocks are shared at biennial albacore workshops involving scientists of these organisations. These workshops provide a means of coordinating and conducting future interaction research.

10.   SOUTH PACIFIC ALBACORE

The session on South Pacific albacore was chaired by Dr. Talbot Murray; the vice-chair was Dr. Mark Labelle, and the rapporteur Mr. Albert Caton.

10.1   Components of the Fishery

Larval distribution patterns and apparent physiological barriers for larval and juvenile albacore in equatorial waters suggest that albacore form a discrete stock in the

South Pacific. Separate North Pacific and Indian Ocean stocks are also suggested by low longline catch rates of adults in equatorial waters and in waters south of Tasmania. The South American land mass serves as an eastern boundary between the Pacific and Atlantic stocks of albacore.

Major surface fisheries for juveniles and longline fisheries for adults exploit the resource throughout most of its range. One surface fishery component, a drift gillnet fishery in the Tasman Sea and Sub-tropical Convergence Zone (STCZ), developed rapidly in the mid- to late 1980s, and ceased in 1991. The biology and description of the fisheries for South Pacific albacore is reviewed by Murray (this document), and the potential for interactions among fishery components is discussed by Murray (this document).

The surface fishery began in 1968 in New Zealand waters, where a troll fleet continues to operate off the west coast of the South Island from January to April. The fleet, consisting of over 200 vessels in some years, made catches ranging from less than 1,000 mt to over 4,000 mt of juvenile albacore annually. The drift gillnet fishery, composed of fleets from Taiwan and Japan, expanded rapidly during the mid to late 1980s to an estimated 136 vessels in 1988/89 (Anon., in press).

The estimate of the 1988/89 drift gillnet catch was nearly 22,000 mt; the total surface-fishery catches for the entire South Pacific was almost 31,000 mt for that period. Following this peak season, when the surface fishery catch exceeded the previous high by threefold, the number of vessels in the drift gillnet fleets declined due to international pressure to terminate this method of fishing. Drift gillnet fishing in the South Pacific ended in June 1991 in accordance with United Nations Resolution 44/225. While the drift gillnet fishery was developing, another major surface fishery, the STCZ troll fishery, developed in the same area and season as the drift gillnet fishery. The STCZ troll fishery increased from two vessels in 1985/86 to more than 70 vessels in 1991; the catch of this troll fishery was about 5,400 mt. There have also been indications that during 1991 some pole-and-line vessels fished in the Tasman Sea, in the same area and season as drift gillnet vessels were operating. The target species of this small pole-and-line fleet is unknown, but it is likely that some albacore were caught.

Japanese longlining began in the South Pacific during the early 1950s. Korean longliners joined the fishery in 1958 and Taiwanese vessels in 1967. The annual longline catch of albacore has ranged between 20,000 mt and 40,000 mt. Japanese activity peaked in 1962, and then declined with a change in target species. The catch by Korean longliners has varied from about 6,000 mt to almost 19,000 mt, declining since 1986 when there was a shift in the target species. The Taiwanese longline catch has been less variable (averaging around 13,500 mt each year since 1980) with most of the fleet effort directed to albacore. In recent years, some Taiwanese longline vessels may have shifted from albacore to target other tuna species.

Since the mid-1980s, several South Pacific countries (Australia, Tonga, New Caledonia, and New Zealand) have started to develop domestic longline fisheries. The total catch from these fisheries in 1990 was less than 2,000 mt; most of the fishing took place in waters in, and adjacent to, their respective EEZs. Since 1980, the albacore catch by all tuna fleets operating in the South Pacific ranged from 21,000 mt to 39,000 mt; the average was 29,300 mt.

10.2   Scope of Interactions

The drift gillnet and troll fisheries catch similar sizes of juvenile albacore from the same areas (particularly in the STCZ) and in the same season, suggesting a high potential for interaction. The longline fishery exploits predominantly adult albacore throughout the year over a broader area north of the surface fishery region. The spatial separation of fleets and size differences in the catch would seem to suggest that most surface and longline interactions would occur after some time lag. On the contrary, some longline operations occur immediately north of the STCZ troll fishery, and follow the STCZ fishing season. The proximity of these fisheries and the oceanographic conditions of the region appear to be responsible for the overlap in the albacore size compositions of the two fisheries. These observations suggest there is a potential for some interaction between the surface and longline fisheries without a substantial time lag.

10.3   Information on Interactions

Based on CPUE estimates (Anon., in press) there is no clear correlation between drift gillnet catch and troll CPUE. Drift gillnet CPUE generally increased until 1987/88, but dropped in 1988/89, a period when drift gillnet catch increased sharply. Subsequently, the drift gillnet catch declined abruptly in 1989/90, while the CPUE increased again. Troll fishery CPUE showed a less marked trend during the period and, in fact, declined in 1990/91 when drift gillnet catch was low. Subsequently, as drift gillnet fleets were reduced, catch rates of the remaining drift gillnetters and the troll fleet improved. The USA trollers have reported gear conflict where fishing has been hampered by the risk of troll vessels becoming entangled in drift gillnets. Some fishers also claim that albacore behaviour changes (become less vulnerable) in the vicinity of drift gillnet operations.

Spatial and temporal patterns of drift gillnet damaged albacore subsequently caught by trolling indicate that the highest incidence of net-marked fish occurs in the vicinity of the drift gillnet fleet with declining incidence to the east and low incidence to the west. Similar indicators were evident in the New Zealand troll fishery, where incidence of net-marked fish increased from about 1% to 7% during the 1989/90 season when drift gillnetters operated in the western Tasman Sea (*i.e.*, west of the New Zealand surface fishery). Observations of drift gillnet marked fish in longline catches off New Zealand, indicated net marks occurred on a substantial portion of the catch in the size categories exploited by surface fisheries. In contrast, observations in the Australian Fishery Zone to the west of the Tasman Sea drift gillnet fishing area did not reveal any net-marked fish. These observations suggest a direction of surface albacore movements, and consequently imply a spatial pattern of interaction between surface and longline fisheries in the Tasman Sea and New Zealand area.

There are other indications of surface and longline fishery interaction from longline catch rates. Longline catch rates in sub-equatorial latitudes, where large adult albacore are the main catch component, were fairly stable through the 1980s. Farther south (20°-30°S), the CPUE was more variable and has shown a decline since 1986. Variability in CPUE and the magnitude of the CPUE decline is greatest in the latitudinal band immediately north of the STCZ troll fishery, where juveniles make up a strong component of the longline catch. While the link is rather tenuous, the catch rate in the

30-40°S band declined after 1986. However, the most recent longline CPUE in this region is at a similar level to the period before the development of the surface fishery.

Movement patterns inferred from parasite studies and tagging experiments are consistent with the patterns suggested by net-marked fish and longline fishing patterns. The parasite studies confirmed the tropical origin for recruitment of juveniles to the New Zealand troll fishery; a parasite unique to the tropics was associated with albacore caught in New Zealand waters. Subsequent prevalence of the parasite declined, suggesting movement eastward. Prevalence also declined with increasing albacore size until about the size of female maturity, after which prevalence increased again. This pattern of parasite prevalence is consistent with juveniles remaining in temperate waters until the first spawning migration to the tropics and a subsequent return to temperate waters. So far, only 11 tagged albacore have been recovered from more than 10,000 releases of juveniles in the troll fishery. However, the movements of the 11 recoveries were consistent with either short-term west-to-east movements or movement northward. Nine of the recoveries were made by the longline fishery and times at liberty ranged from a few months to several years. The results provide direct indication of interaction between the surface and longline fishery.

10.4   Scientific Problems and Research Requirements

The low recovery rate of tags from the relatively large number of releases in surface troll fishery (11 recoveries from more than 10,000 tag releases) contrasts sharply with results from similar albacore tagging in other oceans. The low tag returns could indicate substantial non-reporting of tags, high mortality of tagged fish, or low exploitation rates. Several observations suggest that non-reporting of tags may explain the low recovery rates. The use of a proven tagging method which still resulted in low recovery rates, the reports by some individuals in the surface fishery of a general unwillingness to return tags, the non-return of a few tags "seeded" in the catch of the drift gillnet fishery, and the reduction of the longline CPUE, which was predicted if effort continued at the 1988/89 level (Hampton, 1990), suggest that substantial non-reporting of tags may be occurring. Other tagging experiments involving tropical species with higher mortality rates in the South Pacific (*e.g.*, skipjack tuna tagged in the Solomon Islands) resulted in tag recovery rates many times higher than that for albacore, further suggesting that non-reporting may be a major problem.

Available data from longline fisheries was useful. However, higher-resolution catch-and-effort data were needed to monitor size-specific changes in longline CPUE following the 1988/89 surface fishery peak. Better information on fishery targets and other longline operational details was also needed to standardize CPUE data. Work is planned to extend the length-frequency-based growth rate and population-at-age estimation procedure to incorporate a preliminary size-structured stock assessment for South Pacific albacore.

Development of models of albacore fisheries interaction will require incorporation of spatial and temporal movement patterns. Consideration of spatial distribution is complicated by different depth distributions for juveniles and adults and apparent differences in depth distributions of adults in the western, central, and eastern South Pacific. However, with juvenile catches currently occurring at an order of magnitude lower than adult catches, there may be limited scope for the present troll catch to exert a detectable impact on the longline fishery. This observation adds an impetus to monitor the historical impact of the peak surface catch in 1988/89 when the juvenile catch equalled or exceeded the adult catch. A simulation model which assumed similar parameters as albacore populations in other oceans and which incorporated estimates of surface and longline catches, suggested that continued surface catches at 1988/89 level could cause a significant reduction (perhaps 60% in 5 years) in longline catch.

As further increases in surface-fishery catch cannot be ruled out, it is still appropriate to assess the potential impact of a range of surface-fishery catches on the longline fishery. For this reason, further studies of historical drift gillnet catches could provide a valuable benchmark. If estimates of the number of fish escaping drift gillnets were available it might be possible to do a more detailed analysis of the net-mark observations as a form of tag-recapture study to shed more light on the extent of interaction. In general, the impact of interactions is closely linked with exploitation rate; assessment studies in progress or proposed should lead to considerably improved estimates of relative exploitation rates.

10.5   Recommendations

The following research activities are recommended for the South Pacific albacore:

a)   Fishery data collection from all sectors needs to be maintained and, where possible, data coverage should be increased. In particular:

   1) there is a need to improve the resolution of available high seas catch, effort, and size-composition data, especially to take advantage of the opportunity to monitor any size-specific, longline-CPUE changes following the 1988/89 surface fishery catches, and
   2) better information on target species and other longline operational details is also needed to standardise CPUE data for use as abundance indices.

b)   Tagging experiments should be expanded and reasons for low recovery rates identified. Where possible, tag-reporting rates should be estimated for each fishery sector.

c)   Further work should be done to determine the potential for interaction between surface-troll and longline fisheries, particularly with regard to gear-specific yield-per-recruit analyses.

d)   There is a need to determine the sampling regime and sample sizes required to detect a given level of interaction.

10.6   Institutional Arrangements for Future Research

Scientists from the South Pacific and DWFNs interested in albacore formed an informal "South Pacific Albacore Research Working Group" (SPAR) in 1986, with the objective of determining the appropriate aggregate yields of surface and longline fisheries. The need arose because of interest in the scope for development of surface fishing when longline fisheries were already regarded as operating close to MSY. Consequently, SPAR's focus was on stock assessment and fisheries interaction. Regular updates of fisheries developments and research results are produced for discussion at each SPAR meeting.

With the rapid expansion of the drift gillnet fishery in 1988, interaction issues were highlighted as a research priority, as was the need for management. The SPAR group was asked by South Pacific and DWFNs fisheries managers to serve as an interim scientific advisory body during South Pacific Albacore Management Consultations between and DWFN and South Pacific Island States. While a core function of SPAR remains that of research coordination and planning, it also produces a summary "Status of the Stock" report in which the potential for fisheries interactions is an important consideration. SPAR, is therefore, in a good position to continue to review developments in South Pacific albacore fisheries and coordinate future research activities addressing interactions and related issues.

10.7 References

Anon. In Press. Report of the Fourth South Pacific Albacore Research Workshop, November, 1991. National Taiwan University, Taipei.

Hampton, J. 1990. Simulations of the South Pacific albacore population: effects of rapid developments in the surface fishery. Third South Pacific Albacore Research Workshop, Information Paper No. 1.

11. PACIFIC BIGEYE TUNA

The session on Pacific bigeye tuna was chaired by Dr. Naozumi Miyabe; vice-chair was Dr. Chris Boggs, and rapporteur Dr. Antony Lewis.

11.1 Fishery Components

Both surface and longline fisheries capture bigeye over wide areas of the Pacific between 45°N and 40°S, with longline catches comprising the great majority of the catch by volume. Japanese longliners take the largest proportion of this catch (75-85%) on a year-round basis in mostly equatorial areas, especially east of 140°W. Korean and Taiwanese longliners also take bigeye in quantity principally in the central South Pacific. A recent development has been the increased activity of small (<20 GRT) longliners in several areas; these vessels supply fresh fish to sashimi markets. This fishery catches adult-sized bigeye.

Baitboats in the eastern and western Pacific and in some Pacific islands areas have a long history of capturing juvenile bigeye, although in relatively minor quantities. The rapid expansion of purse-seine fisheries in the western Pacific in recent years has undoubtedly led to a significant increase in incidental bigeye catches. Knowledge of the species composition in surface catches is improving, but is still incomplete. A variety of gears in Indonesia and the Philippines no doubt takes sizeable quantities of juvenile bigeye.

According to declared statistics, recent bigeye catches have ranged from 110,000 to 150,000 mt annually; however, this is likely to be an underestimate. Production models based on longline time-series data provide MSY estimates of the stock in the range of 100,000 to 170,000 mt. Longline CPUEs are now at approximately 45% of its initial levels, but they have been relatively stable since the mid 1960s.

11.2 Scope of Interactions

The recent expansion of purse-seine fisheries in the western Pacific, and other fisheries in Indonesia and Philippines, suggests the potential for interaction with the widely-distributed, established longline fisheries, even though these fisheries may be spatially separated. In some coastal and island states, adjacent surface and longline fisheries could provide a potential source of interaction.

11.3 Importance of Interactions for Fisheries Management

The increase in surface catches combined with the uncertainty about precise species composition of these catches, may have implications for the more valuable longline fishery. Indeed, excluding skipjack, bigeye catches represent the largest volume and highest total value in the Japanese tuna fishery.

11.4 Methods Applied for Studying Interactions

A Y/R model was applied to the four groups of fisheries, longline, Japanese baitboat in the northwest Pacific, western Pacific surface fishery, and eastern Pacific surface fishery. Two sets of catch levels were assumed, the second accounting for unreported catches and assumed misidentification of juvenile bigeye as yellowfin. The second of these cases, with $F = 0.4$, seemed to fit the situation presently observed in the fisheries, and indicates that any further increase in the catch of very small (0+) fish will be detrimental to the longline fishery. The application of this approach was recommended, and its general utility recognized for other tuna species.

11.5 Information on Interactions

There is no reliable information on interactions.

11.6 Recommendations

The following scientific problems were identified and research recommended:

a)   The biology of bigeye remains less known than that of either yellowfin or skipjack. Further work is essential on selected aspects in order to assist studies of interaction. These include stock identification, movement patterns for all stages of the life history, and growth.

b)   More reliable catch statistics, with greater emphasis on separation and correct identification of bigeye and yellowfin in the catches, are needed. The recently-expanded activity of small longliners directing their catches to the sashimi markets and operating in several areas of the Pacific needs to be better documented.

c) Migration and stock structure of bigeye in the Pacific are not well understood; lack of information is partly due to the fewer tagging experiments conducted on bigeye relative to other tunas. Recently, sonic tagging has confirmed the species' ability to range down to 350 m depth. A single Pacific-wide stock is assumed, based on fishery data and extent of spawning. There is an observed east-west cline in the size of longline-caught fish, and, given the distribution of optimal habitat, it is probable that Pacific-wide mixing is possible over the entire life history of the bigeye.

11.7 <u>Research Required and Priorities</u>

Biological research needs, as identified above, should be pursued with some priority. The present Regional Tuna Tagging Programme (RTTP) work should provide some useful information in this regard, particularly with respect to movement, age-specific migrations, and growth. Continued refinement of the Y/R approach should be encouraged as better estimates of the parameters become available.

11.8 <u>Institutional Arrangements for Future Research</u>

Research on bigeye tuna stocks is likely to be continued by, but not limited to, the following organisations: the National Research Institute for Far Seas Fisheries, Japan (NRIFSF), the South Pacific Commission (SPC), the Inter-American Tropical Tuna Commission (IATTC), the Indo-Pacific Tuna Development and Management Programme (IPTP), and national fisheries agencies in Southeast Asia. Regular communication and cooperation among concerned agencies should be encouraged, with the possibility of forming a working group sometime in the future. The Consultation however, did not consider any such arrangements.

12. NORTHERN BLUEFIN TUNA

The session on northern bluefin tuna was chaired by Dr. William Bayliff; the vice-chair was Mr. Yoshio Ishizuka, and the rapporteur Dr. Alex Wild.

12.1 <u>Fishery Components</u>

The northern bluefin tuna has a complex life history, which results in exposure to a variety of fishing gears in the western, central, and eastern Pacific Ocean. The fish are hatched south of Japan and in the Sea of Japan, and are exploited at age-0 by the troll and trap fisheries in the northwestern Pacific Ocean; these fish are about 15 to 55 cm long.

Some fish in their first or second year of life begin a migration to the eastern Pacific Ocean (EPO) during the fall or winter; others remain in the western Pacific Ocean (WPO). It is possible that there are two stocks that sustain the northern bluefin fisheries, one stock remaining in the WPO and the other migrating to the EPO, but the tagging data offer little support for this hypothesis.

The journey from the WPO to the EPO takes as little as seven months. On the way the fish are exposed to small-mesh drift gillnets used to catch flying squid. Upon arriving in the EPO, the migrants form the basis of the fishery that is carried out by purse seiners and, to a much lesser extent, by other commercial and sport gear. The fish may stay in the EPO for a period varying from possibly less than a year to six years before returning to the WPO. The minimum time between the release of a tagged fish in the EPO and its recapture in the WPO is 674 days, so the return journey may take 1-1/2 or more years. In the central and western Pacific the fish are exposed to baitboat and longline fisheries, and to the large-mesh drift gillnets used for albacore. After returning to the WPO they are exposed to purse-seine, trap, and longline fisheries.

The non-migrants are exposed to troll, gillnet, purse-seine, baitboat, trap, and longline fisheries in the WPO.

Northern bluefin are also caught, to a much lesser extent, by longlining east of the Philippines, northeast of Papua New Guinea, southeast of Australia, and especially near New Zealand. It is not known if these fish were juveniles which migrated directly from the spawning grounds, immature fish which migrated south from Japan, or older fish which migrated south after spawning.

12.2   Scope of Interactions

The factors which determine whether the fish migrate to the EPO or remain in the WPO are probably independent of the fisheries. Nevertheless, for the proportion destined to migrate, the mortality inflicted by the various gears prior to and during the migration reduces the potential yield in the EPO. Similarly, the potential catches and reproduction of the fish which would return to the WPO are reduced by the catches taken in the central and eastern Pacific. The potential for fisheries interactions also exists within different oceanic regions. In the WPO, for example, there is interaction among the different gear types. In the EPO, the fish are caught principally by purse seining, so the only significant interaction involves vessels from Mexico and the USA.

12.3   Importance of Interactions for Fisheries Management

There has been a decline in the catch of northern bluefin in the EPO in recent years. The analysis of catch data and the results of tagging experiments initiated in the EPO and WPO suggest that the decline is due principally to a decline in the proportion of fish that migrate to the EPO. Rough calculations based on catch data suggest that the squid drift gillnet fishery in the central Pacific has only a minor effect on the fishery of the EPO. Elimination of the drift gillnet fishery will end that type of interaction. The interaction in the EPO between purse-seine vessels of Mexico and the USA does not appear to warrant study at this time.

12.4   Methods Applied for Studying Interactions

Tagging experiments have been the principal source of information to evaluate the movements of the fish and the relationships between the fisheries in the western and eastern Pacific. The growth rates obtained from tagging data have also been used to estimate size at age. Together with catch and length-frequency data for the WPO and

EPO, this information has been used to estimate the age compositions of the catches in both regions.

12.5   Research Required

The following research requirements were identified:

a)   Catch and effort data are needed for the various fisheries that capture northern bluefin in the western Pacific. The data should be stratified by gear, area, month, and length frequencies.

b)   Further information is needed on the migrations, growth, mortality, and size-related abundance of the fish.

c)   Whether there are separate migrant and non-migrant stocks of northern bluefin needs to be determined.

d)   The oceanographic conditions which influence the migration to and from the EPO should be determined.

12.6   Recommendations

The following recommendations were made:

a)   Improved data on catch and effort are needed to permit more sophisticated yield-per-recruit and cohort analyses. The Inter-American Tropical Tuna Commission (IATTC) intends to reprocess the available length-frequency data for the EPO fisheries to improve their reliability. The National Research Institute of Far Seas Fisheries (NRIFSF) has estimated the amounts of bluefin in *meiji* catches, and intends to make further improvements to the bluefin catch data, beginning in 1992.

b)   There is a need to standardize the effort directed at northern bluefin in the EPO. Contacts with fishermen are planned to help evaluate the meaning of variations in apparent abundance and possibly to obtain information useful for standardization of effort.

c)   Tagging in the Japanese troll fishery would provide useful information on the migrations, mortality, growth, and size-related abundance of the fish. Consideration should be given to the use of new types of tags that would not affect adversely the growth or mortality of the fish, that would provide maximum retention, and that would discourage detection and removal by non-scientific personnel.

d)   The stock structure of the population should be explored by means other than tagging, such as DNA or electrophoretic studies, in order to test the tentative conclusion that there is only one stock of northern bluefin in the Pacific Ocean.

12.7   Institutional Arrangements for Future Research

Future research will be conducted by the NRIFSF and the IATTC, and through cooperation with other scientific organisations; *e.g.*, one or more of those in Mexico.

13.   SOUTHERN BLUEFIN TUNA

The session on southern bluefin tuna was chaired by Dr. Tom Polacheck; vice-chair was Dr. Talbot Murray, and rapporteurs Mr. Peter Ward and Mr. Albert Caton.

13.1   Fishery Components

Southern bluefin tuna comprise a single stock of circumpolar distribution, generally between 35°-50°S. They are also found in the tropical Indian Ocean, south of Java, where they spawn.

Components of the southern bluefin tuna fishery are described by Caton (this document). Three nations (Australia, Japan and New Zealand) target southern bluefin tuna and take the majority of the catch. Bycatches have been reported or suspected in other fisheries, including longline fisheries of Indonesia, Taiwan and Korea, drift gillnet fisheries of Taiwan and, less commonly, inshore fisheries of South Africa. There are also minor catches by recreational anglers, mainly in Australia. The southern bluefin tuna stock is currently characterised by a major decline in the parental biomass and, in recent years, Australia, Japan, and New Zealand have managed the fishery using restrictive catch limits.

Catches of southern bluefin tuna by the Japanese using longlines peaked at about 78,000 mt in 1961 and then declined. Ages of southern bluefin taken range from 3 to 20 years old. The Australian surface fishery, which has mainly taken juvenile southern bluefin tuna aged 2-4 years, expanded rapidly in later years, reaching a peak catch of 21,000 mt in 1982/83. Catches by all components of the Australian surface fishery have also declined or were reduced, with the New South Wales surface component failing altogether in the early 1980s. This failure and concern over the condition of the stock led to the introduction of catch quotas during the 1980s. The quota agreed to by Australia, Japan and New Zealand for 1991/92 was 11,750 mt.

The Japanese longline catch in 1989 was about 9,200 mt. The 1990/91 Australian surface catch was less than 3,000 mt. The remainder of the Australian quota was being taken by local or joint-venture longline operations. The New Zealand catch in 1990 was about 520 mt, the bulk taken by domestic or joint-venture longline operations. Catches by Taiwanese Indonesian, and Korean vessels were not known with certainty, but in total may have exceeded 1,000 mt in 1989.

13.2   Scope of Interactions

Several features of the fisheries and the biology of southern bluefin tuna lead to concern over interactions between the various components of the fishery:

a)   Southern bluefin tuna are long-lived and slow-growing; some live to be 20 years or more. They do not mature until 8 years of age, but information on their reproductive biology is sketchy.

b)   The species is broadly distributed and is capable of moving rapidly over long distances.

c)   Spawning occurs in tropical waters south of Java, while adult feeding grounds occur in widely-dispersed areas of the Southern Ocean.

d)   A substantial, but unknown, proportion of juvenile southern bluefin tuna (2-4 years old) occurs in surface schools in inshore waters of southern and southeastern Australia.

e)   Fishing activities take southern bluefin tuna at sequential stages of its life-cycle.

f)   The geographic expanse of fishing grounds has become progressively condensed.

g)   Southern bluefin tuna are subject to high rates of exploitation, and their abundance is currently at historically low levels, *e.g.*, the spawning biomass has been reduced to 16-25% of pre-exploitation levels.

h)   Several nations fish for southern bluefin tuna on the high seas and within exclusive economic zones (EEZs).

Despite the critical state of the southern bluefin tuna stock and the potential for interaction problems, no formal, broad-based management arrangement exists for the southern bluefin tuna fisheries.

Polacheck (this document), provided details on the interactions and classified them into six types:

a)   Interactions among the three components of the Australian surface fishery.

b)   Interactions among surface and longline gear in the same area at the same time.

c)   Effect of the Australian surface catch on subsequent catches in the longline fishery for the same cohorts.

d)   Interactions among longliners in different geographic areas.

e)   Effect of longline, driftnet, and surface fisheries on recruitment and, thus, future catches.

f)   Interaction among longline vessels of different nationalities.

13.3 Importance of Interactions for Fisheries Management

The importance of different interaction issues among the various components has changed as a consequence of the declining condition of the stock. Historically, the main interaction issues of concern were the effects of catches in the various surface components on current yields and catch rates in the surface fishery, the effects of the surface fishery on subsequent longline catches, and local effects of longline and surface catches on each other.

In recent years, management has been concerned with rebuilding the parental biomass. Therefore the effects of current catches on spawning biomass and future recruitment are the most critical management issues. The associated need for restraining catches, and the decline of catches in major fishery components have intensified interaction issues. Particular emphasis has been placed on evaluating the relative impacts of various components (*e.g.*, the South Australian surface fishery) on the overall condition of the stock.

Interaction among longline fisheries operating in different areas has been a critical, unresolved issue. A large amount of spatial structuring (*e.g.*, "site fidelity") may exist among adults on different feeding grounds. Concern has been expressed over the possibility of sequential depletion of areas as longline activity shifts and effort becomes spatially concentrated.

Another management issue is the effect of altering the structure (gear or location) of the fishery. For example, the Australian industry is shifting towards supplying the high-value sashimi market by converting to longline operations. This should enhance the spawning-biomass-per-recruit and promote escape from the surface fishery in the long-term. In the short term, however, these structural changes may reduce the current spawning biomass, and there is uncertainty about their immediate impact on recent and current cohorts of sub-adults that have escaped the surface fishery.

13.4 Methods Applied for Studying Interactions

Data used to study interactions in the past are described by Polocheck (this document). Primary data sets for these analyses included catch, effort and size data for major fisheries and tag-recapture information. Observer reports and various observations by fishermen were important sources of supplementary information.

Comparisons and evaluation of changes in the size composition of the catch have been used to assess the potential for interactions among components of the Australian surface fishery. Southern bluefin tuna movement patterns, based on tag-recapture data, have provided further insight into interactions between components of the surface and longline fisheries. For example, recent recaptures of tagged fish have demonstrated extensive and rapid movement from the Tasmanian troll fishery to the longline fishery. Recaptures of tagged southern bluefin tuna also show rapid movement between the South Australian surface fishery and nearby longline activities. Interaction occurred between the surface and longline fisheries within the first year of release.

Two different analytical methods have been used with the tag-recapture data for measuring interaction. Applications of the methods of Majkowski *et al.* (1988) and Hearn and Marzanov (this document) suggested that the level of interaction among the various components of the Australian surface fisheries was minimal during the 1960s. During the mid-1980s, however, results based on these methods indicate that significant and direct interactions occurred between Western Australia and South Australia surface fisheries. Hampton (1989) developed a parametric model to estimate transfer rates between fisheries, natural mortality rates and fisheries-specific catchability coefficients from tagging data collected during the 1960s. Results showed rates of mixing and movement between New South Wales and South Australia/Western Australia and vice-versa in the order of 0.1 to 0.2 yr^{-1}. The estimated transfer rate into the longline fishery was from 0.6 to 1.8 yr^{-1}. However, the robustness of the results from these two analytical approaches is poorly known. Factors of particular concern are the effects of pooling data from different years and cohorts, and the robustness of the results to assumptions about steady-state conditions, tag-reporting rates, and/or structural components of the models. A need exists for improved models and analyses of these tagging data.

Finally, Y/R analyses have been employed to evaluate relative yields from a cohort in relation to a range of levels of effort in the Australian surface fishery and Japanese longline fishery. The Y/R analyses, however, assume a spatially-homogeneous stock and fishery. The large degree of spatial heterogeneity, both in the stock and the fishery, makes Y/R analyses difficult to interpret to some extent. Of particular concern is the uncertainty over the proportion of juveniles that are vulnerable to the surface fisheries. Nevertheless, Y/R analyses indicate the potential for large interactions between the surface and longline fisheries.

13.5 <u>Scientific Problems and Recommendations</u>

The most important research and data needs centre on questions of movement, migration routes, mixing rates and site fidelity, particularly for the component of the stock that is vulnerable to the longline fisheries. There is a need for further analysis of existing data, development of new analytical methods and models, improved collection of catch-and-effort statistics and development of new approaches for assessing the spatial structure of southern bluefin tuna. Recommendations are listed below; the order of the recommendations do not indicate any ranking or priority:

a) Uncertainty over mixing patterns, movement rates, migration routes and stock dynamics (for juvenile and adult components of the stock).

1) continue to tag and release juveniles off Western Australia, South Australia, and Tasmania,
2) improve coverage and geographic resolution of high-seas catch, effort, and length frequency data,
3) develop and apply models which take into account age-specific spatial and temporal dynamics, fishery changes, and environmental factors,
4) tag in fishery components other than the Australia surface fishery, (*e.g.*, explore feasibility of tagging from longline vessels and in South Africa),

    5) conduct detailed studies of southern bluefin tuna, particularly for large fish, using sonic and archival tags, and relate this to feeding and environmental conditions, and

    6) conduct genetic analyses of different sizes and classes of fish caught over a range of geographic locations.

b)   Uncertainty over bycatch by other fisheries.

    1) enhance data collection in all fisheries components with known or suspected by-catches (*i.e.*, Indonesian, Taiwanese, and Korean longline, Taiwanese drift gillnets, and South African inshore fisheries), and

    2) develop port sampling for length-frequency data from all fisheries with bycatch.

c)   Uncertainty over biological parameters.

    1) validate estimates of age, using hard parts, and
    2) conduct studies on reproductive biology.

13.6   Institutional Arrangements for Future Research

Australia, Japan, and New Zealand (nations regulating their catches of southern bluefin tuna) undertake research and collect catch and effort statistics. However, there is a need for additional coordination and collaboration among these institutions, in order to prevent duplication and to achieve the most effective use of the limited resources available for research. Scientists from these organisations review the condition of the stock at informal scientific meetings each year prior to trilateral management meetings. Involvement of scientists from other nations catching southern bluefin tuna has been recommended by scientists at these trilateral meetings.

The Consultation identified several interaction issues involving high-seas components of the fisheries. Currently information on bycatch from vessels that do not target southern bluefin tuna is critical for assessing these issues. Regional fisheries agencies could assist in collecting data on these incidental catches of southern bluefin. In lieu of effective actions by a regional agency in obtaining these data, collection and provision of such data will most likely depend on informal and cooperative arrangements between countries. Indonesia and Australia are currently exploring the possibility of developing a joint sampling and monitoring programme for southern bluefin tuna caught around Indonesia.

13.7   References

The following papers are cited in this section:

Hampton, J. 1989. Population dynamics, stock assessment, and fishery management of the southern bluefin tuna (*Thunnus maccoyii*). Univ. of New South Wales, Australia. PhD Thesis: 273 pp.

Majkowski, J., W.S. Hearn, and R.L. Sandland. 1988. A tag-release-recovery method for predicting the effect of changing the catch of one component of a fishery upon the remaining components. *Can.J.Fish.Aquat.Sci.* 45:675-84.

14. SMALL TUNAS

The session on small tunas was chaired by Mr. Mitsuo Yesaki; vice-chair was Dr. Nurzali Naamin and rapporteurs were Ms. Flerida Arce and Ms. Chee Phaik Ean.

Only the small tunas of Southeast Asia were reviewed at the Consultation. There are fisheries for black skipjack (*Euthynnus lineatus*) and eastern Pacific bonito (*Sarda orientalis*) off Central and South America which were not considered. Also, there is evidence from stomach contents and larval studies, of a large resource of *Auxis* spp. in the eastern Pacific Ocean which has not been exploited to date.

14.1 Fishery Components

Three species or groups of species of small tunas are important in Southeast Asia. These are the longtail tuna (*Thunnus tonggol*), kawakawa (*Euthynnus affinis*), and *Auxis* spp.

The largest catches of longtail tuna in 1989 were made by Thailand (65,900 mt) and Malaysia (5,600 mt). The Philippines landed the highest catches of kawakawa with 57,900 mt, followed by Thailand with 26,000 mt. Approximately 117,000 mt of *Auxis* spp. were landed by the Philippines in 1989. Indonesia caught over 120,000 mt of small tunas. However, Indonesia does not report landings separately for these species.

The fisheries for small tunas off the South China Sea coast of Thailand and Peninsular Malaysia were examined for interactions. In Thailand, two major gears exploit small tunas: gillnets and purse seines. Gillnets historically fished narrow-barred Spanish mackerel; however, a shift to small tunas probably started around 1981. Luring purse seines operated in conjunction with FADs and lights, while one-boat purse seines targetted small pelagic fish and also caught small tunas as by-catches. The tuna purse seine was reportedly introduced in 1981 to specifically target small tunas.

Off the South China Sea coast of Peninsular Malaysia, drift gillnets and the luring purse seines catch small tunas as a bycatch. Troll lines that target small tunas are also an important gear.

14.2 Scope of Interactions

Landings of small tunas from the South China Sea coast of Thailand and Malaysia increased from 19,000 mt in 1980 to 141,000 mt in 1989. In 1989, Thailand accounted for 93% of this total, while the remainder of the landings was from Malaysia.

Small-tuna landings from the South China Sea coast of Thailand increased gradually from 4,000 mt in 1970 to 13,000 mt in 1980. In 1983, landings increased markedly to 82,000 mt and totalled 131,000 mt in 1989. Landings by purse seiners and gillnetters were of equal magnitude up to 1982, but thereafter, purse-seine landings greatly exceeded those of gillnetters. Purse-seine landings have increased by 150% and gillnet landings have decreased by 62% since 1984.

Small-tuna landings from the South China Sea coast of Peninsular Malaysia fluctuated between 13,000 mt in 1981 and 19,000 mt in 1988. In 1989, landings decreased to 10,000 mt. Troll and gillnet landings decreased from 1987, while purse-seine landings decreased from 1988.

Of the five basic fisheries exploiting small tunas in the South China Sea off Thailand and Malaysia, catches have increased in recent years only for the Thai purse-seine fishery. Catches of small tunas have decreased in the Thai gillnet and in the Malaysian troll, purse-seine, and gillnet fisheries.

Although specific studies to identify the relationship of these fisheries to each other have not been undertaken, one obvious hypothesis is that the Thai purse seine fishery is having an impact on the other types of fisheries. Catch rates of small tunas by the Thai purse-seine fishery increased from 11 kg/day in 1973 to 522 kg/day in 1988; catch rates of other gears showed a general decline.

14.3 Importance of Interactions for Fisheries Management

A critical management issue that needs to be addressed is whether the Thai and Malaysian fisheries are harvesting the same stocks. If studies should show homogeneity in the various species of small tunas, the problem of allocation will be of major importance, especially if overfishing is evident.

14.4 Methods for Studying Interactions

An indication of interactions between fisheries exploiting small tunas has been obtained from an analysis of the available catch-and-effort statistics from Thailand and Malaysia. The correlative approach to studying fisheries interactions may be the only method available to study interactions among small tuna fisheries of the South China Sea. More detailed catch-and-effort statistics obtained from the tuna-sampling programme for the Thai tuna purse-seine fishery may help better understand the relationship of this fishery with the others. Also, more detailed data on the species and size compositions of luring and one-boat purse seines are required to better assess the magnitude of incidental catches by these gears.

14.5 Information on Interactions

The preliminary analysis suggests that the Thai purse-seine fishery may be affecting the magnitude, species, and size compositions of small tuna catches made by Thai gillnet and Malaysian gillnet, troll, and purse-seine fisheries.

14.6   <u>Scientific Problems to be Resolved</u>

The study of interactions between fisheries for small tunas in the South China Sea is of lesser importance than the assessment of the status of these stocks. Available evidence suggests that the longtail tuna and kawakawa stocks are being affected by exploitation. Priority should be given to developing methods based on minimal data inputs for evaluating the status of these stocks.

14.7   <u>Recommendations</u>

The following recommendations were made; the list is not given in priority order:

a)   The catch-and-effort data generated by the tuna-sampling programmes in Thailand and Malaysia are essential to adequately assess the status of the stocks. The total landings of small tunas from the South China Sea are still increasing, but evidence indicates that some fisheries and some stocks are being affected by exploitation. Hence, there is a need to continue monitoring the small tuna fisheries of Malaysia and Thailand in order to establish a data base adequate for an assessment of the small tuna stocks.

b)   Additional biological information is needed on the small tunas if the various methods that have been developed to study interactions are to be used.

14.8   <u>Institutional Arrangements</u>

The tuna-sampling programmes which would generate the required information for small tunas will be carried out by the Marine Fisheries Department of Thailand and the Department of Fisheries of Malaysia, with assistance from the Indo-Pacific Tuna Development and Management Programme (IPTP).

15.   INFORMAL SESSIONS

During the Consultation several informal sessions were organized to discuss topics of interest to participants. Included were southeast Asian fisheries interactions, eastern Pacific Ocean fisheries interactions, Pacific Islands perspectives, and E-mail. The following sections summarize these informal discussions.

15.1   <u>Southeast Asian Fisheries Interactions</u>

Mr. Mitsuo Yesaki introduced the topic and summarized the discussion that took place in the special session on southeast Asian fisheries interactions.

The Philippines and Indonesia are concerned about the effect increasing western Pacific skipjack and yellowfin tuna catch may have on the national fisheries for those species. Both species are important to those countries, and the purse-seine fisheries of distant water fishing nations (DWFN) are located in the general proximity. It was pointed out that a great deal of fisheries data exists, as does tagging information; both sets of data need to be analysed further. The tagging data from the Philippines experiment are apparently inadequate for assessing movement; a low tag-return rate of only 1.5% was noted. It was suggested that the tagging programme was focused on small fish and that high, intervening mortality is probably the reason that so few tags are returned. It was noted that only small and large yellowfin are captured in the Philippines, with no intermediate-sized fish present.

There is also concern in the Philippines that fisheries (ring-net or purse-seine) utilizing FADs are having negative impacts on small-scale fisheries. In many cases, the fisheries-interaction effects also require that social and economic issues be addressed. Some consideration is being given to closing areas within some distance from shore (perhaps 15 km) to industrial fisheries, but the data are inadequate to justify such action.

In Indonesia, good time-series data are available for the pole-and-line fisheries. FADs have been introduced to this fishery in recent years. There is interest to determine the effect FADs have had on this fishery and also on small-scale fisheries in neighbouring areas. The distances over which FADs exert an attractive force should be determined in order to assess the potential for interaction. Studies on the number of FADs that can effectively be used is important to local areas to make optimal use of this technology.

In Malaysia, there is concern about the interactions of small-tuna fisheries because the stocks of small tunas are spatially-restricted. The statistics on catch and CPUE from the tuna-sampling programme in Thailand have not been used for analysis of fisheries interaction because of their short duration. Catch trends suggest an interaction between the Thai purse seiners and the traditional troll boats.

The general recommendations from this session included (1) stronger government support to allow the use of newer techniques to address these problems, (2) improved statistics throughout the region, and (3) resolution of stock boundaries.

15.2   Eastern Pacific Ocean Fisheries Interactions

Mr. Guillermo Compean described several issues in the eastern Pacific Ocean purse-seine fisheries for yellowfin and skipjack tuna. These fisheries are changing in terms of the composition of the fleet, as well as the components of the stock fished. Two different size groups are being taken: older, larger yellowfin in porpoise sets, and younger yellowfin and skipjack in other types of sets. There is also a distinct spatial component of the fishery that may result in interactions. A variety of research is needed to study the interactions of all components of the eastern Pacific Ocean (EPO) fleet. Dr. Deriso pointed out that many aspects of this research are currently being carried out by the Inter-American Tropical Tuna Commission (IATTC), which continues to conduct research on the spatial and temporal aspects of tunas and their fisheries in the EPO. The general simulation model being developed by IATTC will address the effects of how changes in one component of the fishery will affect another, particularly relating to different size or age groups of fish. It was concluded that all interested parties should discuss the research needed to address fisheries issues in the EPO and to develop research and appropriate methodologies, while insuring that duplication of effort is kept to a minimum.

15.3 <u>Pacific Island Perspectives</u>

Mr. Noah Idechong discussed the results of an informal session held to discuss Pacific Island perceptions of tuna-interaction issues, and to identify areas deserving of consideration under the Consultation. At an earlier meeting involving a wider group (representatives of developing countries), participants were asked to identify interaction issues in their area. A variety of interaction concerns was identified by the group, with emphasis on artisanal/industrial interactions and spatial-exclusion concerns associated with protecting domestic fisheries. The need to continue improving area-coverage and quality of data was also emphasized.

Noting these concerns, the informal session identified three avenues for further action. These were:

a) Undertake a representative case study in a Pacific-island situation where artisanal, domestic-commercial, and DWFN industrial fleets operate in proximity, and where reasonable data-coverage of all fisheries is available. The case study would include biological and economic analyses.

b) Continue development of general models to analyse fisheries interactions, to further the progress made so far with a movement model.

c) Assemble historical data from purse-seine fleets; increased coverage provided by these data will facilitate interaction studies.

Mr. Idechong noted that the above actions, will provide representative benefits to Pacific island countries in terms of enhancing their capability to address interaction issues, with the assistance and support of regional organisations (South Pacific Commission and Forum Fisheries Agency).

The subsequent discussion considered the best candidates for such a representative case study. Although the group identified no specific case study, it was noted that Palau and Kiribati were candidates because of the characteristics of their fisheries and availability of past tagging data. In the case of Kiribati, for example, over 12,000 fish have been tagged in the past year by the SPC's Regional Tuna Tagging Programme (RTTP). Some concern was voiced about the problems of non-reporting in artisanal fisheries, but this problem has been improved with better publicity of programmes; it has been evident in the past year that these efforts by the RTTP are paying off.

Questions were also raised about the adequacy of existing methodology, such as the Sibert/Fournier model, to address the interactions in such an island setting. It was revealed that the current modelling effort has been directed at skipjack tuna, and that modification to consider yellowfin tuna in these island settings would involve refinements to the existing model. It was also pointed out that an interaction study has been proposed for the domestic-pelagic fisheries around the main Hawaiian islands. Although the types of fisheries in the latter case are less diverse than those desired for the study outlined above, it was noted that this study and its methods should be coordinated with any subsequent study developed around other Pacific islands.

15.4 <u>E-mail</u>

Dr. Kleiber chaired the special session on e-mail, convened to introduce participants to electronic communication and data-file transfer. He discussed the systems currently in use, both commercial (fee-based) and organisational (often free or charging a nominal connection fee). He pointed out that Mr. Les Allinson of the Forum Fisheries Agency is developing a proposal to have an e-mail and data-transfer system specifically tailored to fisheries and fisheries officers in the Pacific; the proposed system would utilize the PEACESAT system. It was noted that the timetable for the PEACESAT proposal implementation is uncertain. It is possible however, to use PEACESAT individually at present, for voice and data transmission at many locations throughout the Pacific. A coordinated system targetted directly at fisheries is the intent of the proposal, and this will take at least several months to implement. It was noted that there is no charge for the use of PEACESAT lines.

Dr. Kleiber noted that a good example of an e-mail system that goes to relatively-undeveloped regions is the ORSTOM system. Dr. Fonteneau stated that ORSTOM would gladly assist in developing this type of system. ORSTOM experts can help with development problems. This assistance, as well as providing access to the current system, would probably be free in countries where ORSTOM is currently based. In other countries, there could be assistance given, but probably at some cost. All telephone-line costs within the ORSTOM e-mail system are charged to each user, but these are minimal because of the reduced costs of data transmission. Dr. Kleiber suggested that it would be useful to widely distribute the document describing the ORSTOM e-mail system, and suggested that each TUNET member should receive a copy.

16. GENERAL DISCUSSION

This session was convened to stimulate a general discussion among the participants and to elicit thoughts and observations arising from the preceding sessions. As such, the comments in the following text do not necessarily represent the consensus opinion of the participants. The session began with a series of general observations by the Chair Mr. Richard Shomura. He noted that, although he was concerned with the limited scope of discussion during some of the sessions, he was pleased with the written material provided by the working groups. As an example, he noted that little time was spent on fisheries interactions where the available data are lacking or only partially available, such as in the Philippines. He also noted that the general sessions spent too little time on anticipated future interactions. There is a lot taking place in tuna-resource development in the Pacific, and he stated that this group is perhaps best suited to anticipate data needs and potential interaction problems. It was noted that fisheries interaction problems in the future will most likely occur in those stocks that are currently exploited at a near-optimal level. The assumption is that these fisheries will eventually be subject to over-exploitation, so potential interaction problems should be explored before that occurs.

Mr. Shomura noted that in any large group with diverse scientific disciplines and expertise, developing a dialogue is difficult due to differing interest levels in the range of topics at hand. He suggested that this may have hindered discussions at this Consultation. As an example, he pointed out that modellers should recognize that many participants were not capable of evaluating the mathematics of the models and that they should therefore be careful to communicate their findings and approach in a language easily understood. Likewise, biologists should take great pains to effectively communicate their results to modellers to ensure the maximum use of biological or fisheries data.

In addition to improving communication between biologists and modellers, it was pointed out that good fisheries management adopts a comprehensive point of view, not simply concentrating on any component parts; informed decisions require a variety of input. One must deal with the "trade-off" of gaining further types of information and determining whether the added cost is justified. This should be a common approach in deciding where to put limited research funds. If the benefits are small relative to the research costs, we should carefully examine the need for that research. The consensus on this issue seemed to be that informed decisions require input from a range of disciplines and expertise and that communication among the people involved is therefore crucial.

Mr. Shomura covered several issues related to methodology. Most modelling work requires movement data (typically from tagging), catch-and- effort data, and biological data. He noted specifically that despite 40 years of research and data gathering, tuna scientists still seek improvement in these sources of information.

In tagging data, there remain problems with tagging mortality, tag shedding, and non-reporting and/or incomplete reporting (returning tags with incomplete recapture data). It was pointed out that methods which do not rely on some level of reporting are of real importance to understanding the interaction problem. Although this generated some dissenting views, most agreed that modelling efforts must estimate non-reporting in some manner, and take into account the impacts of this problem. Various means of minimizing the problem were discussed in detail, including tag seeding, improved port sampling, rewards, and observer coverage. It was also pointed out that the improvements in tagging over the last decade in the Pacific have been significant. The group was also reminded that alternative methods for assessing movement need to be discussed, including archival tags, remote-sensing methodology, and other high-technology applications that may advance the quality and quantity of information presently generated in tagging studies. Finally, the difference between estimation and forecasting of tuna movements was noted. Tagging experiments are good for estimating movement, but one can question their use for forecasting it; the issue is extrapolation beyond one's data, and this will apply to generalities about interactions, as well. If exploitation increases, are data on movement patterns collected earlier still useful? In the eastern Pacific Ocean, experience has shown that tagging experiments in different years can give contrasting estimates of movement, resulting in a bias depending upon when the work was done. The need to understand the biological rationale for movement is important, *i.e.*, to answer the question "why do the fish move?" Understanding the causes of variation in movement is inherently important, and must be related to physical, chemical, and biological conditions in the ocean.

Fisheries catch and effort data were then discussed, and the continued need for improved data was identified. In the Pacific, there is no ocean-wide database that insures that good catch-and-effort data are available. Development of methods to standardize effort based on gear type, fishing power, effect of FADs, has lagged, possibly due, at least partly, to lack of such a data base. The overall improvement of measures of effort was briefly discussed, and all agreed that such improvements are needed. It was suggested that a common data apparatus would help solve many data- related problems; this has been effective in other oceans.

In the area of biological data, tuna scientists have been collecting these for years, but even many basic types of information are not available in some of the areas where they are needed. Current data on maturation, spawning, growth, and natural mortality in tunas are insufficient in some areas. In terms of stock structure, morphometric and meristic data have been studied for years and are now being reconsidered, indicating the importance of spatial structure to the modelling efforts currently underway.

Models of fisheries interaction integrate all of these sources of information. General simulation models are quite useful, so that "what-if" questions can be asked, particularly as they relate to impacts on the stock. The question of fleet dynamics, however, must also be addressed. If fleets are dislocated by management, potential changes in interaction will likely occur and should be taken into account.

A final issue raised concerned interaction effects near islands as a subset of the more global interaction effects. In the Philippines, for example, local and distant water fishery interactions must be considered; results obtained there may be applicable to other locations. The Philippines situation may not be unique if the same range of sampling types was used in other island areas, but the apparent magnitude of yellowfin tuna nursery grounds in the Philippines is probably unique. In Hawaii, the longline fishery takes large yellowfin that are not evident in many other island areas. The lack of large yellowfin tuna near other islands does not necessarily mean that they are not there, but perhaps that they are not subject to a fishery. Finally, compared to the open ocean, nearshore waters of islands appear to be characterized by high levels of spawning by several species of tunas, *e.g.*, yellowfin. Additional biological research will be needed to be incorporated into any interaction studies around islands. It was agreed that several appropriate sites may exist for local-interaction studies. The local interactions issue has an additional benefit in that adaptive management approaches may be used; manipulative experiments can address many questions about fisheries interaction. Research on local interaction studies may assist progress on larger-scale interactions.

17.   RECOMMENDATIONS

The Consultation reviewed a summary of the technical recommendations made in the session reports and a general recommendation of possible follow-up action.

17.1   <u>Technical Recommendations</u>

During the course of the Consultation, extensive discussions were held on "state of the art" methodologies to study tuna interactions. It became apparent that substantially more work needs to be done to develop models, especially simulation models, that incorporate fisheries information, resource assessment information, and biological information.

The specific recommendations resulting from the Consultation can be found in the appropriate sections of the session reports. The majority of these specific recommendations can be grouped into six major categories as follows:

17.1.1　　　　Methodology

Effort should be increased in developing models, especially simulation models, that integrate fisheries statistics, movement information, assessment information, and biological information.

17.1.2　　　　Resource assessment

Effort should be increased to obtain resource assessment information which can be incorporated into tuna interaction studies. The list includes horizontal and vertical spatial movements and age-structure information.

17.1.3　　　　Fisheries statistics

Effort should be drastically increased to improve the quality and quantity of fisheries statistics needed to understand tuna interaction issues and the tuna resources.

17.1.4　　　　Tagging

Effort should be increased while carrying out tagging experiments to provide information on tuna movement and population assessment studies. An important aspect of these studies should be the defining of statistical properties of the tagging results.

17.1.5　　　　Biological and ecological

Effort should be increased to undertake biological and ecological studies of tunas that relate to the interaction issues. Included in the needed research is some basic information on life history parameters, *e.g.*, age and growth, maturation and spawning, mortality and time-and-space movements. Particular attention should be paid to the tuna-environment relationship as it affects all aspects of tuna science.

17.1.6　　　　Management

The "decision theory" methodology (Medley, this document) should be considered in fisheries management where choices need to made in addressing interaction issues.

17.2　General Recommendations

The Consultation recognized the importance of tuna-fisheries interactions, particularly for developing countries, and the possible funding limitations to adequately address all fisheries-interaction issues. The Consultation recommended that:

a)　　FAO review and update guidelines for proposals to seek funding for interaction studies and distribute these through the TUNET network.

b)　　FAO develop procedures for the review and establishment of priorities for research proposals related to studies of interaction issues.

c)   FAO ensure that adequate funding is available to conduct approved research proposals on interaction issues.

d)   If funds available through the FAO Trust Fund Project (Cooperative Research on Interaction of Pacific Tuna Fisheries) are insufficient to implement projects judged worthy of funding, FAO should seek additional resources adequate to meet identified needs.

18.   CLOSING SESSION

Before the Consultation was closed during the afternoon of 11 December 1991, Dr. Majkowski thanked SPC and ORSTOM for hosting the Consultation and thanked all individuals for their active participation and contribution during the Consultation. Mr. Shomura in closing the Consultation extended his personal thanks to participants for bearing with him as Chair and concluding, in his opinion, a very successful and fruitful meeting.

49

APPENDIX A. Provisional Agenda.

FAO EXPERT CONSULTATION ON INTERACTIONS
OF PACIFIC TUNA FISHERIES

Noumea, New Caledonia
3-11 December 1991

Provisional Agenda

3 December 1991 (Tuesday)

08.30 - 09.00	Registration (SPC conference room foyer)
09.00 - 10.00	Preliminaries (SPC conference room)
10.30 - 12.00	Opening of the Consultation (ORSTOM)
12.00 - 01.30	Lunch break
Afternoon	Types of Tuna Fisheries Interaction and Related Scientific Problems
	Scientific Methods for Studying Tuna Fisheries Interactions
18.30 - 20.30	Cocktail Party

4 December 1991 (Wednesday)

Morning	Scientific Methods for Studying Fisheries Interactions (cont.)
Afternoon	Interactions of Pacific Skipjack Fisheries
Evening	Informal discussions

5 December 1991 (Thursday)

Morning	Interactions of Eastern Pacific Yellowfin Tuna Fisheries
Afternoon	Interactions of Western Pacific Yellowfin Tuna Fisheries
Evening	Informal discussions

6 December 1991 (Friday)

Morning	Interactions of North Pacific Albacore Fisheries
Morning/Afternoon	Interactions of South Pacific Albacore Fisheries
Afternoon	Interactions of Pacific Bigeye Tuna (PBT) Fisheries
Evening	Informal discussions

7 December 1991 (Saturday)

All day	Additional sessions if required, or excursions and free time.

8 December 1991 (Sunday)

Morning	Demonstration of Computer Software Related to Studying Tuna Fisheries Interactions (Informal Session)
Afternoon	Scientific Methods for Studying Fisheries Interactions (Informal Session)

9 December 1991 (Monday)

Morning	Interactions of Pacific Northern Bluefin Tuna Fisheries
Morning/Afternoon	Interactions of Southern Bluefin Tuna Fisheries
Afternoon	Session on Interactions of Pacific Small Tuna Fisheries
Evening	Informal discussions

10 December 1991 (Tuesday)

Morning	Reports from Informal Sessions and General Discussions
Afternoon	Formulation of Recommendations
19.00 - 22.00	Barbecue at SPC

11 December 1991 (Wednesday)

Morning	Additional session if required
Afternoon	Distribution of the Report
	Summary of the Consultation by the Consultation and Session Chairmen

APPENDIX B. List of Participants.

FAO EXPERT CONSULTATION ON INTERACTIONS
OF PACIFIC TUNA FISHERIES

Noumea, New Caledonia
3-11 December 1991

List of Participants
(Current known address as of October 1993)

Ms. Flerida Arce
Bureau of Fisheries & Aquatic Resources
Department of Agriculture
860 Quezon Avenue, Quezon City
Metro Manila 3008 PHILIPPINES
Tel: (632) 965498/965428
Fax: (632) 988517
Tlx: 23149/64111

Dr. Norman Bartoo
Southwest Fisheries Science Center
National Marine Fisheries Service
La Jolla, California 92038 USA
Tel: (619) 5467000
Fax: (619) 54677003
Tlx: 3734386 UC WWD SIO SDG (USA)
E-Mail net: OMNET
E-Mail add: N.BARTOO

Mr. Noel Barut
Bureau of Fisheries & Aquatic Resources
Research Division
860 Quezon Avenue, Quezon City
PHILIPPINES
Tel: (632) 991249/965428
Fax: (632) 988517

Dr. William Bayliff
Inter-American Tropical Tuna Commission
8604 La Jolla Shores Drive
La Jolla, California 92037 USA
Tel: (619) 5467025
Fax: (619) 5467133
Tlx: 697 115 TUNACOM
E-Mail net: BITNET
E-Mail add: WBAYLIFF 8604 UCSD

Dr. George Boehlert
Southwest Fisheries Science Center
National Marine Fisheries Service
P.O. Box 831
Monterey, California 93942 USA
Tel: (408) 656-4689 or 656-3311
Fax: (408) 656-3319
E-Mail net: OMNET
E-Mail add: G.BOEHLERT,
Internet gboehler@honlab.nmfs.hawaii.edu

Dr. Chris Boggs
Honolulu Laboratory
Southwest Fisheries Science Center
National Marine Fisheries Service
2570 Dole Street
Honolulu, Hawaii 96822-2396 USA
Tel: (808) 943-1222
Fax: (808) 943-1290
E-Mail net: BITNET
E-Mail add: T110700@UHCCMVS

Mr. John Bungitak
Marshall Islands Marine Resources
  Authority (MIMRA)
Ministry of Resources & Development
P.O. Box 860
Majuro, MARSHALL ISLANDS 96960
Tel: (692) 3262, 3202, 3206
Fax: (692) 3685
Tlx: 730-0927

Mr. Albert Caton
Bureau of Rural Resources
Department of Primary Industries &
 Energy
P.O. Box E11 Queen Victoria Tce
Parkes ACT 2600 AUSTRALIA
Tel: (616) 272-5287
Fax: (616) 272-4013

Mr. Atilio Coan
Southwest Fisheries Science Center
National Marine Fisheries Service
P.O. Box 271
La Jolla, California 92038 USA
Tel: (619) 546-7000
Fax: (619) 546-7003
Tlx: 3734386
E-Mail net: SCIENCEnet
E-Mail add: A.Coan

Mr. Guillermo Compean
Universidad Autonoma de Nuevo Leon
Facultad de Ciencias Biologicas
A.P. 16-F
Ciudad Universitaria
66450 San Nicolas de Los Garza
N.L. MEXICO
Tel: (52 83) 522139
Fax: (52 83) 762813

Dr. Richard Deriso
Inter-American Tropical Tuna Commission
8604 La Jolla Shores Drive
La Jolla, California 92037 USA
Tel: (619) 546-7100
Fax: (619) 546-7133
Tlx: 697 115 TUNACOM
E-Mail net: BITNET
E-Mail add: RDERISO@UCSD.EDU

Mr. Sylvester Diake
Fisheries Division
Ministry of Natural Resources
P.O. Box G24
Honiara, SOLOMON ISLANDS
Tel: (677) 30107
Fax: (677) 30256
Tlx: 66465 SOLFISH

Ms. Chee Phaik Ean
Fisheries Research Institute
Department of Fisheries
Ministry of Agriculture
11700 Glugor
Penang, MALAYSIA
Tel: (04) 872777, 873150
Fax: (604) 876388
Tlx: MA 28157

Dr. Alain Fonteneau
Institut Senegalais de Recherches
 Oceanographiques
Centre de Recherches Oceanographiques
B.P. 2241, Dakar, SENEGAL
Tel: (32) 340534/340462
Fax: (221) 324307 ORSTOM DAKAR
Tlx: 468 ORSTOM SG
E-Mail net: FNET
E-Mail add: FONTENEA%ORSTOM.FR

Dr. David Fournier
Otter Research Ltd.
Box 265, Station A
Nanaimo, B.C. V9R 5K9 CANADA
Tel: (604) 756-0956
Fax: (604) 2482531
E-Mail net: INTERNET
E-Mail add: 72730
223@COMPUSERVE.COM

Mr. Reuben Ganaden
Bureau of Fisheries & Aquatic
Resources, Department of Agriculture
Arcadia Building
860 Quezon Avenue, Quezon City
3008 PHILIPPINES
Tel: (632) 991249/987075
Fax: (632) 988517/965444

Ms. Sofia Ortega Garcia
Centro Interdisciplinario de
 Ciencias Marinas
Apartado Postal 592
La Paz, Baja California Sur
MEXICO
Tel: (91 682) 25366/25322/25344
Fax: (91 682) 25322

Mr. Robert Gillett
Ministry of Fisheries
P.O. Box 871
Nuku'alofa, KINGDOM OF TONGA
Tel: (676) 21399
Fax: (676) 23891

Dr. John Hampton
South Pacific Commission
B.P.D5, Noumea, NEW CALEDONIA
Tel: (687) 262000
Fax: (687) 263818
Tlx: SOPACOM 3139 NM
E-Mail net: BITNET
E-Mail add: DASNET, USER,
    MP 1 SH 66/USA

Mr. Victor Gomez Munoz
Centro Interdisciplinario
 de Ciencias Marinas
Apartado, Postal 592
La Paz, Baja California Sur
MEXICO
Tel: (682) 25322/25344/25366
Fax: (682) 25322

Dr. Bill Hearn
Division of Fisheries
CSIRO Marine Laboratories
GPO Box 1538
Hobart, Tasmania 7001 AUSTRALIA
Tel: (6102) 20-6222
Fax: (6102) 24-0530

Mr. Noah Idechong
Chief Marine Resources Division
Ministry of National Resources
P.O. Box 100
Koror, PALAU 96940
Tel: (680 9) 488-2266
Fax: (680 9) 488-1475, 488-1725

Mr. Yoshio Ishizuka
National Research Institute of Far Seas
 Fisheries
7-1, Orido 5-Chome
Shimizu-shi 424 JAPAN
Tel: (0543) 340715
Fax: (0543) 359642
Tlx: 03965689FARSEA J
E-Mail net: OMNET
E-Mail add: J.SUZUKI

Mr. Erwan Josse
ORSTOM
B.P. 529
Papeete, Tahiti
FRENCH POLYNESIA
Tel: (689) 43 98 87
Fax: (689) 42 95 55
E-Mail net: FNET
E-Mail add: ORSTOM.ORSTOM.FR
 JOSSE@

Dr. Pierre Kleiber
Southwest Fisheries Science Center
National Marine Fisheries Service
P.O. Box 271
La Jolla, California 92038 USA
Tel: (619) 546-7076
Fax: (619) 546-7003
Tlx: SOPACOM 3139 NM
E-Mail: PKLEIBER@UCSD.EDU

Dr. Marc Labelle
South Pacific Commission
B.P. D5
Noumea Cedex
NEW CALEDONIA
Tel: (687) 262-000
Fax: (687) 263-818
Tlx: SOPACOM 3139 NM
E-Mail net: bitnet
E-Mail add: DANET.USER/MEP.1
SM66/USA

Mr. Philip Langford
Department of Marine and Wildlife
 Resources
P.O. Box 3730
Pago Pago, AMERICAN SAMOA 96799
Tel: (684) 633-4456
Fax: (684) 633-5944

Mr. Viliami Langi
RDA International
P.O. Box 83
Vavau, KINDGOM OF TONGA
Tel: (676) 70055
Fax: (676) 70130

Dr. Antony D. Lewis
Tuna and Billfish Assessment Programme
South Pacific Commission
B.P. D5
Noumea Cedex, NEW CALEDONIA
Tel: (687) 262000
Fax: (687) 263818
Tlx: SOPACOM 3139 NM

Ms. Dorothy Lowman
Western Pacific Regional Fishery
 Management Council
1164 Bishop Street #1405
Honolulu, Hawaii 96813 USA
Tel: (808) 523-1368
Fax: (808) 526-0824

Dr. Jacek Majkowski
NF 512, FIRM
Food and Agriculture Organization of the
 United Nations
Via delle Terme di Caracalla
00100 Rome ITALY
Tel: (396) 52256656
Fax: (396) 52253020 or 52256656
Tlx: 600181 FAO I
E-Mail net: INTERNET
E-Mail add: Jacek.Majkowski@
 FAO.ORG@INTERNET

Dr. Naozumi Miyabe
National Research Institute of Far Seas
 Fisheries
7-1, Orido 5-Chome
Shimizu-shi 424 JAPAN
Tel: (543) 340715
Fax: (543) 359642
Tlx: 03965689FARSEAJ

Mr. Arturo Muhlia-Melo
Biological Research Center
KM 2 Carr San Juan de La Costa
EL COMITAN Apdo Postal 128
La Paz, Baja Califonia Sur
MEXICO CP 23000
Tel: 52 (682) 53633/53626
Fax: 52 (682) 53625
Tlx: CIBPME 52223

Dr. Paul Medley
Fisheries Department
Turks & Caicos Islands
WEST INDIES
Tel/Fax: (809) 946 2970

Dr. Talbot Murray
Research Centre
MAFF Fisheries
P.O. Box 287
Wellington, NEW ZEALAND
Tel: (644) 386-1029
Fax: (644) 386-1299
Tlx: MAFFCC NZ 30049
E-Mail net: USENET
E-Mail add: TEM%FRCMAFGOVTNZ

Dr. Nurzali Naamin
Research Institute for Marine Fisheries
Komplek Pelabuhan
Perikanan Samudera
JI, Muara Baru Ujing
Jakarta Utara INDONESIA
Fax: (6221) 570-9158

Mr. Paul Nichols
Forum Fisheries Agency
P.O. Box 629
Honiara, SOLOMON ISLANDS
Tel: (677) 21124
Fax: (677) 23995

Dr. Yeong Chull Park
Deep-Sea Resources Division
National Fisheries Research &
 Development Agency
Shirang-ri, Kijang-up, Yangsan-gun
Kyoungsangnam-do 626-900
REPUBLIC OF KOREA
Tel: (051) 465 0091
    (0523) 361 8052
Fax: (0523) 361 8076
Tlx: K 52647

Dr. Tom Polacheck
Division of Fisheries
CSIRO Marine Laboratories
G.P.O. Box 1538
Hobart, Tasmania 7001, AUSTRALIA
Tel: (61) 02 206222
Fax: (61) 02 240530
Tlx: AA 57182

Mr. Renaud Pianet
ORSTOM
B.P. A5
Noumea Cedex
NEW CALEDONIA
Tel: (687) 261000
Fax: (687) 264326
Tlx: ORSTOM 3193 NM
E-Mail net: FNET
E-Mail add: IANET%ORSTOM.FR

Mr. Andrew Richards
Research & Survey Branch
Department of Fisheries and
  Marine Resources
P.O. Box 165
Konedobu, PAPUA NEW GUINEA
Tel: (675) 214522
Fax: (675) 214369

Mr. Carlos Salvadó
Southwest Fisheries Science Center
National Marine Fisheries Service
P.O. Box 271
La Jolla, CA 92038 USA
Tel: (619) 546-7000
Fax: (619) 546-7003
Tlx: 3734386 UC WWD STO SDG (USA)
E-Mail: CSALVADO@UCSD.EDU

Mr. Richard Shomura
School of Ocean and Earth Science
Hawaii Institute of Marine Biology
University of Hawaii, Manoa
1000 Pope Road
Honolulu, Hawaii 96813 USA
Tel: (808) 956-6715
Fax: (808) 956-6751

Dr. John Sibert
Pelagic Fisheries Research Program
JIMAR, University of Hawaii
1000 Pope Road
Honolulu, Hawaii 96822 USA
Tel: (808) 856-4109
Fax: (808) 956-4104
E-Mail net: INTERNET
E-Mail add: jsibert@soest.hawaii.edu.

Dr. Ziro Suzuki
National Research Institute
of Far Seas Fisheries
Shimizu-shi 424 JAPAN
Tel: (543) 340715
Fax: (543) 359642
Tlx: 03965689FARSEAJ
E-Mail net: OMNET
E-Mail add: J.SUZUKI

Mr. Jacobus C.B. Uktolseja
Tuna Research and Programme Division
Research Institute for Marine Fisheries
J1. Muara Baru Ujung
Komplek Pelabuhan Perikanan Samudera
(Jakarta Fishing Port)
Jakarta 14440 INDONESIA
Tel: (6221) 6602044
Fax: (6221) 661-2137

Mr. Hugh Walton
South Pacific Commission
B.P.D5
Noumea, NEW CALEDONIA
Tel: (687) 262000
Fax: (687) 263818
Tlx: SOPACOM 3139 NM

Mr. Peter Ward
Fisheries Resources Branch
Bureau of Rural Resources
P.O. Box E11
Queen Victoria Terrace
Parkes ACT 2600 AUSTRALIA
Tel: (61) 6272 5534
Fax: (61) 6272 4014

Dr. Alex Wild
Inter-American Tropical Tuna Commission
8604 La Jolla Shores Drive
La Jolla, California 92037 USA
Tel: (619) 546-7100
Fax: (619) 546-7133
Tlx: 697 115 TUNACOM
E-Mail net: BITNET
E-Mail add: AWILD%UCSD.EDU

Mr. Mitsuo Yesaki
6180 Steveston Highway
Richmond, B.C. CANADA V7E 2K8

APPENDIX C. Organisation of Consultation.

FAO EXPERT CONSULTATION ON INTERACTIONS
OF PACIFIC TUNA FISHERIES

Noumea, New Caledonia
3-11 December 1991

Organisation of Consultation

Consultation Chair:	Mr. Richard S. Shomura Hawaii Institute of Marine Biology School of Ocean and Earth Science and Technology University of Hawaii, Manoa
Vice-Chair:	Dr. Ziro Suzuki National Research Institute of Far Seas Fisheries Japan Fisheries Agency
Convenor and Technical Secretary:	Dr. Jacek Majkowski Marine Resources Services Fishery Resources and Environment Division Food and Agriculture Organization of the United Nations
Consultation Secretaries	Mr. Robert Gillett FAO/UNDP South Pacific Regional Fisheries Programme Mr. Hugh Walton FAO/UNDP South Pacific Regional Fisheries Programme
Local Organisers:	Dr. Antony Lewis Tuna and Billfish Assessment Programme South Pacific Commission Mr. Renaud Pianet Institut Francais de Recherche pour le Developpement en Cooperation (ORSTOM)

APPENDIX D. List of Documents.

FAO EXPERT CONSULTATION ON INTERACTIONS
OF PACIFIC TUNA FISHERIES

Noumea, New Caledonia
3-11 December 1991

List of Documents

- Evaluation of albacore tuna fishery interaction by use of a simulation model (P. Kleiber and B. Baker)

- Interactions between two fisheries: a tag- recapture method for estimating the effects on catches of changing fishing intensity (W. Hearn and A. Mazanov)

- Interaction between tuna fisheries in the Federated States of Micronesia (J. Sibert)

- A review of skipjack tuna biology and fisheries in the Pacific Ocean (A. Wild and J. Hampton)

- A review of yellowfin tuna biology and fisheries in the eastern Pacific Ocean (A. Wild)

- A review of the biology and fisheries for yellowfin tuna, *Thunnus albacares*, in the western and central Pacific Ocean (Z. Suzuki)

- A synopsis of the biology and fisheries for North Pacific albacore, *Thunnus alalunga* (N. Bartoo and T. Foreman)

- A review of the biology and fisheries of albacore tuna in the southern Pacific Ocean (T. Murray)

- General review for Pacific bigeye tuna (N. Miyabe)

- A review of information on the biology of northern bluefin tuna, *Thunnus thynnus*, and the fisheries for this species in the Pacific Ocean (W. Bayliff)

- Review of aspects of southern bluefin tuna biology, populations, and fisheries (A.E. Caton, editor)

- A review of the *Auxis* fisheries of the Philippines and some aspects of the biology of frigate (*A. thazard*) and bullet (*A. rochei*) tunas in the Indo-Pacific region (M. Yesaki and F. Arce)

- A review of the biology and fisheries for longtail tuna (*Thunnus tonggol*) in the Indo-Pacific region (M. Yesaki)

- A review of the biology and fisheries for kawakawa (*Euthynnus affinis*) in the Indo-Pacific region (M. Yesaki)

- A tag release-recovery method for predicting the effect of changing the catch of one component of a fishery upon the remaining components (J. Majkowski, W. Hearn and R. Sandland)

- Population field theory for tag analysis and fisheries modeling: the empirical propagator (C. Salvadó)

- Status of Korean tuna longline and purse-seine fisheries in the Pacific Ocean (Y.C. Park, W.S. Yang and T.I. Kim)

- The Mexican tuna fishery (A. Muhlia-Melo)

- Relationship between sea surface temperature and dolphin-associated fishing activities by the Mexican tuna fleet (V.M. Gomez-Monoz and S. Ortega-Garcia)

- Types of fishery interaction in the Pacific Ocean and methods of assessing interaction (P. Kleiber)

- Discrete population field theory for modeling fisheries interaction based on tagging data (C. Salvadó)

- Assessment of skipjack fishery interaction in the eastern tropical Atlantic using tagging data (P. Kleiber, and A. Fonteneau)

- Evaluation of diffusion advection equations for the estimation of movement patterns from tag-recapture data (J. Sibert and D. Fournier)

- Methods of analysing interaction in Hawaii's pelagic fisheries (C. Boggs)

- A review of skipjack fishery interaction in the western and central Pacific Ocean (J. Hampton)

- Skipjack tuna fishery interaction in the Solomon Islands: an example of small scale spatial interaction with FADs as an influencing factor (P. Kleiber and J. Hampton)

- A brief review of interaction between purse-seine and longline on yellowfin tuna *Thunnus albacares*, in the western and central Pacific Ocean (Z. Suzuki)

- U.S. distant water and artisanal fisheries for yellowfin tuna in the central and western Pacific (A. Coan)

- Taiwanese yellowfin fisheries in the Pacific Ocean (C. Wang)

- The interactions of yellowfin tuna fisheries in the eastern part of Indonesian waters (N. Naamin and E. Bahar)

- Estimating the impact of purse seine catches on longline (P. Medley)

- Evidence of interactions between high seas driftnet fisheries and the North American troll fishery for albacore (N. Bartoo and D. Holts)

- Examples of interactions among fisheries exploiting the southern Pacific albacore stock (T. Murray)

- Application of SSA method in South Pacific albacore, 1971-1989 (C. Wang)

- A brief analysis of fishery interaction between fisheries for Pacific bigeye tuna, *Thunnus obesus* in the Pacific Ocean (N. Miyabe)

- Interactions among fisheries for northern bluefin tuna, *Thunnus thynnus*, in the Pacific Ocean (W. Bayliff)

- Commercial and recreational components of the southern bluefin tuna (*Thunnus maccoyii*) fishery (A.E. Caton)

- Interactions among fisheries for southern bluefin tuna (T. Polacheck)

- Interactions between fisheries for small tunas off the South China Sea coast of Thailand and Malaysia (M. Yesaki)

- Skipjack pole-and-line fishing in east Indonesia: effect of location upon catch size and catch variability, with and without FADs, and other interactive effects (J.K. McElroy and J.C.B. Uktolseja)

- A review of tuna fishery interaction in the western and central Pacific Ocean (J. Hampton)

TYPES OF TUNA FISHERY INTERACTION IN THE PACIFIC OCEAN AND METHODS OF ASSESSING INTERACTION

Pierre Kleiber
Southwest Fisheries Science Center
National Marine Fisheries Service, NOAA
La Jolla, California 92038 U.S.A.

ABSTRACT

Fishery interaction can be classified by the causal mechanisms by which one fishery can affect another. This paper focuses on mechanisms of direct competition for the same fish stock. This type of interaction can occur between two fisheries operating in overlapping or separated fishing grounds and between fisheries harvesting the same or different size classes or life stages of the target species. Characteristics of tuna that pertain to the mechanisms of competitive interaction are their migratory nature, fast growth, and high fecundity. Relevant characteristics of the fisheries are the range of gear types and the geographic and temporal deployment of fleets.

A second way to classify fishery interaction is by the type of index used to measure it. There are proxy indices, and also indices that attempt to directly measure the impact of one fishery on another. The choice of index can affect the results of the assessment. A simple model presented in this paper gives an example of how with increasing effort in one fishery, the measured impact on another can increase in severity according to one type of index and can decrease in severity according to another.

Various strategies for estimating interaction indices are reviewed, including analysis of fishery statistics, analysis of tag data, and use of models. Analytical models applied to data such as fishery statistics and tagging data can be used to estimate relevant parameters of population, harvest and movement dynamics. Simulation models can be used to conduct experimental manipulations aimed at testing for fishery interaction — manipulations that are not feasible in the real world.

1. INTRODUCTION

The tuna fisheries in the Pacific are a complex mixture of several gear types and fleet characteristics. Gear types include hand line, troll, pole and line, purse seine, gill net, and longline. Fleet characteristics range from local artisanal fisheries to local industrial fisheries to large fleets from several distant water fishing nations (DWFNs). The characteristics of the tuna stocks are also complex. There are several economically important tuna species in the tropical and temperate waters of the Pacific. All are pelagic animals, roaming freely in the oceanic environment, but the nature of their wandering varies considerably from species to species.

In the central and western tropical Pacific, where islands are plentiful, the ocean surface is covered by a mosaic of economic zones of the island countries, most of which are home to small-scale, artisanal fisheries, and some of which have developed local, industrial-scale fleets. Superimposed on the mosaic and on the international waters are broad and partially overlapping fishing grounds of DWFN fleets. Most of the gear types listed above can be found in both local and DWFN fleets. Some fleets make extensive use of fish aggregating devices (FADs) which attract concentrations of tuna. Some fleets are aided in locating concentrations of tuna by aircraft, radar, sonar and other electronic gear, and for other fleets such assistance is unavailable or infeasible. The situation is rife with possibilities for interaction among various combinations of the different fisheries.

This paper is a review of the types of fishery interaction that might occur among tuna fisheries in the Pacific Ocean. The purpose is to set the stage for the work of the Expert Consultation on Interactions of Pacific Tuna Fisheries (hereafter "the consultation") in considering methods to assess fishery interaction and in investigating specific instances of actual or suspected fishery interaction. What constitutes a "fishery", we will purposefully leave vague. It could be a fleet, a management unit, an industrial unit, or even a sub-grouping within a fleet — any delineation of a group of fishermen who might be concerned about impacts on them resulting from activities of another group.

We will examine two basic ways to classify fishery interaction. One way considers the biological and fishery-related mechanisms by which one fishery may affect another. Because our focus is on tuna fisheries in the Pacific, we will review general features of tuna biology and behaviour, and of the Pacific tuna fisheries, and we will note those features that could help, or hinder, the effects of one fishery being visited upon another. Details of specific fish stocks, regions, and fisheries will be presented in other working papers.

The second classification of fishery interaction considers the type of metric we choose to assess interaction. To have an objective assessment, we need to measure something, and it turns out that there is a variety of possible types of interaction index to choose from. We shall see that different indices have different properties which in some cases can lead to different assessments of the seriousness of interaction. We will consider indices of interaction not only from the point of view of fisheries scientists but also of those who might make use of the results of an interaction assessment, for example, fisheries administrators making management decisions or conducting negotiations with other fisheries administrators.

The discussion of what to measure naturally leads to discussion of how to measure. We will review some of the methods that have been used. For an exhaustive review of methods, see Anonymous (1988).

2. CLASSIFICATION OF FISHERY INTERACTION MECHANISMS

The mechanisms of fishery interaction can be loosely categorized into impacts of one fishery on another mediated by effects on a common resource stock and impacts unrelated to the resource stock. We will dispense with the latter category quickly because it will probably get little attention in the consultation. It includes non-harvest competition such as competition in marketing and a variety of other types of interaction not mediated by effects on the resource population — gear interference, for example.

Resource-mediated interaction comprises direct competition between fisheries harvesting the same or overlapping resources and secondary effects of one fishery on another's resource. Secondary effects include impact by one fishery on the environment of the target species of a second fishery. For example, harvesting the food resource of another fishery's target species might have a significant effect on that species. Such an interaction is theoretically possible but is not known to be significant in tuna fisheries. Another secondary effect would be over-harvest of spawning stock (known as "recruitment over-fishing") by one fishery which could depress the resource of another fishery (and presumably of the first fishery as well). Except possibly for southern bluefin tuna, recruitment over-fishing has never been thoroughly documented in tuna fisheries.

It is the category of interaction through harvest competition that will likely draw most of our attention in this consultation, and from now on this is what we will mean by the unmodified term "competition". This is where one fishery harvests animals that would otherwise be destined to be harvested by another fishery. The fish stock in question does not need to be the principal target of both fisheries. The target stock of one fishery may well be by-catch of another, and that by-catch may have a significant impact on the stock, thereby affecting the fishery that targets it.

A general property of this kind of interaction is that the overall exploitation rate affects the severity of interaction. In the case of lightly exploited stocks, most fish are destined to die naturally. Therefore a particular fish caught in one fishery is unlikely to have been otherwise destined to be harvested by another fishery. Thus competition between fisheries for lightly exploited stocks would be less than for heavily exploited stocks.

It is not necessary for two fisheries to target the same size range or life stage of the fish for competitive interaction to occur. If fish are exposed to a series of fisheries as they grow in size or pass through stages of their life cycle, we can have consecutive interaction where fisheries harvesting early stages may have an effect on fisheries harvesting later stages, but where the converse is less likely to be true. An exception would be recruitment overfishing by a later-stage fishery, but that would be a secondary effect rather than direct competition in our classification. If two fisheries harvest the same size classes or life cycle stages, we have concurrent interaction where the effects can go both ways. Of course there can be intermediate situations where size distributions of the catch in different fisheries overlap, but are not completely congruent.

It is furthermore not necessary for competing fisheries to operate in the same geographic area. Certainly we would expect that fisheries could interact with each other if they operate in congruent or proximal fishing grounds (coincident interaction), but given that fish can migrate from one area to another or exchange diffusively between two regions, it is entirely possible for fisheries located at some distance from each other to interact competitively (remote interaction). With diffusive movement we would expect the severity of remote interaction to diminish with distance between the fisheries because the exchange rate between the fishing areas would decrease. However, with advective movement, that would not necessarily be so.

3. MECHANISMS OF TUNA FISHERY INTERACTION IN THE PACIFIC

Having laid out a general classification of interaction mechanisms, it is useful to review the characteristics of tuna and their fisheries in the Pacific as they pertain to those mechanisms. We will focus on competitive interaction which, as we have seen, can come in different varieties depending on the pattern and rate of fish movement between the fishing grounds and on whether the fisheries harvest the same or different life stages of the fish. The latter factor depends in turn on how vulnerability to different gears changes through the life cycle of the fish.

Individual tagged tuna have been known to migrate over long distances. Tag-recovery data for some tunas, such as skipjack, imply that the movement is diffusive in character, but some tunas, such as albacore, undergo extensive annual migration cycles that are highly advective, and others, such as bluefin, appear to conduct annual cycles of homing to localized spawning areas followed by dispersal over broad ocean regions. Given the migratory nature of these animals, we have the potential for remote interaction, perhaps over long distances.

On the other hand, the highly migratory nature of some tunas has been questioned (Hilborn and Sibert, 1986). Some tropical species appear to congregate near islands, reefs, and seamounts which occur in the tropical Pacific. The attraction to islands might mean that the bulk of these tuna populations does not move very much in spite of the odd individual that makes a long journey. Albacore, a temperate species, also congregate, not around islands, but on oceanic frontal structures that meander along the interface between temperate and sub-polar water masses (Laurs, 1983; Laurs and Lynn, 1977). Rather than inhibitors of movement, these fronts may serve as conduits for the annual trans-Pacific migrations of these fish. It is clear that various features of the environment (including FADs) modify tuna movement-behaviour, but the details are poorly known. Therefore there is a crucial need for more information about the movement-behaviour of the various tuna species and the effect of the environment on that movement.

We might expect with diffusive exchange of fish between two fishing grounds, that interaction would be balanced, with approximately the same potential impact in each direction. The actual impacts would depend on the fishing intensity in the two fisheries. This might be the case with two neighbouring fisheries having similarly sized fishing grounds. Another geographical arrangement would be a local island fishery operating near shore and a DWFN fishery operating offshore in a broad area surrounding the island. In this case we could expect the interaction to be unbalanced and perhaps more severe (for the local fishery) than is the case in neighbouring fisheries. Both these geographic arrangements exist in the Pacific, and evaluation of interaction in these situations is aided by knowledge of movement-behaviour of the fish.

Most tunas grow quickly from larval size to a size of a few kilograms, at which point they become vulnerable to gear specializing in smaller size classes. Then, as they continue to increase in size and perhaps change behaviour (swimming deeper, for example), many tunas become vulnerable to other fishing gear which may lead to consecutive interaction between a series of gear types. In some cases the fish can be exposed to a series of fisheries simply because the fisheries are located at different positions along the migratory path of the fish. Therefore, we may find cases of consecutive interaction between similar or different gear types located in distantly separated fishing grounds.

If we were managing a particular fishery and were concerned about potential impacts of other fisheries outside our area, a critical question would be the origin of the fish targeted by our fishery. The inputs to that stock can be immigrants that are exposed, or potentially exposed, to other fisheries before reaching our fishing grounds, or they can be local recruits for which the first exposure of their lives is to our fishery. Obviously, the greater the proportion of inputs due to immigration of already recruited fish, the greater our concern must be. Therefore knowledge of the geographic and seasonal pattern of recruitment would be useful for judging which instances of potential interaction might be significant. But that information is sketchy for most of the tuna species. We do know that the pattern of spawning is variable amongst the tuna species. Some spawn over broad areas, and some are localized. Therefore we can expect that the pattern of recruitment will also vary among species, and we may find instances of localized recruitment where one or a few fisheries have exclusive first access to each cohort, or other instances of recruitment over a broad area where first access to each cohort is shared by many fisheries.

The fast growth and high fecundity of some tuna species mean that there is a high turnover rate in the population. The fish have a high natural death rate, but they are also replaced at a high rate. This is a situation that can mitigate fishery interaction because for a given catch rate, a higher turnover means a lesser exploitation rate, which would lessen the degree of interaction.

4. CLASSIFICATION OF INTERACTION INDICES

In addition to the types of interaction mechanisms, it is useful for us to review the types of indicators that could be used to evaluate interaction. The choice of which indices we want to evaluate is critical. It affects the kind of data that we need to collect. It also affects the interpretations that we can safely draw from our results because different indicators have different properties.

One of the approaches that has been used to assess interaction is to look for cross-correlations between time series of various fishery statistics in two potentially competing fisheries. The statistical techniques could range from simple to complex. Such an interaction index, then, is a correlation coefficient or perhaps more sophisticated indicator of statistical relationship between various series of fishery data in the two fisheries.

Another index of potential interaction (Kleiber *et al.*, 1984) is a measure of the rate of input of animals to the area of one fishery (fishery-y) due to immigration from the area of another fishery (fishery-x). The index is the estimated proportion of the total rate of input (local recruitment plus total immigration) to fishery-y's area.

The above are really only proxy indices for the root measurement that needs to be made, which is how much the actual performance of one fishery is affected by the activity of another fishery. To develop such a direct index of interaction, we need to specify just what parameter we want as a measure of performance in the affected fishery and what parameter we want as a measure of activity in the affecting fishery. There is a wide choice ranging from strictly fishing-related parameters (such as catch, catch per effort, and effort) to economic parameters (such as profits) to socioeconomic parameters (such as employment or availability of protein). Some parameters (such as catch per effort) may be appropriate for performance, some (such as effort) may be appropriate for activity, and some (such as catch) may be appropriate for both.

In asking about the effect of fishery-x on fishery-y, we are presuming that performance of fishery-y is a function of activity in fishery-x. Fishery-y certainly also affects itself. Therefore we should be able to write

$$P_y = P_y(A_x, A_y, \ldots) \tag{1}$$

where P and A are respectively the parameters of performance and activity that we have chosen, and the subscripts, x and y, indicate the fishery the parameter refers to. The dots indicate that P_y, is a function of factors other than just current activity levels. We will neglect these other factors for the moment, but we should not forget them.

If we could measure function, P_y, at many different values of A_x, and A_y, it could be plotted as a 3-dimensional surface as in the contour plot in Figure 1. The data in this figure come from a model which predicts steady-state catch in fishery-y when it and fishery-x exploit a common stock. In each fishing ground, the local population dynamics in the model follow the simple Schaefer model (logistic dynamics with harvest) and there is a diffusional exchange of fish between the two areas. If fishery-x is inactive, $A_x = 0$, then fishery-y has an optimum activity, A_y, somewhat less than 1.0, but with increasing activity in fishery-x, the optimum for activity in fishery-y decreases as does the maximum catch. The blank area in the upper right of the figure is where the total activity in the two fisheries is enough to wipe out the population; that is, the model predicts negative population levels in this area.

Information such as in Figure 1 would be ideal in assessing the effect of fishery-x on fishery-y. The current effort levels of both fisheries define a position on the plot, and a change in effort in fishery-x would move that position parallel to the A_x axis. We can then read the resulting change in the catch of fishery-y. The rate at which the performance changes in one fishery with small changes in the activity of another has been proposed as a measure of interaction (Sibert, 1984). This is the marginal effect of one fishery on another. In our example, it is the slope of the surface in Figure 1 parallel to the A_x, axis. A mathematical definition of marginal interaction is the partial derivative of P_y, with respect to A_x, that is,

$$marginal\ interaction = I^m_{x \rightarrow y} = \frac{\partial}{\partial A_x} P_y(A_x, A_y) \tag{2}$$

where the subscript, $x \rightarrow y$, indicates the effect of fishery-x on fishery-y as distinct from the reverse. Another possible measure of interaction could be the absolute effect

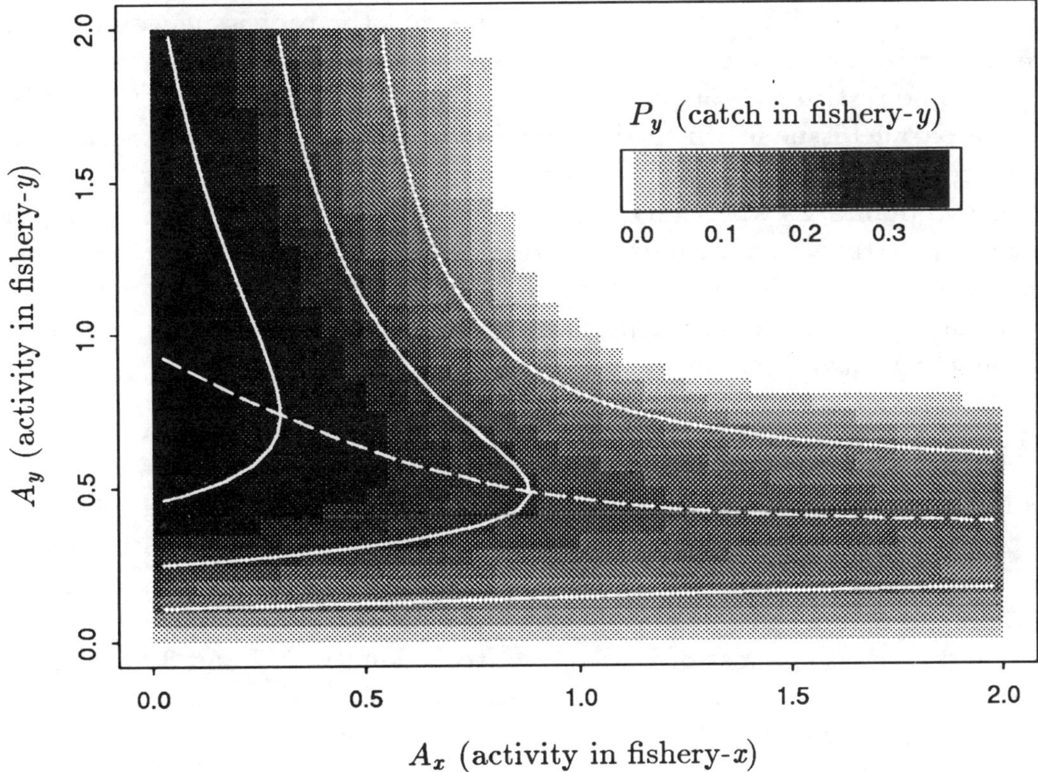

Figure 1. Performance at steady-state of fishery-y as a function of both its own fishing activity, A_y, and activity, A_x, of fishery-x. Performance is measured by catch rate and is predicted by a simulation model in which the two fisheries interact by diffusive exchange of fish between their respective fishing grounds and in which the dynamics internal to the fishing grounds follow the Schaefer model. Activities are measured by fishing effort. Dashed line shows optimum level of A_y as a function of A_x.

of one fishery on another, which would be the way the performance of fishery-y is altered as a result of the fact that fishery-x exists at all, that is,

$$absolute\ interaction = I^a_{x \to y} = P_y(A_x, A_y) - P_y(0, A_y) \qquad (3)$$

The above variations of interaction indices are all contingent on the current activity level in the affected fishery, A_y, but we could alternatively have an interaction parameter that is independent of A_y, by looking at the effect of A_x on the maximum catch obtainable by fishery-y if it is allowed to adjust its effort to an optimum level. We could call this "opportunity" interaction in that it is the effect of fishery-x on the opportunity available to fishery-y for optimizing catch. In this case in Figure 1 we would be concerned either with the slope of the surface along the ridge indicated by the dashed line (if we want marginal interaction) or with the height of the ridge at current A_x and at $A_x = 0$ (if we want absolute interaction).

In any serious official inquiry as to the existence of significant interaction, or negotiation to resolve an interaction dispute, where fishery-x is the affecting fishery and fishery-y is the affected fishery, we can imagine that the root issue would centre on asking fishery-x to sacrifice some of its activity for the benefit of fishery-y or otherwise provide some restitution for past or present impacts on fishery-y. Of course there may be a parallel issue in which the affecting and affector roles are reversed. But in any case, in such a negotiation, information about the response of P_y to A_x at just one or two points, as with marginal and absolute interaction indices, may not be very satisfactory. However, if we had information over the range of values that A_x might be asked to take, we can be sure that such information would be considered with great attention. In this case our measure of interaction is not a single index but a function, namely P_y as a function of A_x, over some range of values.

5. CHOICE OF INTERACTION INDEX AFFECTS ASSESSMENT

The point in emphasizing the multiplicity of choices for a measure of interaction is that the choice can influence the outcome of an assessment. I will illustrate this with a comparison of a marginal with an absolute interaction index using the data in Figure 1. With the model that produced those data, it was also possible to calculate the marginal and absolute indices of interaction from Equations 2 and 3 at points on the A_x, A_y plane where the steady-state population is greater than zero. Contour plots of absolute and marginal interaction are given in Figures 2 and 3 respectively.

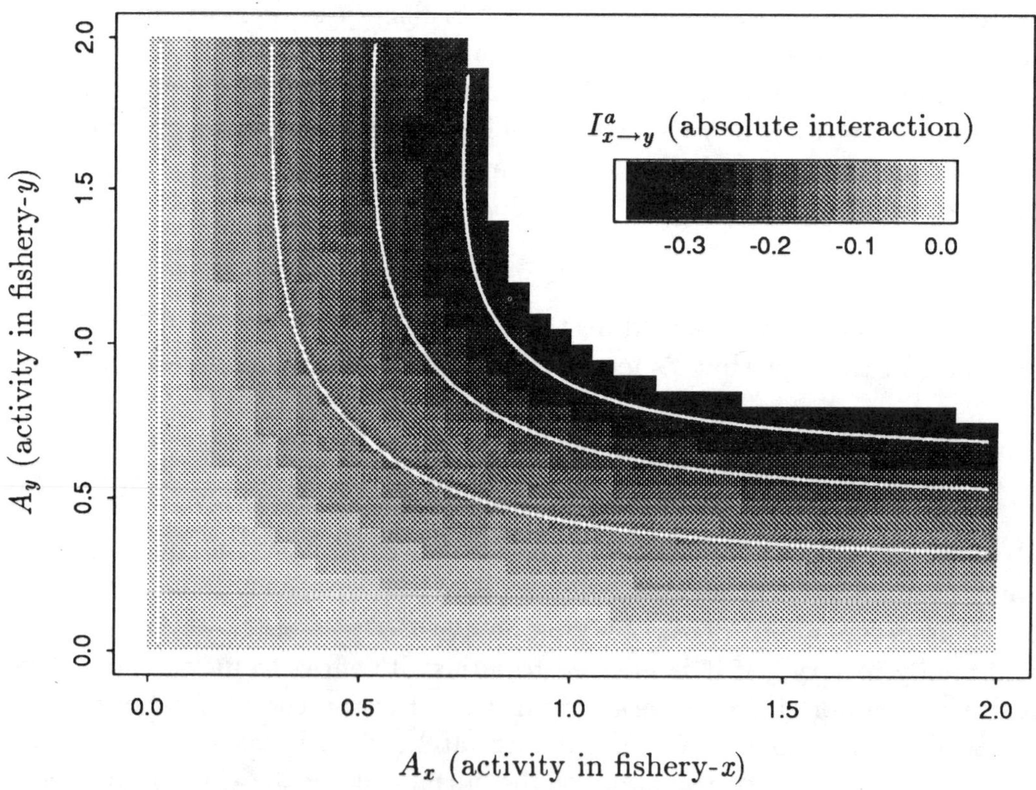

Figure 2. Absolute effect of fishery-x on fishery-y predicted by same model giving results in Figure 1, calculated as the level of the surface in Figure 1 at point A_X, A_y minus the level at point $0, A_y$.

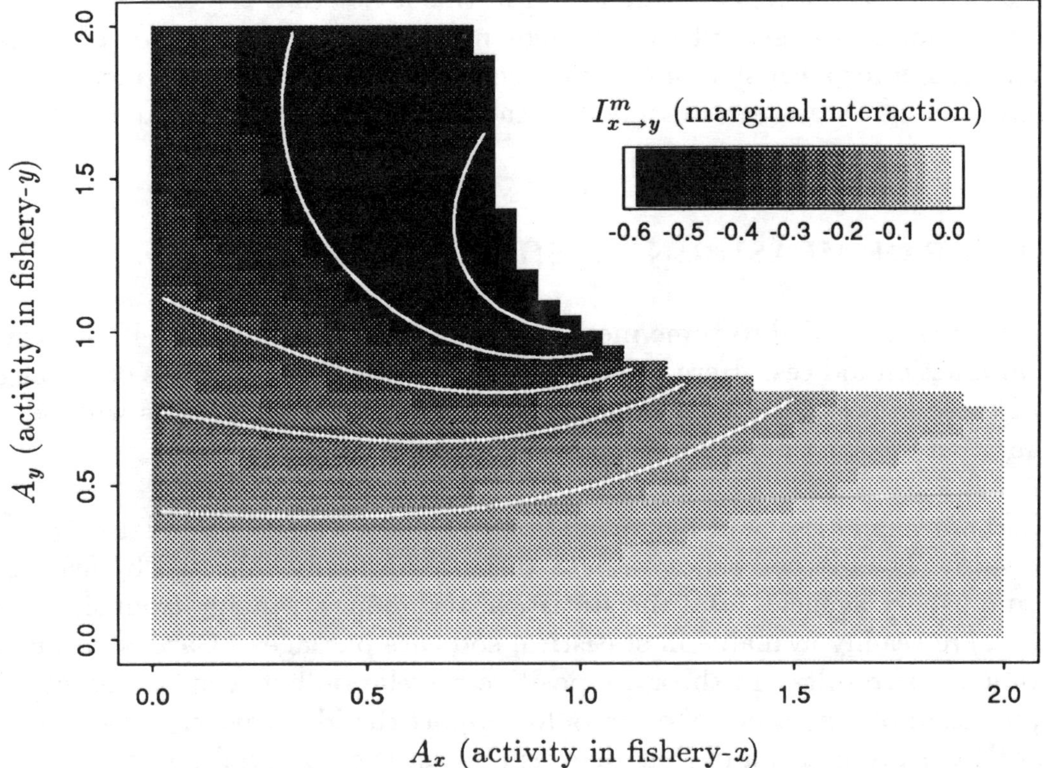

Figure 3. Marginal effect of fishery-x on fishery-y predicted by same model giving results in Figure 1, calculated as the slope of the surface in Figure 1 at point A_X, A_y parallel to the A_x axis.

As expected, there is a tendency with both types of interaction index for the severity of interaction to increase with increasing activity in either or both fisheries. For absolute interaction (Figure 2), this is always the case. For any given level of activity in fishery-y, A_y, if fishery-x increases its activity (moves right, parallel to the A_x axis) the measure of absolute interaction increases. This implies that a manager of fishery-y should be more concerned the more fishery-x increases its activity. However, in the case of marginal interaction (Figure 3), there is a region in the A_x, A_y plane where the measured effect of fishery-x on fishery-y decreases as A_x increases. Does this imply that the manager of fishery-y should become less concerned about the activity of fishery-x as that activity increases?

The possibly counter-intuitive result of this modelling exercise stems from the combination of two factors. One is the movement dynamics becoming overwhelmed by the local dynamics in the fishing grounds as exploitation rate increases, which effectively decouples the two fisheries. The other factor has to do with the fact that even if the two fisheries were so tightly coupled as to be in effect one fishery, when effort is adjusted for maximum catch, the marginal change in catch per change in effort is zero.

There is no reason to expect that the effect we have just seen is limited to the particular model used here to demonstrate it. In general we can expect that at high exploitation rates remotely interacting fisheries will tend to become decoupled and marginal interaction will decrease. The lesson in this is not that the concept of marginal interaction is useless, but that a single measurement of any index of interaction is not enough for a satisfactory assessment of interaction. Such a measurement should be embedded in a wider assessment of the status of the fisheries and of the resource stock.

6. METHODS OF MEASURING FISHERY INTERACTION

We have already alluded to some methods for measuring interaction in our classification of interaction indices. Here we will confront some of the realities of getting estimates of those indices. Detailed examples of most of these techniques will be presented in other working papers of the consultation.

The mechanics of assessment by correlation of fishery statistics is relatively straight forward. The method relies on data that are normally collected by fishing industries and fishery agencies, and the statistical techniques, ranging from simple to sophisticated, are readily available in statistical software packages. Care is needed in interpretation of the results. In theory, a positive correlation between stock-status indicators, for example catch per effort, would support the idea that the resources of the two fisheries are linked and that there is at least the potential for interaction. On the other hand, a negative correlation with certain combinations of statistics, for example catch in one fishery (fishery-x) and catch per effort in another (fishery-y), could indicate that fishery-x is having a deleterious effect on fishery-y. The interpretation of the results could easily be clouded by environmental effects that could cause fluctuations in the two fisheries that are either positively or negatively correlated with each other or uncorrelated regardless of whether the actual or potential interaction is strong or weak.

The index of potential interaction (Kleiber *et al.*, 1984) requires tagging data, which is less often available. In this case, tagged fish should be released and recoveries collected in the areas of both fishery-x and fishery-y. Again, the index is easy to calculate, but interpretation can be problematical. If this index is large in the direction to fishery-y from fishery-x, then fishery-y should clearly be concerned about any plans that fishery-x might have to increase its catch. On the other hand, if the index is low, the interpretation is ambiguous without additional information. The index may be low because there is naturally little migration from x to y, in which case there is no cause for concern, or it may be low in spite of a high natural migration rate because fishery-x harvests most would-be migrants before they have a chance to depart, in which case there is great cause for concern.

For any of the indices of direct interaction, we have said that data such as in Figure 1 would be ideal. Unfortunately, empirically determined information such as that would never be available in a real situation. At best, we would see only glimpses of that surface at points on the A_x, A_y plane where actual activity levels have prevailed, but what we really need is a glimpse of the surface in Figure 2 or 3 (depending on whether we want absolute or marginal interaction) at the current A_x, A_y position. A simple measure of performance, P_y, at this point is not enough.

With tagging, there is a way to measure the current level of marginal interaction (Majkowski *et al.*, 1988). In simplified terms, tagged fish are released in the area of the affecting fishery, fishery-x, those that are caught in fishery-x are re-released, and recoveries of those re-releases in fishery-y are noted. These recovered fish are representatives of the fish which would have been caught in fishery-y if they had not been caught in fishery-x. The proportion of re-released fish that are recovered is an estimate of marginal impact of fishery-x on fishery-y where performance and activity are both measured by catch.

Other than the above tagging method, the only direct way to estimate marginal or absolute interaction is to observe performance at other than current activity levels. A small change, ΔA_x, would allow us to approximate marginal interaction from

$$I^m_{x \to y} \approx \frac{P_y(A_x + \Delta A_x, A_y) - P_y(A_x, A_y)}{\Delta A_x} \tag{4}$$

and a large change (to zero) would allow us to estimate absolute interaction from Equation 2. But there are serious problems with this approach. It is unlikely that we could order up experimental changes in fishing activity; so we would probably have to be content with a historical series of observed P_y at various fishing activity levels. But even if we had the power to manipulate activity levels, or if we had a suitable history of performance and activity data, we would still be in difficulty because of the extra arguments to function, P_y, in Equation 1. In addition to fishing activities, P_y is bound to be sensitive to a variety of environmental factors which can vary in time, and to further complicate matters, it is sensitive to the history of those environmental factors and also the history of fishing activity. It is actually a dynamic variable that at any time is in a transient state unless all the factors that affect it have been held constant for some time. Under some circumstances, fishery managers, or delegates at interaction negotiations, might be interested in information about transient states, but presumably for setting or negotiating long-term policy, the desired information would be expected steady-state performance under a set of nominal environmental conditions and fishing activity levels. Attempting to measure the expected steady-state of a dynamic variable by observing transient states of that variable while it is buffeted by changing conditions is, to say the least, very difficult — much like trying to aim at an erratically moving target with loose sights on one's gun.

This is not to say that historical series of fishery data are useless. Observation of the response of fisheries to changing conditions, particularly in concert with other sources of information (fecundity studies, growth studies, or tagging data for example) helps to inform us about the underlying dynamics of the fisheries and their resource populations. Armed with such understanding, we can build dynamic models that predict fishery performance in response to various factors including activity in other fisheries. With such models, we can control confounding factors that cannot be controlled in the real world, and we can perform experimental manipulations of activity levels at will while observing the behaviour of P_y. We can readily generate P_y as a function of A_x, and in fact we can generate the same kind of data as in Figures 1, 2, and 3. To establish steady-state results, we can hold conditions constant as long as we wish, but if we are also interested in transient effects, we can investigate those as well. With the same model, or minor modification, we could turn the question around and deal with the effect of fishery-y on fishery-x, and we could incorporate multiple fisheries if appropriate.

To implement an interaction model suitable formulations and parameter values need to be found for all the processes to be incorporated in the model, including at minimum the population dynamics of the fish stocks, the harvest of fish by the relevant fisheries, and the movement of fish between the fisheries. Examples of interaction models of Pacific tuna fisheries that explicitly incorporate those processes are described by Sibert (1984), and Kleiber and Baker (1987).

The process of fish movement has been slow to be incorporated into fishery models in general partly because of computational overhead, and partly because of ignorance of relevant movement parameters (exchange rates, advective velocities, diffusivities, and the like). Tag data are probably the best source of movement information, but when the recovery effort is non-uniform in time and space (as is the case with most tagging experiments), the analysis of tag data to estimate movement parameters is not straight forward. Computational techniques to overcome these difficulties and thereby estimate parameters of movement models from tag recovery data have been developed only recently, to a large extent using tag data from Pacific tuna (Ishii, 1979; Sibert, 1984; Hilborn, 1990). These techniques are still being refined and improved, and new techniques are being developed, both with tag data and with other sources of information, such as genetic data. Since fish movement is a central issue in assessing interaction for tuna fisheries, many of the methodological papers in this consultation will focus mostly, or entirely, on techniques for estimating fish-movement parameters.

7. CONCLUSION

Our review of types of fishery interaction mechanisms has illustrated the many potentialities for interaction between tuna fisheries in the Pacific. Our review of the variety of indices for evaluating interaction has underscored the need for care in choosing interaction indicators and in interpreting them in a way that addresses the particular concerns of a given situation. Of the methods for measuring fishery interaction, the most flexible for addressing a variety of interaction concerns is the construction of suitable models which can serve as test beds for various experimental manipulations aimed at revealing the effect of one fishery on another. Much of the methodology development needed for assessment of tuna-fishery interaction is then the development of methods to gain the underlying knowledge and understanding necessary to produce such models. Thus what is needed is more knowledge of the life processes of tunas, tuna movement in particular, and better understanding of the way those processes are affected by fishing activities and by environmental changes. These are fundamental issues for any approach to effective assessment of fishery interaction as well as to a panoply of other kinds of stock assessment and fishery management questions.

8. REFERENCES CITED

Anonymous. 1988. Methods of studying fishery interaction. Paper presented at 20th Regional Technical Meeting on Fisheries, S.Pac.Comm., Noumea, New Caledonia, 1–5 August 1988, WP/5:14 p.

Hilborn, R. 1990. Determination of fish movement patterns from tag recoveries using maximum likelihood estimators. *Can.J.Fish.Aquat.Sci.*, 47(3):635–43.

Hilborn, R. and J.R. Sibert. 1986. Is international management of tuna necessary? *Fish.Newsl., S.Pac.Comm.*, (39):31–40.

Ishii, T. 1979. Attempt to estimate migration of a fish population with survival parameters from tagging experiment data by the simulation method. *Inv.Pesq.*, 43:301–17.

Kleiber, P. and B. Baker. 1987. Assessment of interaction between north Pacific albacore, *Thunnus alalunga*, fisheries by use of a simulation model. *Fish.Bull.NOAA-NMFS*, 85:703–11.

Kleiber, P., A.W. Argue, J.R. Sibert, and L.S. Hammond. 1984. A parameter for estimating potential interaction between fisheries for skipjack tuna (*Katsuwonus pelamis*) in the western Pacific. *Tech.Rep.Tuna Billfish Assess.Programme, S.Pac.Comm.*, (12):11 p.

Laurs, R.M. 1983. The North Pacific Albacore — An Important Visitor To California Current Waters. *Rep.CCOFI*, 24:99–106.

Laurs, R.M., and R.J. Lynn. 1977. Seasonal migration of North Pacific albacore, *Thunnus alaunga*, into North American coastal waters: Distribution, relative abundance, and association with Transition Zone waters. *Fish.Bull.NOAA-NMFS*, 75:795–822.

Majkowski, J., W.S. Hearn, and R.K. Sandland. 1988. A tag-release/recovery method for predicting the effect of changing the catch of one component of a fishery upon the remaining components. *Can.J.Fish.Aquat.Sci.*, 45:675–84.

Sibert, J.R. 1984. A two-fishery tag attrition model for the analysis of mortality, recruitment and fishery interaction. *Tech.Rep.Tuna Billfish Assess.Programme, S.Pac.Comm.*, (13):27 p.

METHODS FOR ANALYZING INTERACTIONS
OF LIMITED-RANGE FISHERIES: HAWAII'S PELAGIC FISHERIES

Christofer H. Boggs
Honolulu Laboratory
Southwest Fisheries Science Center
National Marine Fisheries Service, NOAA
Honolulu, Hawaii, USA

ABSTRACT

A review of previous studies indicated that local fishing intensity can affect local pelagic fish abundance, but overall abundance on a wider scale may have a greater local effect. Interactions of limited-range fisheries were explored using a theoretical model in which local catch per unit effort (CPUE) declines as the catch approaches or exceeds the rates of fish immigration and recruitment in a limited area. The model has the local catch increase with effort to an asymptotic level. This simple model was used to simulate local CPUE in relation to varying rates of immigration and local fishing. Simulated data were used to test an analytical approach--the regression of CPUE on catch--which was then applied to real data from Hawaii's troll fishery. Analyzing CPUE in relation to catch rather than effort was convenient because fishing effort is poorly documented in Hawaii and includes effort by diverse methods. The analysis indicated that the CPUE of yellowfin tuna (*Thunnus albacares*) in Hawaii's troll fishery was not affected significantly by total yellowfin tuna catch in Hawaii in 1987-90, but was directly related to an index of abundance for surface-caught yellowfin tuna in the western Pacific.

## 1.	INTRODUCTION

In Hawaii and in many Pacific island nations, locally-caught pelagic fishes constitute a small fraction of stocks that extend far beyond the range of the local fisheries. Primary concern over the status of these stocks is appropriately focused on abundance and production throughout their range (Suzuki, 1989). However, the rate of replacement of fish within any area is finite. Theoretically, if fishing mortality in an area increases greatly in relation to net immigration and recruitment, local catch per unit effort (CPUE) will decline. If there are several fisheries in such an area, fishery interactions may occur.

Fishery production models, which typically are used to estimate maximum sustainable yields for pelagic species (Suzuki, 1989; IATTC, 1992), are not very useful for detecting fishery interactions. Nor are they useful for estimating the optimal level of fishing effort in localized fisheries that are too small to significantly affect the size of the stock or its level of production (Sathiendrakumar and Tisdell, 1987). In many areas of the Pacific, including Hawaii, tropical tuna and billfish production may be limited mostly by immigration from surrounding areas rather than by the reproduction and growth of resident fish. For such fisheries, a model in which the local catch increases with effort towards an asymptote was proposed by Sathiendrakumar and Tisdell (1987).

In this paper, I develop a simple model for the relationship between catch and effort in a limited-range fishery on a highly mobile pelagic stock. The model is similar to that of Sathiendrakumar and Tisdell (1987) but is explicitly formulated as a function of immigration, emigration, natural mortality, and catchability. I use the model to simulate data for a fishery in which immigration, emigration, and fishing effort are seasonal. The simulated data are used to test whether the effect of fishing intensity on local catch rates can be adequately quantified by regressing CPUE on catch rather than on effort. Using catch rather than effort to quantify fishing intensity is helpful if total fishing effort is poorly documented and derived from widely different fishing methods, as is the case in Hawaii. Finally, the CPUE in Hawaii's troll fishery for yellowfin tuna (*Thunnus albacares*) is analyzed in relation to the total catch by troll, handline, and longline fisheries to determine whether the troll CPUE has been affected by total fishing intensity.

2. REGULATING LOCAL FISHING EFFORT IN HAWAII

Local fishery managers in the USA Exclusive Economic Zones (EEZs) of the central and western Pacific can do little about the distant-water foreign fisheries, which operate outside the EEZs and harvest the majority of pelagic fish caught in the region (NMFS, 1991). The primary USA fishery-management objective is to prevent recruitment overfishing (NMFS, 1989), defined in relation to pelagic populations which are mostly exploited beyond USA jurisdiction. This objective is dysfunctional, since there exists no international management organization through which stock-wide objectives can be achieved (NMFS, 1991). Therefore the USA Western Pacific Regional Fishery Management Council (WPRFMC) has focused its objectives on "equitable domestic utilization of the resources. . ." among user groups (WPRFMC, 1991). Recognizing that the domestic catch within the EEZs of the USA may be large enough to reduce local CPUE and cause local fishery interactions, the WPRFMC has attempted to regulate fishing effort in Hawaii's EEZ (WPRFMC, 1991).

Recent increases in the total catch of pelagic species in the USA Pacific EEZs have mostly been due to an expanding domestic longline fishery in Hawaii (Ito, 1991; WPRFMC, 1991). The less mobile, small-vessel troll and handline fisheries in Hawaii have experienced declines in total catch and CPUE (Boggs, 1991) as the longline fishery has expanded. Concern by troll, handline, and longline fishermen over reduced CPUE, together with gear conflicts, dangerous confrontations between fishermen, and overcrowding of dock facilities, induced the WPRFMC to pass regulations halting the entry of additional vessels into the USA domestic longline fishery in Hawaii in 1990. To prevent gear conflicts, the WPRFMC in 1991 also passed regulations closing nearshore areas (<50-70 miles, depending on location) to longline fishermen.

Scientific evidence in support of limiting longline-fishery participation in Hawaii has been sparse. Prior to 1986 the WPRFMC focused on the impacts of foreign longline fishing within Hawaii's EEZ on catch rates for pelagic management unit species (PMUS) which included billfish, mahimahi (*Coryphaena hippurus*), wahoo (*Acanthocybium solandri*), and pelagic sharks, but not tunas (Lovejoy, 1977, 1981; Wetherall and Yong, 1983; Skillman and Kamer, 1992). The possible impacts of foreign fishing in the EEZ were regulated by the original Pelagic Fisheries Management Plan (WPRFMC, 1986) but became irrelevant as no foreign longliners exercised the option to fish legally in Hawaii's EEZ after 1980. However, the expanding domestic longline fishery in Hawaii now catches more fish than the foreign longliners did before 1980 within the EEZ (WPRFMC, 1991). This catch increase, plus the inclusion of tunas as a Pacific Pelagic Management Unit Species (PPMUS) in 1992, provided some impetus to regulate the domestic fisheries.

Regulating fishing effort rationally requires a quantification of the effect of local fishing pressure on local CPUE for pelagic species, and a means of choosing a desirable level of fishing effort. An optimal level of fishing effort can be estimated if the cost of effort and the value of catch are defined and the relationship between catch and effort is quantified as an asymptotic curve (Sathiendrakumar and Tisdell, 1987). In the asymptotic catch model, although catch never declines because of overfishing, CPUE does decline as effort increases. Thus, detecting and quantifying an asymptotic relationship between local catch and fishing effort would help indicate whether regulating the local fishing effort is justified.

The optimization of fishing effort, in relation to the appropriate costs and benefits, depends on the relative mix of the fisheries comprising the total effort. Thus, the allocation of fishing effort among interacting fisheries is an additional problem. Changing the composition of effort by different gear types will alter the optimum. Furthermore, the decision as to which fishery to regulate, and how, may be strongly influenced by historical, social, or logistical considerations. For example, the least-mobile fisheries are most vulnerable to local declines in CPUE, and this could be a reason to limit the effort of the more mobile longliners in some areas rather than the effort of all the fisheries in those areas. In any case, evidence that local fishing pressure affects local CPUE should be the basis of any scheme for regulating fishing effort.

3. REVIEW OF PREVIOUS RESEARCH

Several previous studies suggest that localized fishing effort can reduce catch rates (CPUE) for wide-ranging pelagic stocks in a local area (Lovejoy, 1977, 1981; Wetherall and Yong, 1983; Squire and Au, 1990; Skillman and Kamer, 1992). Lovejoy (1977, 1981) simulated the pelagic fisheries in Hawaii's EEZ and predicted that small increases in the domestic catch of blue and marlin (*Makaira mazara*) and striped marlin (*Tetrapturus audax*) would result if foreign fishing in the EEZ was eliminated. Wetherall and Yong (1983) modelled blue marlin catch rates near Hawaii as a function of mid-Pacific abundance, recruitment, and fishing effort, and found that increases in local, adjacent, and mid-Pacific effort had negative impacts on CPUE near Hawaii. Skillman and Kamer (1992) found significant negative correlations between foreign longline effort and Hawaii's longline catch rates for blue and striped marlins. Most recently, Squire and Au (1990) found that local longline and troll catch rates of striped marlin rebounded when longline fishing was temporarily excluded from an area off Mexico. All of these studies found that local fishing effort affected local CPUE, but they lacked the appropriate model or sufficient data to accurately quantify the relationship, and thus provided no means of choosing an optimal intensity for local fishing effort.

The model of Lovejoy (1977) demonstrated that short-term abundance of transient populations was reduced by increasing local fishing effort, but the relative magnitude of the reduction was dependent on the actual (but unknown) number of fish in the area. Lovejoy (1977) used monthly Japanese longline CPUE data (1962-75) to model the spatial distribution, abundance, and catch of marlin in 27 subareas of Hawaii's EEZ and a single

Pacific (pooled) area. Fish movements through the EEZ were simulated to match geographic changes in Japanese CPUE in the 27 Hawaiian areas, assuming general north-south movements for blue marlin and northwest-southeast movements for striped marlin. However, the estimates of abundance and catchability were essentially guesses, and the catches predicted by the model were very sensitive to these parameters. This marlin-fishery simulation (Lovejoy, 1977) was repeated with the removal of the fishing mortality caused by longline fisheries in the EEZ (*i.e.*, set to zero). When Japanese longline fishing in the EEZ was eliminated, the changes in the fish abundance were small in areas fished by the small-vessel trollers, as were the effects of changes in abundance on troll catches of blue and striped marlins (2 and 5%, respectively). The simulated increases in troll catches were larger when domestic longline fishing also was eliminated (5 and 21%, respectively). Though smaller than the Japanese fishery in the EEZ at that time, the domestic longline fishery had a greater simulated impact because it had a greater geographic overlap with the troll fishery.

The sensitivity of the Lovejoy (1977) simulations to parameter estimates was illustrated when the estimated abundances of the marlin stocks were altered (Lovejoy, 1981). When the number of marlins moving from the Pacific-pooled area to Hawaii's EEZ was halved (Lovejoy, 1981) and catchability was increased to simulate the same catches as in the original model, eliminating all longline-fishing mortality increased the simulated troll catches of blue marlin by 13% and striped marlin by 45%, respectively. In contrast, when abundances were doubled, eliminating all longline fishing increased the troll catch of blue marlin by only 1% and striped marlin by 3%.

Exogenous factors may overwhelm the influence of local fishing effort on local catch rates of highly-mobile pelagic fish. Indices of abundance (CPUE) in local areas, for example, have been found to be significantly correlated with CPUE over a much wider range. Using Japanese longline catch-and-effort statistics for 1962-79 to compute estimates of abundance, Wetherall and Yong (1983) found that variation in blue marlin catch rates at the beginning of a year in a mid-Pacific area explained 80% of the annual variation in peak third-quarter catch rates in a 5x10 degree (latitude x longitude) area around the main Hawaiian Islands. No other statistically-significant predictors of local catch rates were found, but by including the variables for a recruitment trend and the foreign fishing effort in local, adjacent, and mid-Pacific areas, Wetherall and Yong's (1983) regression model increased the amount of accountable variation from 80% to 95%. Their analysis suggests that the impact of local effort on Hawaii's blue marlin fishery was small compared with the impact of abundance on a wider scale, under the conditions prevailing in 1962-79.

Pronounced seasonal cycles characterize the local, apparent abundance of large, tropical, pelagic species, especially towards the higher latitudes. The effect of local effort on CPUE may well depend on the season. Wetherall and Yong (1983) and Squire and Au (1990) accounted for seasonal effects by eliminating all but the data for the season of peak CPUE. Another method used by Skillman and Kamer (1992) was to decompose the seasonal and nonseasonal components of quarterly and monthly time series and examine the nonseasonal component. At a quarterly resolution, deseasonalized, local, domestic longline CPUE statistics for blue and striped marlins in Hawaii (1962-78) were significantly negatively correlated with foreign longline effort in local and adjacent areas ($P < 0.05$), whereas no significant correlation was found at the annual resolution

(Skillman and Kamer, 1992). The correlation coefficients were low (-0.26 to -0.32, nonparametric Spearman coefficients), and no attempt was made to quantify the relationships. Domestic longline CPUE statistics were most negatively correlated with Japanese longline effort in the local area, again suggesting a stronger relationship with increased proximity.

An unintended "experiment" on the effect of local fishing effort on local marlin CPUE was performed by the government of Mexico in 1977 when it enforced regulations against foreign longline fishing within its 200-mile exclusive economic zone. The enforcement caused a major decrease in fishing effort, allowing an interesting comparison to be made between catch rates before and after 1977 (Squire and Au, 1990). Striped marlin catch rates by troll fishermen in the area west of Mazatlan and around the tip of Baja California doubled during 1977-80 (Squire and Au, 1990). Joint-venture longline operations beginning in 1979-80 in the area also experienced catch rates twice as high as those in 1976. This series of events suggests a much stronger effect of local fishing effort on CPUE than that indicated for Hawaii's fisheries (Lovejoy, 1977; Wetherall and Yong, 1983; Skillman and Kamer, 1992). The opportunity to make a comparison like that of Squire and Au (1990) between CPUE statistics before and after the abatement of foreign longline fishing in Hawaii's EEZ (1980) was given as a reason for regulating foreign longline fishing (WPRFMC, 1986). Unfortunately this analysis was never undertaken, partly because of a severe decline in reporting of Hawaii's fishery statistics beginning around 1979 (S. Pooley, Honolulu Laboratory, National Marine Fisheries Service, Honolulu, HI 96822-2396, unpubl. manuscr.).

Squire and Au (1990) described the Mexico striped marlin fishery as operating in a "core area" in which fish naturally aggregate in concentrations that are much higher than throughout the population range. When they analyzed catch rates of the joint-venture fishery and the foreign longline fishery off Mexico in relation to fishing effort during 1962-84, they found no clear quantitative relationship. This was largely because the CPUE in the core area declined in 1981-84, despite relatively low levels of effort. Squire and Au (1990) suggested that a reduction in core-area fishing effort disproportionately raises the local CPUE, because core-area CPUE depends more on the formation and fishing down of "hot spots" than on stock production. Hypothetically, the relationship between fishing effort and CPUE in "hot spots" could be different from that in a larger area. Such a relationship could be demonstrated only if the data had very fine geographic resolution.

4. ASYMPTOTIC MODEL FOR LOCAL CATCH

A model for a local fishery in which the catches reach an asymptote as fishing effort increases was proposed by Sathiendrakumar and Tisdell (1987) to determine the optimal level of local fishing effort. I propose a similar model except that the parameters are defined in relation to immigration, emigration, mortality, and catchability. Emigration is treated as if it were analogous to natural mortality, so the two are combined. In the simplest case of this model (*i.e.*, assuming steady-state equilibrium and constant "natural mortality" and catchability), the asymptotic maximum catch is governed by the level of immigration (Figure 1). The parameters and derivation of the model are as follows (all fluxes--immigration, catch, and mortality--are annual, and numbered expressions are considered axiomatic):

I = local immigration (metric tons)--analogous to biological production but independent of local biomass;

C_i = catch (metric tons) at effort level i;

f_i = effort (10^3 hooks) at level i;

q = catchability ($1/10^3$ hooks) [assumed constant];

F_i = fishing mortality $[F_i = qf_i]$; (1)

M' = "natural mortality" including emigration [assumed constant],

Z'_i = total "mortality" including emigration $[Z'_i = M' + qf_i]$; (2)

B_i = equilibrium biomass (metric tons) in the local area $[B_i = I/Z'_i]$; (3)

$$C_i = B_i F_i \text{ (metric tons);} \qquad (4)$$

$$C_i = \frac{IF_i}{Z'_i} \text{ (metric tons)} \qquad \text{[from (3)];}$$

and

$$C_i = \frac{Iqf_i}{M' + qf_i} \text{ (metric tons)} \qquad \text{[from (2)].}$$

  The example shown (Figure 1) uses arbitrary values: I = 5,000, 10,000, and 20,000 metric tons, q = 0.0001/1,000 hooks, and M' = 0.06. When fitting the model to data where I, q, and M' are unknown, the model can be simplified:

$$\text{If} \quad a = \frac{Iq}{M'} \quad \text{and} \quad b = \frac{q}{M'} \,,$$

$$\text{then} \quad C_i = a\left(\frac{f_i}{1+bf_i}\right) \text{ (metric tons).}$$

Algebraic manipulation of the model equation yields the following linear relationship between CPUE and catch (Figure 2):

$$\frac{C_i}{f_i} = a - bC_i \quad \text{(metric tons/10}^3\text{ hooks),}$$

which makes the model easy to fit.

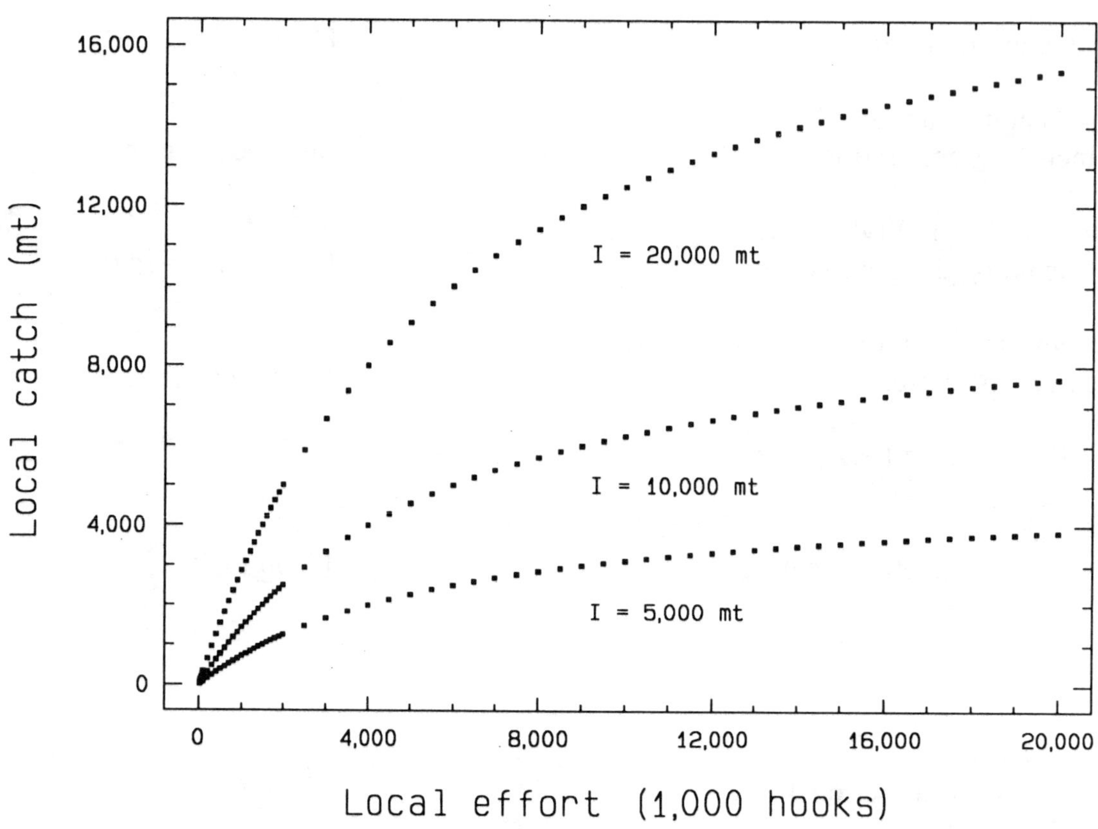

Figure 1.   Asymptotic catch model for a limited-range pelagic fishery that catches a small fraction of the stock of a highly-mobile pelagic species. Natural mortality, emigration, and catchability are constant, and the local biomass of fish is at equilibrium. Annual immigration (I) determines the annual asymptotic catch in metric tons (mt).

Unfortunately, in reality, immigration, emigration, and effort vary seasonally, and equilibrium is not achieved instantaneously. A more realistic model was created by making immigration, natural mortality (plus emigration = M'), and fishing effort vary according to annual cycles, with effort also increasing over time. Simulated data were obtained by calculating the catch, mortality, and the resulting nonequilibrium local biomass at 10 time steps per day for 4 years, summarized by month (Figure 3). The annual cycles of immigration and emigration were chosen to produce a cycle of local biomass roughly resembling the pattern seen in CPUE data from Hawaii's yellowfin tuna fisheries.

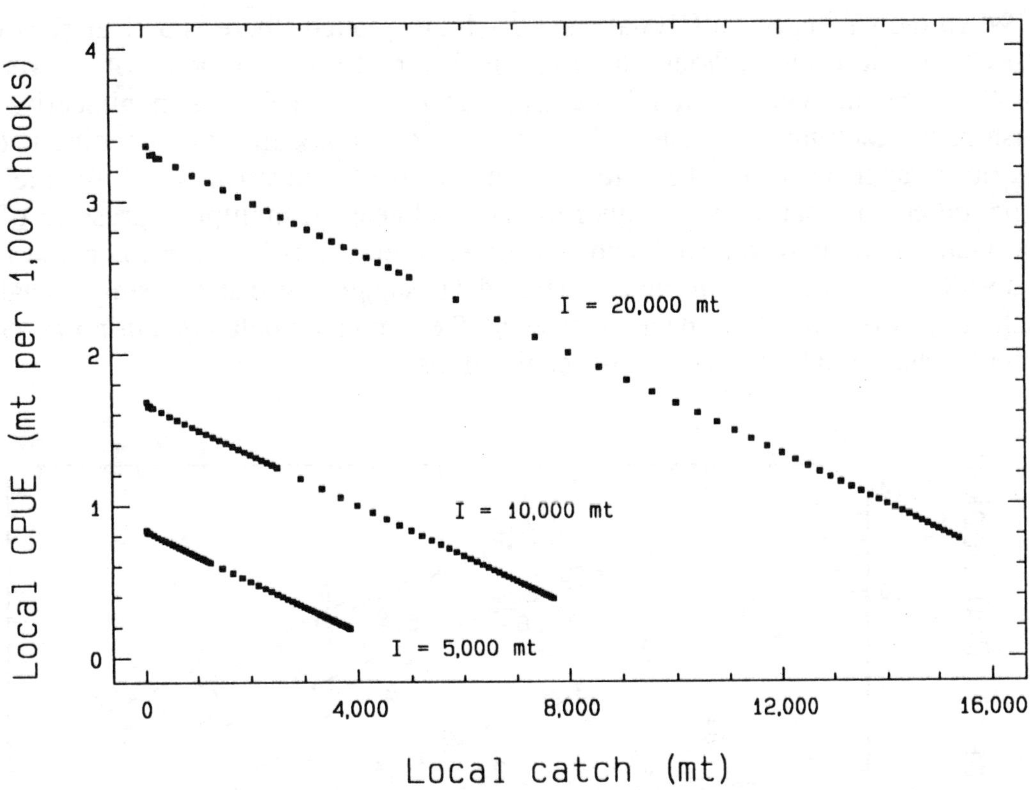

Figure 2.   Relationship between local CPUE and total local catch in metric tons (mt) according to the asymptotic catch model in Figure 1.

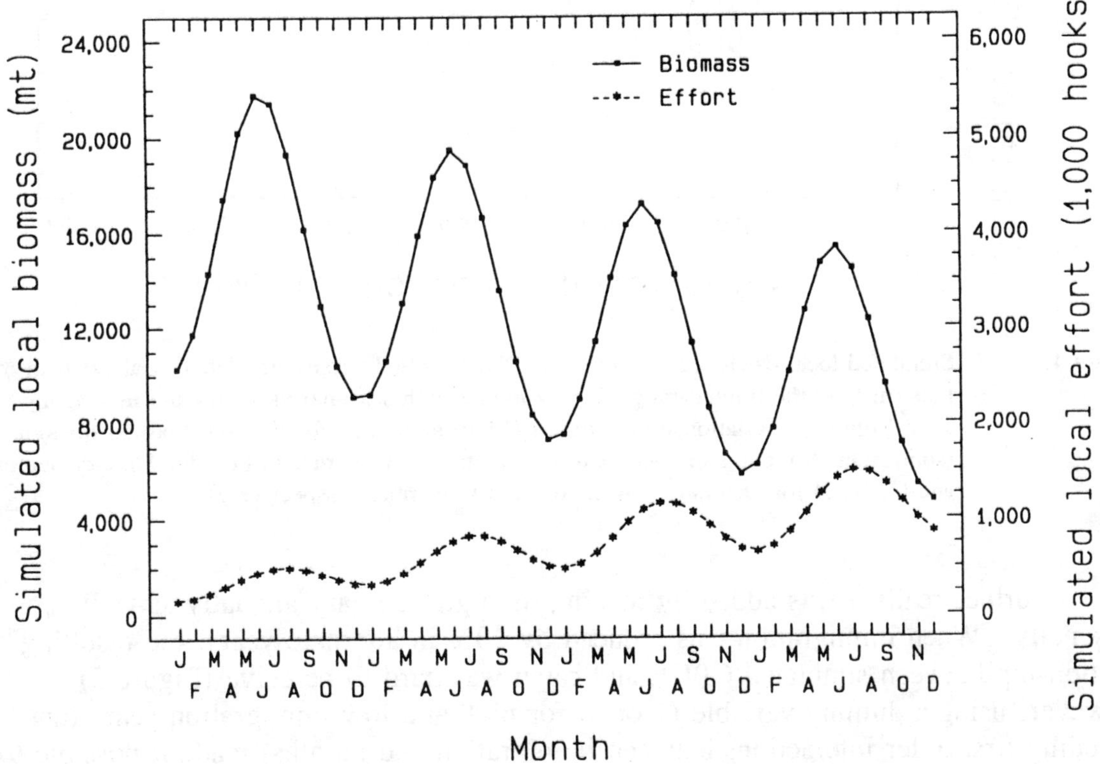

Figure 3.   Simulated monthly biomass in metric tons (mt) and fishing effort in a limited-range fishery like that in Hawaii. Immigration, emigration, and effort vary in an annual cycle, and effort increases threefold over 4 years. Local biomass is not in equilibrium.

When the simulated CPUE data and catch are plotted, there is a clear relationship between CPUE and catch, although the slope is different for each month of the year (Figure 4) in keeping with the monthly changes in *I*, *M'*, and *F*. The nonlinearity of relationships for each month is due to lag effects under nonequilibrium conditions (the nonlinearity disappears when CPUE for each month is plotted *versus* catch for the 5-month period ending that month). Ignoring the nonlinearity, multiple regression of seasonally-adjusted CPUE on catch and first-order interactions between catch and months explains 96% of the variance in the simulated data, suggesting that this simple analytical approach could work with real data. However, the simulated data contain no random components which might obscure weak relationships.

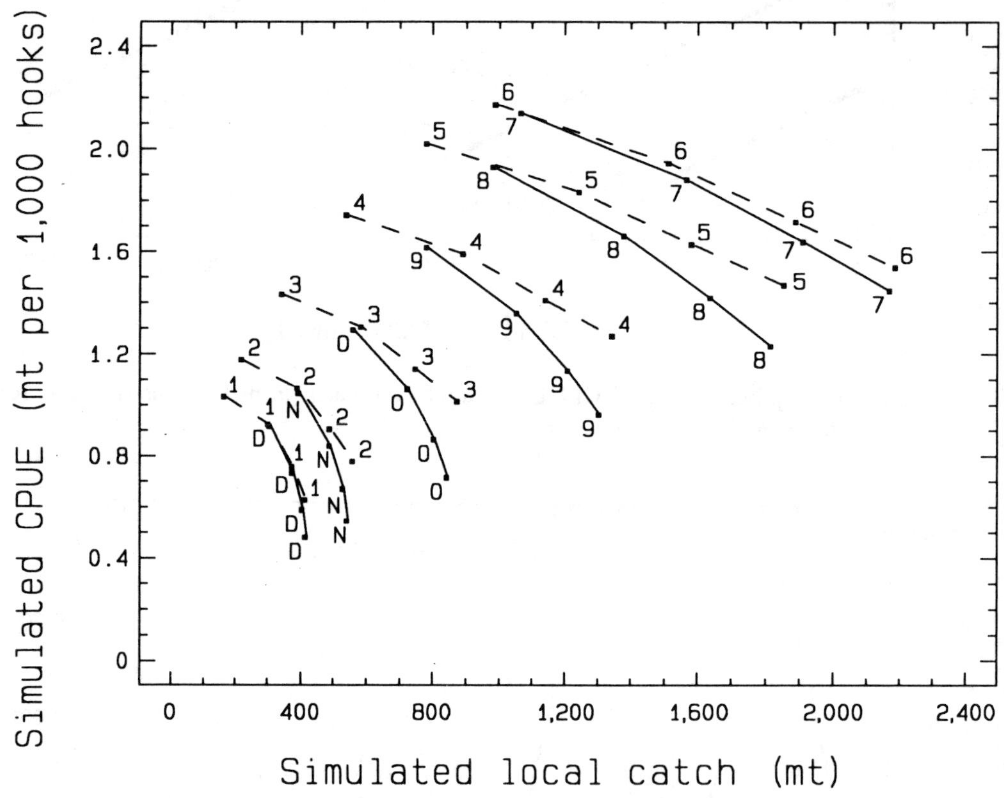

Figure 4.   Simulated local catch per unit effort (CPUE) plotted versus simulated local catch in metric tons (mt) for the limited-range fishery model with seasonal variation and increasing effort as in Figure 3. Relationships between CPUE and catch for the first (dashed lines) and last (solid lines) 6 months of each year are identified by characters (1-9 for January-September, and O, N, D for October, November, and December, respectively).

Further realism was added by making immigration vary annually as well as seasonally. When immigration was reduced by 40% in alternate years, the resulting relationship between simulated CPUE and catch was hard to perceive (Figure 5). However, using a dummy variable (1 or 0) for high and low immigration years (and including first-order interactions between immigration and months) made it possible to use multiple regression to describe the slopes and intercepts of the CPUE- *versus*-catch relationships for each month and level of immigration (Figure 5). These slopes and intercepts were then used to compute the asymptotic relationships between catch and effort (Figure 6).

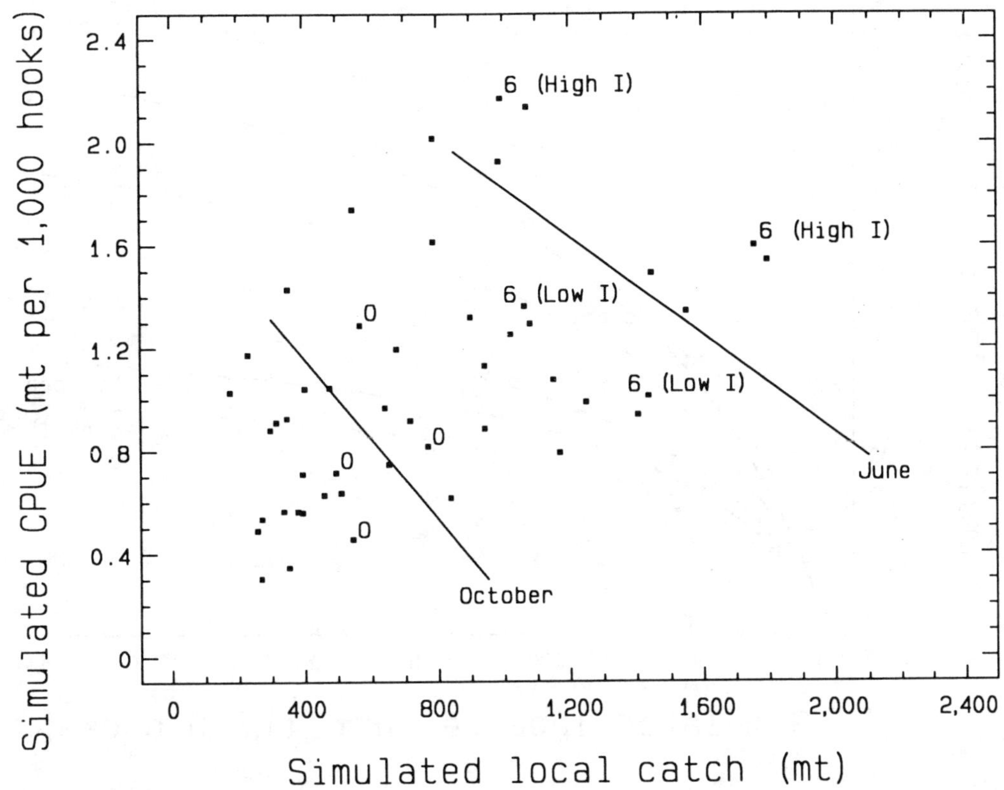

Figure 5.   Simulated local catch per unit effort (CPUE) plotted versus simulated local catch in metric tons (mt) for the limited-range fishery model with immigration (I) alternating from high to low (40% less than high) in alternate years. Relationships for an average amount of immigration, calculated with the multiple regression analysis are shown for June (points labelled 6) and October (points labelled O).

In reality, local immigration is an unknown quantity, but the apparent abundance of fish over a larger area can be used as an index of immigration in a multiple regression analysis of real data. It is logical that more fish would immigrate in years when they are more abundant in the surrounding area, and this behaviour would be consistent with work showing strong correspondence between the apparent abundance of pelagic fish in local areas and abundance on a wider scale (Wetherall and Yong, 1983; Squire and Au, 1990; Skillman and Kamer, 1992). Another mechanism that would explain the correspondence in apparent abundance would be widespread changes in catchability. In contrast, changes in immigration or local catchability due to localized environmental conditions would not be consistent with the observed correspondence in apparent abundance, and thus the local environment may represent an additional source of variation.

The simulations illustrate how relationships between catch and effort within a limited area may differ seasonally and annually. The optimal level of fishing effort may be highly dynamic, making the rational management of fishing effort difficult. Analyzing CPUE data *versus* catch should at least be useful for detecting an impact of local fishing pressure on local CPUE, even when that impact differs seasonally and annually. Detecting such an impact however may require quantifying variation due to other factors.

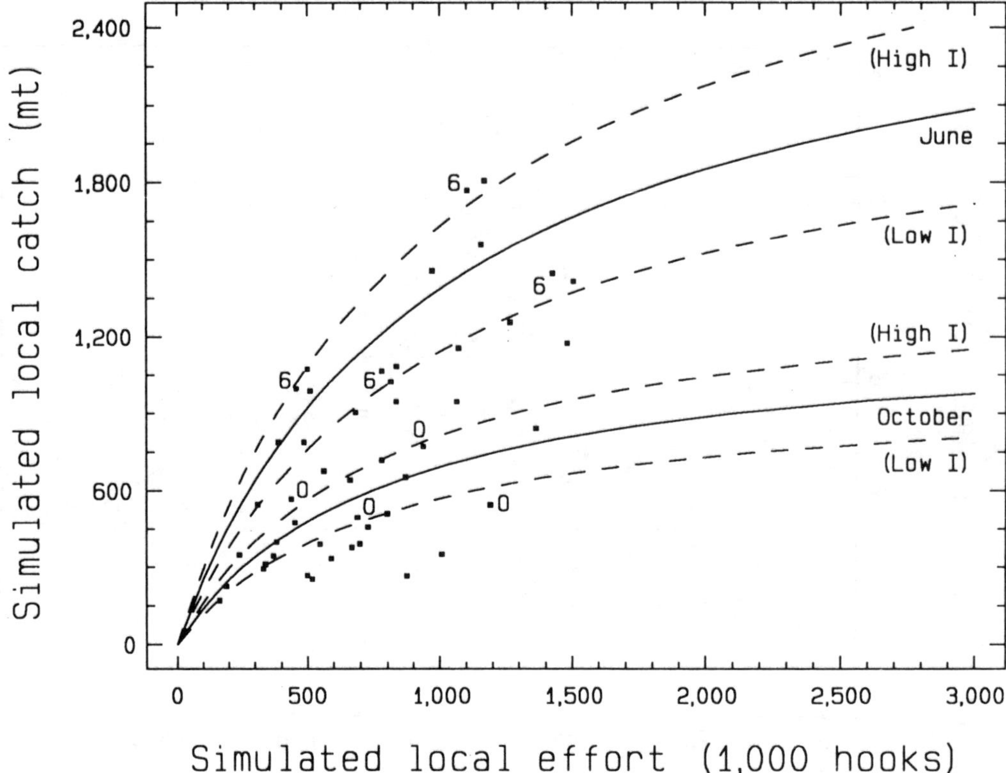

Figure 6. Asymptotic catch relationships for monthly local catches in metric tons (mt) as a function of monthly local effort in a limited-range fishery from the analysis of simulated data shown in Figure 5. The relationships for high and low immigration (I) years (dashed lines) differ from the average relationship (solid lines) for each month (June data labelled 6; October data labelled O).

5. YELLOWFIN TUNA CPUE IN HAWAII

In a preliminary study (Boggs, 1991), troll and handline CPUE data (pounds per trip) were plotted *versus* longline catch to determine whether relationships appeared that might be indicative of fishery interactions. This graphical inspection of the CPUE *versus* catch data was applied to yellowfin tuna, bigeye tuna (*Thunnus obesus*), blue and striped marlins, mahimahi, and wahoo. In the preliminary study (Boggs, 1991), the analyses covered data for January 1987-June 1990. The present study extends the analysis through the end of 1990 for yellowfin tuna troll CPUE (pounds per trip) and compares troll CPUE with the total catch (by all fisheries) rather than just the longline catch.

The data used to calculate troll catch per trip (CPUE) were provided by the Hawaii Division of Aquatic Resources (HDAR) as summaries of commercial catch (pounds of fish) by year (1983-90) and month (1987-90) along with summaries of the total number of fishing trips per year or month, respectively. Annual and monthly catch rates were calculated as the ratio of total catch (in pounds) to the number of trips. No geographic categorization of the data was used, but most of the troll fishing was conducted within 50 miles of shore around the eight main Hawaiian Islands. The HDAR catch data did not contain reports of trips; rather, each date for each vessel in the records was counted as a trip if any PMUS or tuna was reported caught. The assumption of 1-day trips is fairly realistic for the small-vessel troll and handline fisheries.

Catch per trip may not be a good measure of yellowfin tuna abundance since important operational changes and improvements in trolling methods have undoubtedly occurred over the years. The catch per trip index contains no data from trips with zero catches, no standardization of trips as a unit of effort, no estimate or correction for underreporting, and no estimate or correction for changes in reporting over time. Any of these factors could bias trends or relationships in the data, or give the appearance of a trend or relationship where none exists.

Despite these potential problems, catch per trip as an abundance index can provide some indication of changes in the local abundance or availability of fish to trollers and handliners in Hawaii. Catch per trip indices based on HDAR data often mirror patterns seen in more sophisticated CPUE indices, such as catch per hook or catch per set, from nearby fisheries (Wetherall and Yong, 1983; Skillman and Kamer, 1992). When the data from several different sources show a similar pattern, those data probably indicate a true pattern of apparent abundance unless some unknown bias affects several sources of data similarly. In any case, catch per trip is the only available measure of CPUE for Hawaii's troll and handline fisheries. More definitive examinations of trends in CPUE will require data that more accurately specify fishing effort.

Annual troll CPUE (pounds per trip) for yellowfin tuna declined from a relatively high level in 1979 to low levels in 1982-84 (WPRFMC, 1991). Troll CPUE for yellowfin tuna returned to high levels in 1987, dropped again and remained low through 1989, and showed some recovery in 1990 (Boggs, 1991). The decline in troll CPUE for yellowfin tuna in 1987-1989 corresponded with a period of dramatic expansion of Hawaii's domestic longline fishery (Boggs, 1991; Ito, 1991; WPRFMC, 1991), suggesting that some fishery interactions may be occurring. The low troll CPUE seen in 1982-84, however, occurred before the domestic longline fishery expanded and after the foreign longliners ceased fishing in Hawaii's EEZ, suggesting that periods of low troll CPUE may be unrelated to longline fishing effort.

The total longline catch, as estimated by the National Marine Fisheries Service (NMFS) shoreside monitoring programme, showed increasing monthly variation and an upward trend from 1987 through 1989, levelling off somewhat (annual average) in 1990 (Figure 7). The longline catch estimates were for the entire range of the longline fishery, mostly within the Hawaii EEZ but extending beyond it. The range of the troll and handline fisheries is much smaller and is roughly centred within the area fished by the longline fishery (these data precede the establishment of nearshore area closures for longline fishing).

The NMFS data rather than the HDAR data were used to estimate longline catch because they cover most of the longline catch in Hawaii, whereas the HDAR data cover only a small fraction of that catch. Conversely, the NMFS market sample data cover only a fraction of the troll and handline catch and do not distinguish between these two gear types, which have markedly different catch rates. Therefore to obtain total catch, the HDAR data on the catch by gear other than longline (mostly troll and handline) were combined with NMFS estimates of the longline catch. A potentially large component of

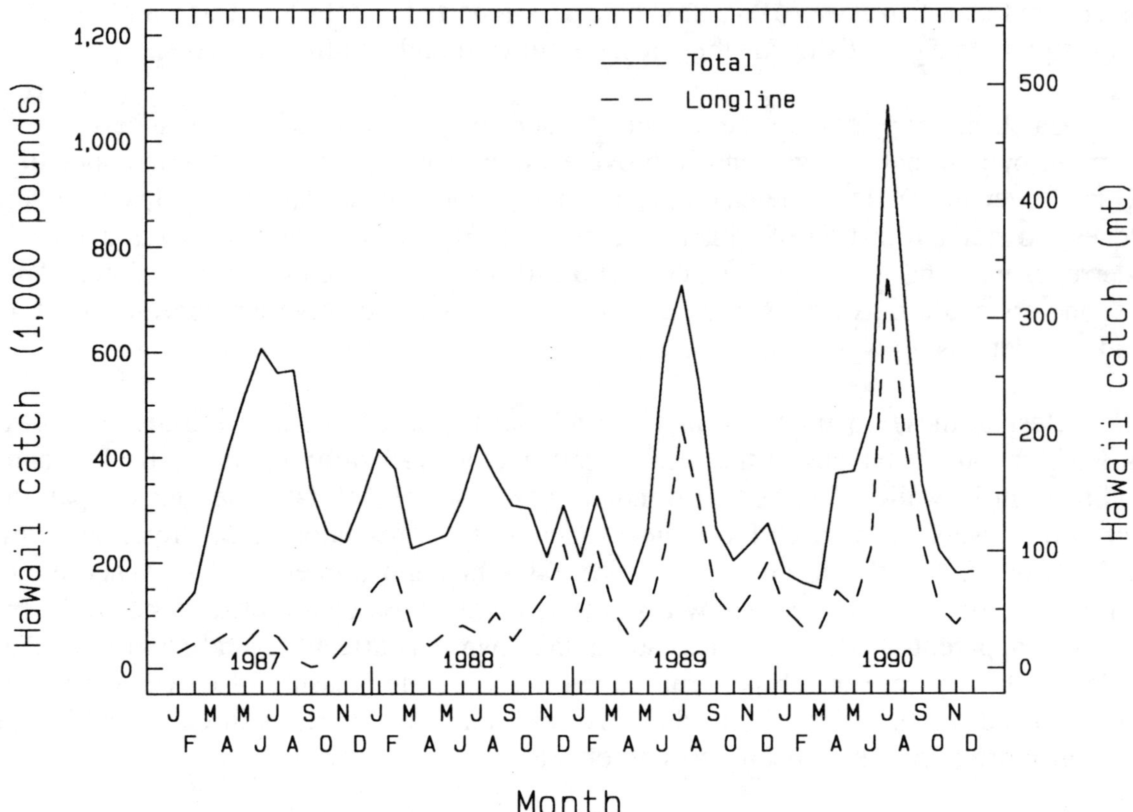

Figure 7. Total catch (all gear types combined, in pounds and metric tons) and longline catch of yellowfin tuna in Hawaii each month in 1987-90. Longline catch is estimated from the shoreside monitoring programme of the National Marine Fisheries Service, and the remainder of the total is from the non-longline catches in the commercial catch reports of the Hawaii Division of Aquatic Resources. The recreational catch is not known.

the total catch, namely the recreational catch, remains unquantified, and there are no good estimates of under-reporting by commercial troll and handline fishermen. Thus, total catch is known with less certainty than longline catch, which was one reason Boggs (1991) used only longline catch.

The total catch of yellowfin tuna by all gear types (Figure 7) varied monthly in 1987-90 but did not show much of an annual trend. As catch by the longline fishery increased, catch by the other fisheries decreased. Fishermen have suggested that increased longline catch might be causing the decline in troll and handline catch; this is why troll and handline CPUE was plotted *versus* longline catch in Boggs (1991). However, examination of CPUE in relation to total catch is more appropriate since the catch by trollers and handliners would also be expected to contribute to any local reduction in biomass.

Hawaii's trollers, handliners, and longliners depend on the same size range of fish to provide the bulk of their yellowfin tuna catch, so interactions of these fisheries are possible. Even though the troll and handline size-frequency distribution (combined) includes more small (<50 lb) fish than the longline size-frequency distribution (Ito, 1991), the small fish add relatively little to the weight of troll and handline landings (Figure 8). The combined troll and handline data (Figure 8) are mostly from Oahu where trolling predominates. Handline-caught fish tend to be even more similar in size to longline-caught fish. The reason for the decreasing catch of small fish (Figure 8) is not known.

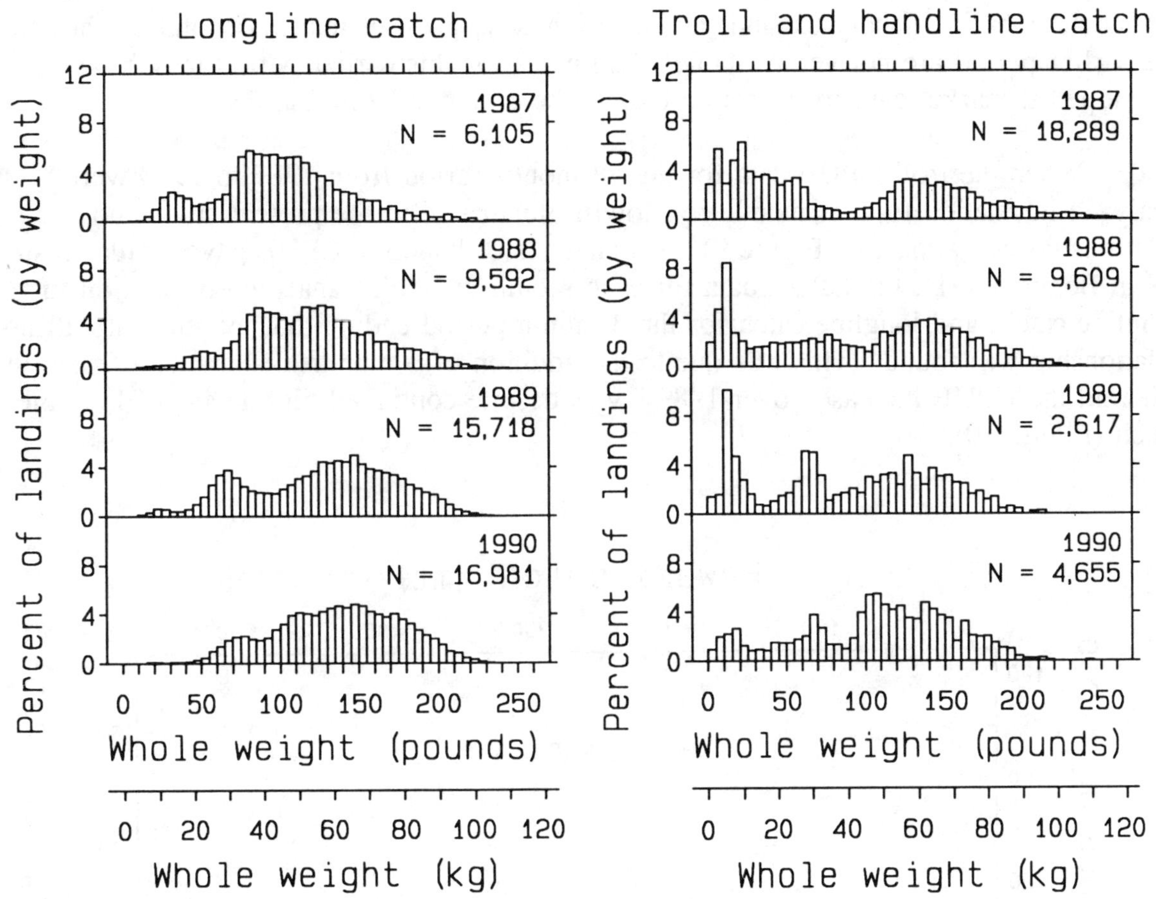

Figure 8. The proportional landings (percent of total weight landed) by weight category (pounds and kg) in Hawaii's longline and combined troll and handline fisheries. Weights of fish were obtained by the National Marine Fisheries Service shoreside monitoring programme (N = number of fish). The combined troll and handline samples are from Oahu, where trolling is the predominant small-vessel pelagic fishing method.

Changing operational characteristics in recent years make longline trips a poor measure of effective effort. For the same reason, longline-catch per trip is a poor measure of yellowfin tuna abundance and thus has not been examined. Troll and handline trips differ from each other and from longline trips in their relative effectiveness, making total effective effort difficult to estimate. Thus, total catch, rather than effort, was used as a more convenient estimate of fishing pressure.

In the preliminary analysis by Boggs (1991) of the monthly data from 1987 to June 1990, six species were examined, but only yellowfin tuna had CPUE values that appeared to be negatively related to longline catch. The relationship was clearest when monthly CPUE was plotted *versus* longline catch for the 3-month period ending the same month. Months were categorized by Boggs (1991) as "in-season" and "nonseason" to account for seasonal effects, whereas the present study used multiple regression analysis of seasonally-adjusted CPUE data on catch with interaction terms for each month.

The apparent relationship found in the preliminary study may have been due to chance, since very low troll and handline catch rates were also observed in 1982-84, well before the longline fishery expanded. Unfortunately, good estimates of total catches for the earlier period are not yet available. Estimates for this earlier period may be developed if market data from this period can be obtained from fish dealers.

When the troll CPUE data for the 48-month period from 1987 to 1990 were analyzed in relation to total monthly yellowfin tuna catch in the present study, no relationship was apparent (Figure 9). An analysis of handline CPUE gave similar results. As in Boggs (1991), the CPUE data for each month were also analyzed in relation to longline catch, and longline catch for the 3-month period ending that month, but still no relationship was found. This was due to the addition of 6 months of new data from 1990, wherein the CPUE increased over 1989 levels despite continued high levels of longline catch (Figure 10).

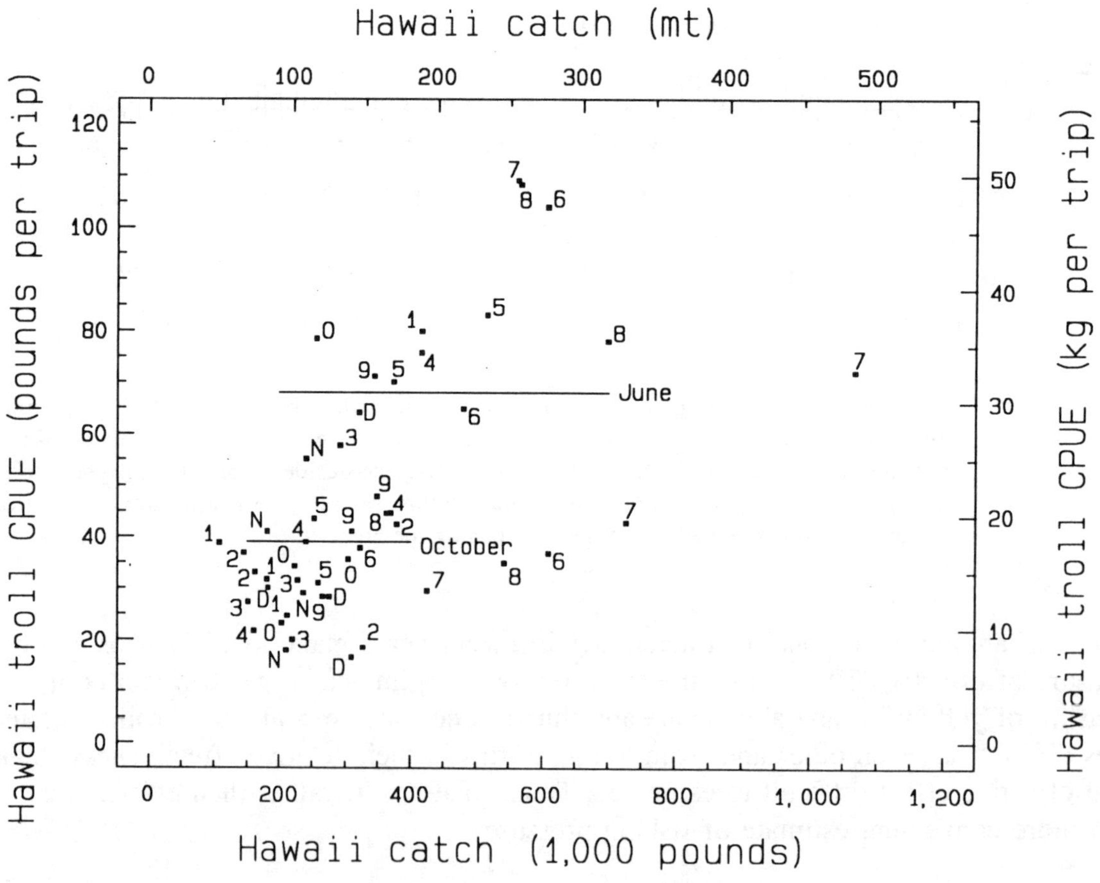

Figure 9. Observed monthly yellowfin tuna catch per unit effort (CPUE) (pounds per trip and kg per trip) in Hawaii's troll fishery, versus total monthly yellowfin tuna catch by all fishing gears (labelled as in Figure 4) for 1987-90. The lack of any CPUE versus catch relationships (slopes $\approx$ 0) based on multiple regression analysis is shown for June and October.

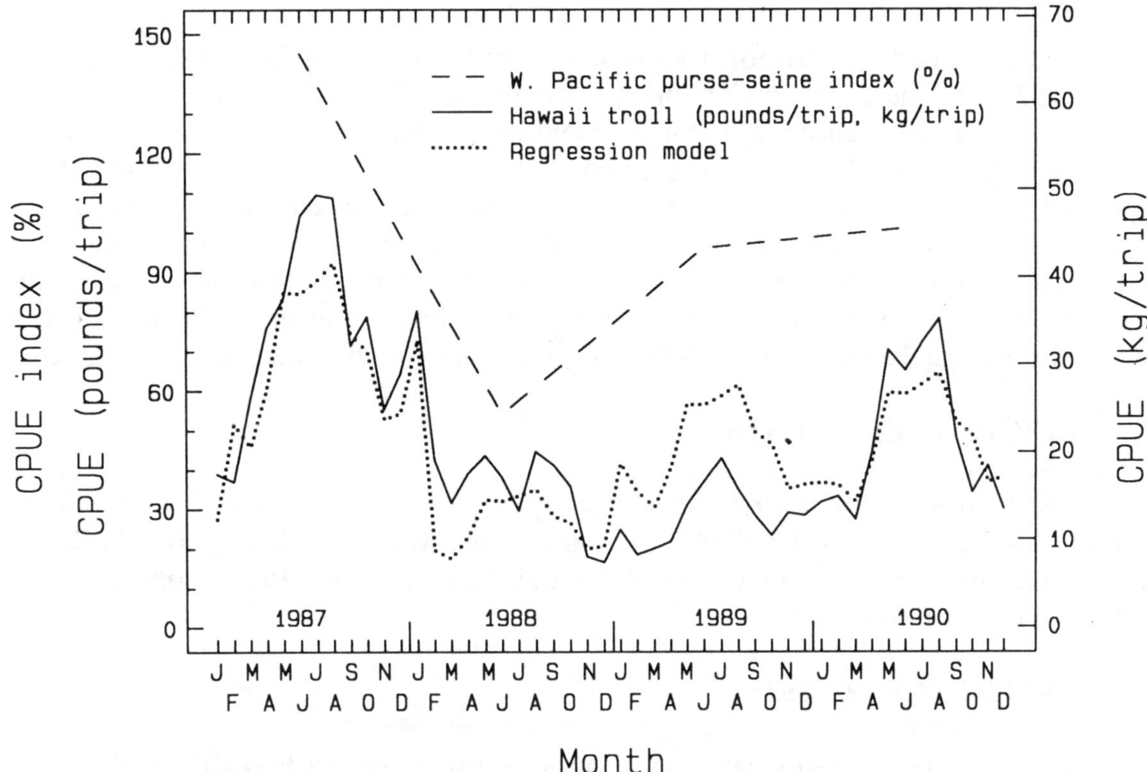

Figure 10. Observed catch per unit effort (CPUE) (pounds per trip and kg per trip) for yellowfin tuna by Hawaii's troll fishery for each month in 1987-90, and the annual index of wide-scale abundance of yellowfin tuna from the USA and Japan purse-seine fisheries (percent of average CPUE). Predicted local CPUE from a regression of seasonally-adjusted CPUE on the purse-seine index is also shown.

An index of wide-scale surface yellowfin tuna abundance from purse-seine fisheries in the western Pacific (Figure 10) was added to the regression analysis to see whether removing some of the interannual variation in immigration would reveal an underlying relationship between local CPUE and fishing intensity. The purse-seine index was calculated as the geometric mean of CPUE from the USA (Coan, 1993) and Japan (Suzuki, 1992) purse-seine fisheries in the western Pacific, expressed as a percent of mean CPUE (Figure 10). This index was the most significant factor ($P < 0.0001$) in a multiple regression of seasonally-adjusted local CPUE on total catch, the purse-seine index, and first order interactions by month.

Catch was not a significant factor (slope $\approx$ 0, Figure 9). The predicted CPUE, based on the seasonal adjustment and regression analysis, followed the pattern of observed CPUE fairly well (Figure 10). It seems that seasonality and the exogenous supply (*i.e.*, the immigration rate) of yellowfin tuna are the dominant factors affecting local CPUE, and the available evidence suggests that local CPUE was independent of local levels of exploitation in 1987-90.

It should also be noted that the purse-seine index may not necessarily represent increases in wide-scale biomass or increased local immigration. Instead it may reflect a widespread change in catchability caused by environmentally-induced changes in behaviour and vulnerability to fishing gear.

There is clearly a need for more research on the local distribution dynamics of tropical pelagic species. Simulation models could be much more helpful if more was known about true abundance and fish-movement dynamics. Tagging studies would be a good way to get such information, if they coincide with improved fisheries data collection emphasizing the geographic distribution of fishing effort and catch. The results of previous studies make it clear that local fisheries data alone may not provide evidence of existing fisheries interactions, because of the important effects of exogenous factors such as changes in stock-wide abundance and catchability. Environmental influences on fish movements and catchability may also play an important role that must be investigated.

6. ACKNOWLEDGMENTS

The NMFS data summaries for this study were provided by Sam Pooley (NMFS Honolulu Laboratory), and the HDAR data summaries were provided by Reggie Kokubun (State of Hawaii Department of Land and Natural Resources) and Reese Tokunaga (NMFS Honolulu Laboratory).

7. REFERENCES CITED

Boggs, C.H. 1991. A preliminary examination of catch rates in Hawaii's troll and handline fisheries over a period of domestic longline fishery expansion. *NOAA Admin.Rep.NMFS-SWFC, Honolulu,* H-91-05:62 p.

Coan, A. 1993. U.S. distant-water and artisanal fisheries for yellowfin tuna in the central and western Pacific. *In* Proceedings of the First FAO Expert Consultation on Interactions of Pacific Tuna Fisheries, edited by R.S. Shomura, J. Majkowski, and S. Langi, 3-11 December 1991, Noumea, New Caledonia. [See this document.]

IATTC. 1992. Annual report of the Inter-American Tropical Tuna Commission, 1990. *Annu.Rep.I-ATTC,* (1990):261 p.

Ito, R.Y. 1991. Western Pacific pelagic fisheries in 1990. *NOAA Admin.Rep.NMFS-SWFC, Honolulu,* H-91-10:43 p.

Lovejoy, W.S. 1977. A BFISH population dynamics analysis of the impact of several alternative fishery management policies in the Hawaiian Fishery Conservation Zone (FCZ) on the Pacific stocks and Hawaiian sport fishing yields of blue and striped marlins. *Western Pac.Reg.Fish.Mgt.Council,* Honolulu, Contract Report, WP-77-107:33 p.

Lovejoy, W.S. 1981. BFISH revisited. *Western Pac.Reg.Fish.Mgt. Council,* Honolulu, Contract Report:37 p.

NMFS. 1989. Guidelines for fishery management plans. *U.S. Dep. Commer., NOAA, NMFS,* 50 CFR, Part 602, 96 p.

NMFS. 1991. Status of Pacific oceanic living marine resources of interest to the USA for 1991. *NOAA Tech.Memo.NMFS-SWFC,* (165):78 p.

Sathiendrakumar, R., and C.A. Tisdell. 1987. Optimal economic fishery effort in the Maldivian tuna fishery: an appropriate model. *Mar.Res.Econ.,* 4:15-44.

Skillman, R.A., and G.L. Kamer. 1992. A correlation analysis of Hawaii and foreign fisheries statistics for billfishes, mahimahi, wahoo, and pelagic sharks, 1962-78. *NOAA Admin.Rep.NMFS-SWFC, Honolulu,* H-92-05:44 p.

Squire, J., and D. Au. 1990. A case for local depletion and core area management. *In* Planning the future of billfishes, research and management in the 90's and beyond. Part 2. Contributed papers, edited by R.H. Stroud. *Mar.Rec.Fish., Natl. Coalition Mar. Conserv.,* 13:199-214.

Suzuki, Z. 1989. Catch and fishing effort relationships for striped marlin, blue marlin, and black marlin in the Pacific Ocean, 1952 to 1985. *In* Planning the future of billfishes, research and management in the 90's and beyond. Part 1. Fisheries and stock synopses, data needs, and management, edited by R.H. Stroud. *Mar.Rec.Fish., Natl. Coalition Mar. Conserv.,* 13:165-77.

Suzuki, Z. 1992. A brief review of interaction between purse seine and longline on yellowfin tuna *Thunnus albacares,* in the western and central Pacific ocean. *In* Proceedings of the First FAO Expert Consultation on Interactions of Pacific Tuna Fisheries, edited by R.S. Shomura, J. Majkowski, and S. Langi, 3-11 December 1991, Noumea, New Caledonia. [See this document.]

Wetherall, J.A., and M.Y.Y. Yong. 1983. An analysis of some factors affecting the abundance of blue marlin in Hawaiian waters. *NOAA Admin.Rep. NMFS-SWFC, Honolulu,* H-83-16:33 p.

WPRFMC. 1986. Fishery management plan for the pelagic fisheries of the Western Pacific Region. *Western Pac.Reg.Fish.Mgt.Council,* Honolulu.

WPRFMC. 1991. Pelagic fisheries of the Western Pacific Region 1990 -- Annual Report. *Western Pac.Reg.Fish.Mgt.Council,* Honolulu, 98 p.

INTERACTIONS AMONG FISHERIES: TAG-RECAPTURE METHODS FOR ESTIMATING THE EFFECT ON CATCHES OF CHANGING FISHING INTENSITY[1]

William S. Hearn
CSIRO Division of Fisheries, Marine Laboratories
G.P.O. Box 1538, Hobart, Tasmania, 7001, Australia

and

Alexander Mazanov[2]
School of Biological Science
University of New South Wales
P.O. Box 1, Kenningston
New South Wales, 2033, Australia

ABSTRACT

A tag-recapture experiment is suggested for determining the effect that a change in the fishing intensity of one fishery will have on its own catch and on the catch of another. The results are expressed as size-based formulae in terms of one control variable: the proportional change in the fishing intensity of the altered fishery. There is no requirement to develop a complex parametric model of the dynamics of the population and fisheries or to collect comprehensive effort data. For the year-class being studied, the ideal experimental design requires that a representative sample of the fish caught by one fishery be tagged, and all tagged fish recaptured by the same fishery be re-released. In the second fishery (or fisheries), no fish need be tagged or re-released. Alternative equivalent procedures to re-releasing tagged fish are described. The fish population and fisheries are assumed to be independent of the sizes of the population and catch, i.e. a linear dynamic system is postulated. The formulae are proven for two population models for which, (i) two fisheries catch fish from a common, homogeneously mixed population, and (ii) two fisheries catch fish from separate grounds (fish may migrate between grounds). Experimental design and formulae are developed to assess interactions when the fishing intensities of two or more fisheries are to be changed or a new fishery is proposed. The results are compared with those from another experiment that requires a representative sample of the population be tagged before they are old enough to be recruited to the fisheries.

---

[1] Paper submitted to ICES J.Mar.Sci. for publication.

[2] Died shortly after preparation of an early draft of the manuscript.

A TAG-RELEASE/RECOVERY METHOD FOR PREDICTING THE EFFECT OF CHANGING THE CATCH OF ONE COMPONENT OF A FISHERY UPON THE REMAINING COMPONENTS[1]

Jacek Majkowski and William S. Hearn
CSIRO Division of Fisheries, Marine Laboratories
G.P.O. Box 1538, Hobart, Tasmania, 7001, Australia

and

Ronald L. Sandland
CSIRO Division of Mathematics and Statistics
P.O. Box 218, Lindfield, NSW, 2070, Australia

ABSTRACT

We describe (and illustrate, using data on southern bluefin tuna, *Thunnus maccoyii*) a tag-release/recovery experiment designed to predict the effect of changing the catch of one component of a fishery upon the remaining components for which fishing patterns and intensities are fixed. Formulae are developed for estimating the predicted changes and their standard errors. We also demonstrate how to determine the number of fish that should be tagged to achieve a desired accuracy of results from such an experiment. The proposed approach requires neither comprehensive knowledge of mechanisms governing the fish population and fishery being considered nor extensive historical catch and fishing effort data, but it will reflect all aspects of the dynamics of the system if the tag-release/recovery experiment is properly implemented. Consequently, it can be applied to many complex fisheries systems for which this knowledge is lacking. However, the proposed approach does not enable predictions to be made when the change in the catch significantly alters the number of recruits to the fishable stock, changes the fish behaviour, or changes the rates of natural mortality, migration, or growth.

---

[1] Published: Majkowski, J., W.S. Hearn, and R.L. Sandland. 1988. A tag-release/recovery method for predicting the effect of changing the catch of one component of a fishery upon the remaining components. *Can.J.Fish.Aquat.Sci.*, 45:675-84.

ASSESSMENT OF SKIPJACK FISHERY INTERACTION IN THE EASTERN TROPICAL ATLANTIC USING TAGGING DATA

Pierre Kleiber
La Jolla Laboratory
Southwest Fisheries Science Center
National Marine Fisheries Service, NOAA
La Jolla, California 92038 U.S.A.

and

Alain Fonteneau
O.R.S.T.O.M
Centre de Recherches Océanographiques
Dakar, Senegal

ABSTRACT

Using tagging data and catch-and-effort data from the eastern tropical Atlantic skipjack fisheries, two models were constructed for the purpose of assessing fishery interaction among skipjack fisheries in that area.

The first model deals with fish harvest, natural mortality, diffusive movement, and seasonal advective movement of tagged fish. Fitting this model to tagging data gave estimates of catchability, natural mortality, and parameters of movement. The results concerning movement imply that diffusive movement and advective movement are of about equal importance. The amplitude of annual cyclic advective displacement in the model is of the order of 1,000 km, and the average diffusive displacement in 6 months is approximately the same.

The parameter estimates obtained from the first model were incorporated into a similar model for untagged fish with a recruitment sub-model added in. By fitting this second model to catch-and-effort data, we were able to estimate parameters of the recruitment sub-model. We could then investigate interaction with this model by running it under various hypothetical regimes of fishing effort to determine how changing the effort in one region would affect the catch of fisheries operating in other regions. The interaction was small. Doubling the effort in one zone caused at most a 5% decrease in the catch of other zones.

1. INTRODUCTION

Although this work concerns the Atlantic Ocean, rather than the Pacific, it is presented here as an exercise in assessing interaction among fisheries operating in separate geographic zones. The degree of interaction between fisheries depends on the exploitation rate, that is, the fraction of total population turnover that is due to fishing mortality. Interaction between geographically separated fisheries depends additionally on movement of fish from one fishing zone to another. Tagging data contain powerful information on both turnover and movement. Our purpose, therefore, was to use skipjack tagging data as well as fishery data from the eastern tropical Atlantic (ETA) to estimate parameters of a simulation model of fish population dynamics, harvest, and movement. In the fitting procedure we utilized two similar models — one dealing with tagged fish to get movement and turnover information from tagging data, and one dealing with untagged fish to get recruitment information from catch data. The tag data we used are the results of the International Skipjack Year Programme conducted by the International Commission for the Conservation of Atlantic Tuna (ICCAT), for which tagged skipjack were released in various zones and various months during 1980 and 1981.

After estimating the parameters, we adjusted the untagged fish model to list predicted catch under the existing (or nominal) regime of effort in the fisheries. Then we compared the nominal catch with catch under various altered regimes of effort to see to what degree changes in the effort in one fishery can affect the catches in other fisheries.

The skipjack fisheries in the ETA are located along the coast of west Africa from Senegal to Angola (Figure 1). They consist of two gear types, purse-seine and pole-and-line. Previous analyses of data from the International Skipjack Year have indicated that skipjack in the ETA are under-exploited (Bard, 1986; Cayré *et al.*, 1986; Kleiber *et al.*, 1984). As such, little interaction among fisheries is to be expected. These assessments were not conducted with the benefit of quantitative information on skipjack movement within the region. They therefore dealt with the region as a whole and did not exclude the possibility of local areas within the region having higher than average exploitation rates and thereby having a deleterious effect on other local areas. The present assessment of interaction among fisheries in the region deals explicitly and quantitatively with movement.

2. FIRST MODEL

2.1 Description of the Model

The catch-and-effort data for ETA skipjack fisheries were aggregated into five zones along the west African coast (Figure 1). Little fishing effort, and no tag returns, have occurred outside these zones. Accordingly, we decided to simplify matters by making a movement model in one dimension (along the coast) rather than in two dimensions. Movement of fish away from the coast was therefore confounded with natural mortality, and our model has five spatial compartments corresponding to the five fishing zones. Skipjack do range beyond the northern and southern limits of these fishing zones. We therefore added five extra compartments to the north and five to the south so as not to impose artificial restrictions on skipjack movement due to the boundaries of the model.

The change in the number of tagged fish at large in any zone, T_i, depends on the population of tagged fish in zone i and in neighboring zones as follows:

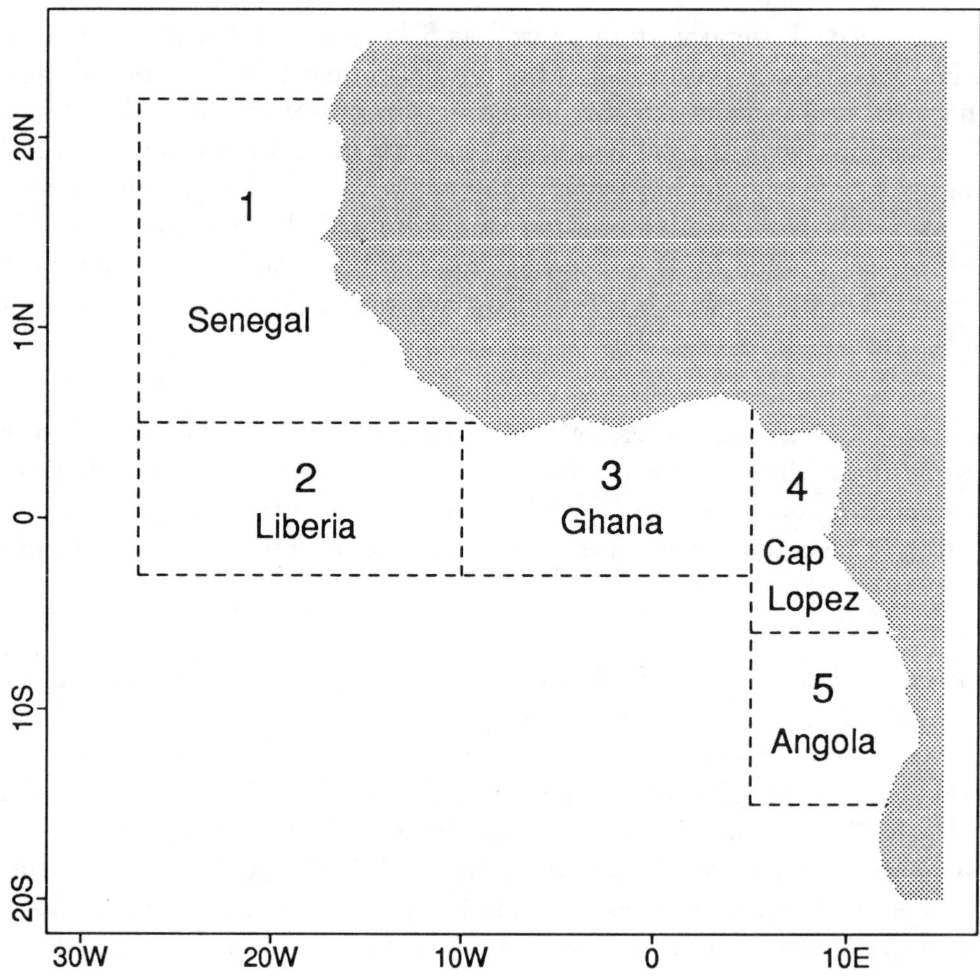

Figure 1. Five fishing zones of eastern tropical Atlantic (ETA) skipjack fisheries. These are the zones used in the model.

$$\frac{dT_{t,i}}{dt} = D(T_{t,i-1} - 2T_{t,i} + T_{t,i-1}) - V_{\max}\left|\cos\left(\omega(t-\phi)\right)\right|(T_{t,i} - T_{t,i\pm1})$$
$$- MT_{t,i} - \hat{r}_{t,i} \tag{1}$$

where

$$\hat{r}_{t,i} = qe_{t,i}T_{t,i} \tag{2}$$

is the predicted returns in time slot t and zone i, and where the initial conditions at some release time slot, τ, are zero everywhere except in the release zone, j. $T_{\tau,j}$ is then the effective number of releases.

The first term in Equation 1 manages diffusive movement with diffusivity parameter D. The second term manages variable advective movement with maximum velocity $V_{\max}$. The cosine function induces a north-south pattern of advective movement that cycles in intensity and direction. We did this because in the catch-per-unit-effort (CPUE) data there is a hint of annual north-south movement (Figure 2a). The subscript, $i \pm 1$, points to the neighbouring zone that is upstream in the movement. Plus or minus in that subscript is chosen appropriately, depending on the sign of the cosine function.

The third term in Equation 1 is natural mortality with parameter M, and the fourth term is capture by the fishery. The number of captures in any time slot, t, and zone, i, is given by Equation 2, where q is the catchability and $e_{t,i}$ is the fishing effort.

2.2 Fitting the First Model to Tag Data

We aggregated the tag-release data into sets consisting of releases in one zone within one month, and for each release set we eliminated releases, and returns therefrom, of fish that seemed to be in a different cohort from the bulk of the fish in the set. Six of the resulting release sets had 1999 or more releases. Two of those occurred within a few months of the end of the available fishery data, and we chose to concentrate on the remaining four sets. The zones and months of release are indicated in Figures 3a–d, along with plots of the number of returns by zone and month.

The effective number of releases is less than the actual number released by a factor, α, that is

$$T_{\tau,j} = \alpha R_{\tau,j} \tag{3}$$

where α is the proportion of fish that survive and do not shed their tags shortly after tagging multiplied by the proportion of recaptured tags that are returned by fishermen with useful information on time and location of recapture. In working with the International Skipjack Year tagging data, Bard (1986) estimated that α is between 0.48 and 0.72. We chose to use a value of 0.6, which is the middle of the range.

Because there are two gear types involved in the ETA skipjack fisheries, we had to deal with returns from both types of gear. Rather than explicitly providing for more than one gear type in our models, we chose to combine the purse-seine and pole-and-line catch, effort, and recovery data. To combine the effort data, we estimated a conversion factor for converting pole-and-line effort into purse-seine equivalents by comparing CPUE for the two gears for the three years covering the period of recovery (1980–1982). For the middle zone (Ghana), where most of the pole-and-line effort is exerted, the conversion procedure was more involved because the size distribution in the catch in this zone was markedly different for the two gear types (Figure 4). Because the skipjack grow rapidly (approximately 1 cm per month at the size at which they were tagged), the conversion from effort in one gear-type to equivalents in the other changes significantly during the time the fish are at large. Therefore for each tag set we generated a time series of effort in the Ghana zone by noting the average size of tagged fish at time of release and advancing the presumed size of the cohort of tagged fish by 1 cm per month. We then assumed that the ratio of pole-and-line effectiveness to purse-seine effectiveness for each month was the ratio of the pole-and-line and the purse-seine selectivity line in Figure 4 at the presumed average fish size for that month.

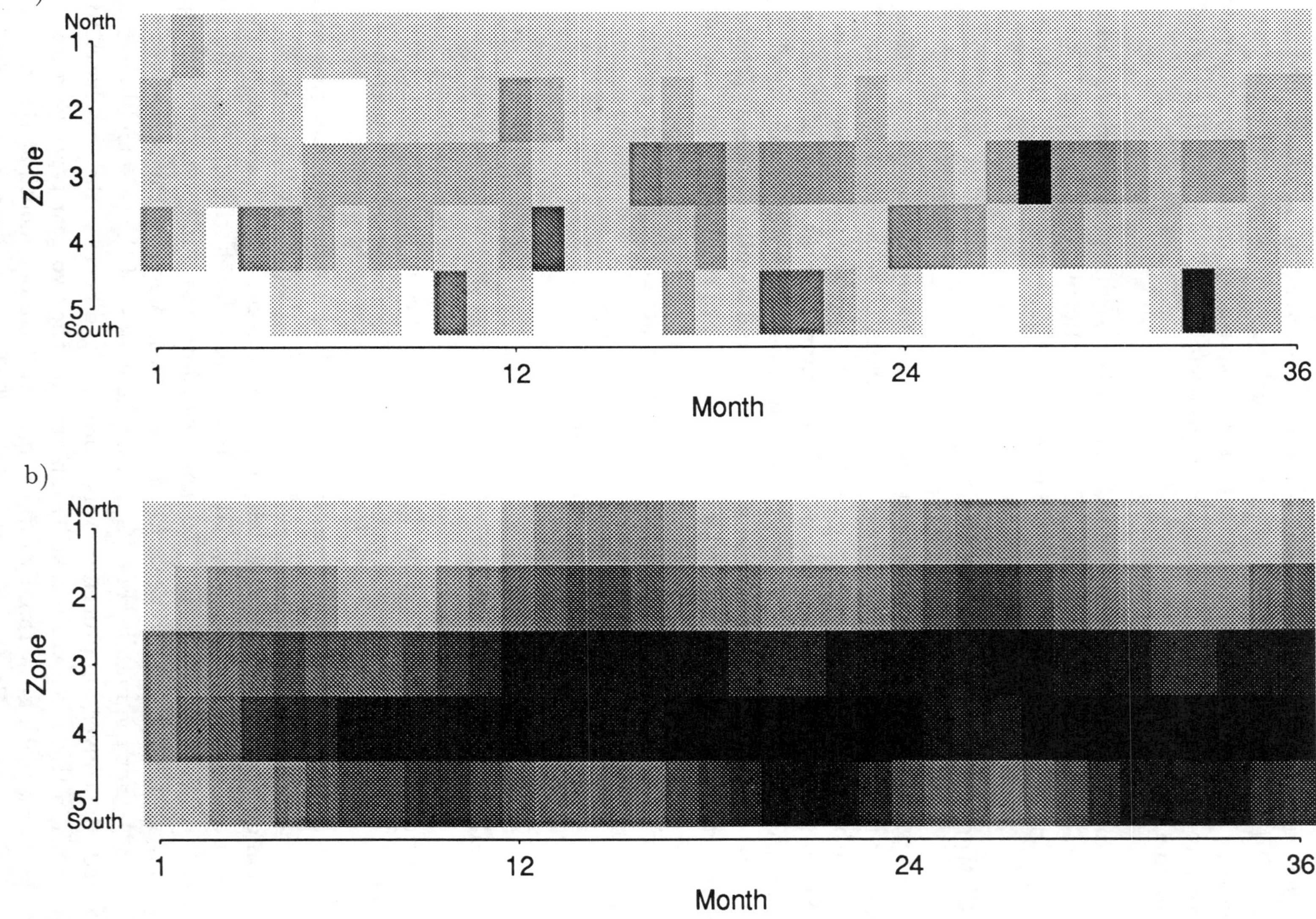

Figure 2. **a)** Catch per unit effort by month in ETA fishing zones for 36 months from January 1980 to December 1982. Blank areas indicate missing data due to lack of fishing effort. **b)** Predicted population abundance by zone and month from the second model for 36 months of simulation with nominal fishing effort in all zones.

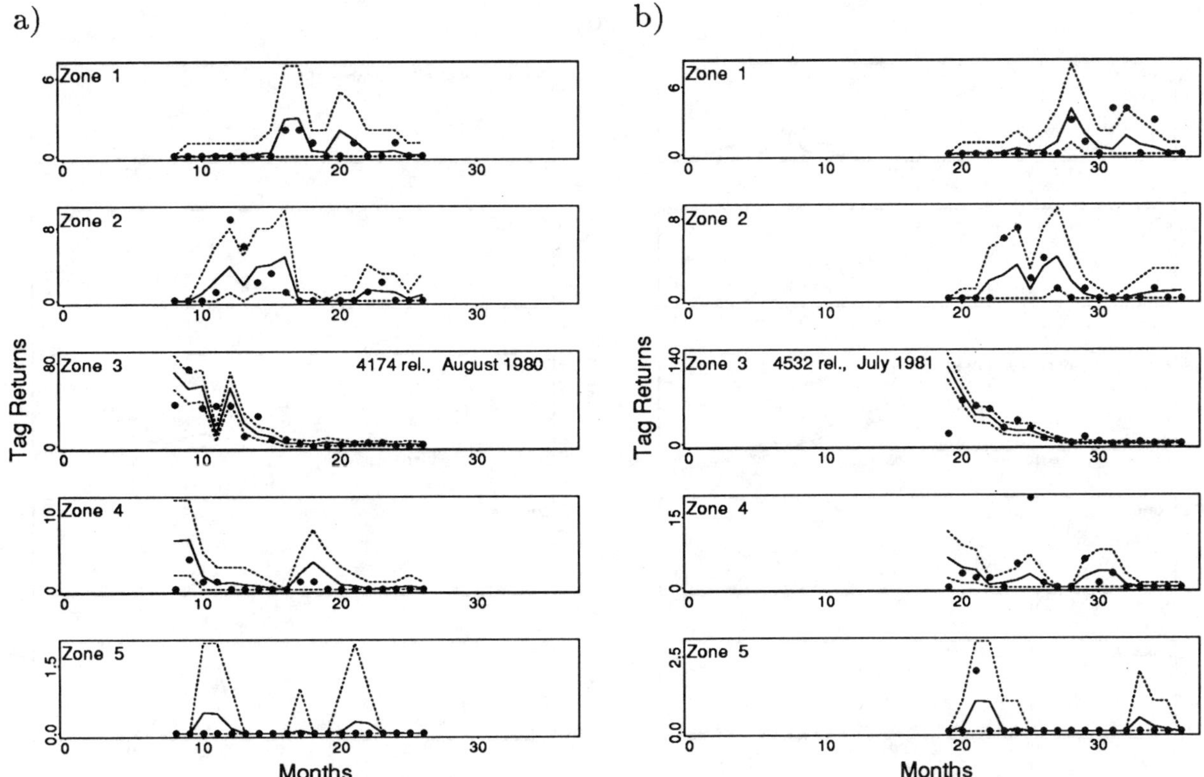

Figure 3 a–b. Predicted and actual tag return data. The "•" marks show real tag returns, and the solid lines show predicted returns from the first model with first month returns and zone 1 releases excluded from parameter estimation. The dotted lines show the 95% confidence band around the predicted values based on a poisson distribution. **a)** Returns from 4174 tagged skipjack released in zone 3 during August 1980. **b)** Returns from 4532 tagged skipjack released in zone 3 during July 1981.

To fit the model to the tag and effort data, we followed Hilborn's (1990) approach of using a model of movement and population-dynamics to predict the probability of recovery in any time slot t and zone i for a tag released in time slot τ and zone j. That probability is

$$p(t, i \mid \mathbf{B}, \tau, j) = \frac{\hat{r}(\mathbf{B})_{t,i,s}}{\alpha R_s} \tag{4}$$

where $\mathbf{B}$ is the vector of parameter values, and $\hat{r}(\mathbf{B})_{t,i,s}$ is the number of returns predicted by the model in time slot t and zone i for tag set s in which R_s tagged fish were released in time slot τ and zone j. From such probabilities we can calculate a likelihood value corresponding to the set of parameter values, $\mathbf{B}$, used to generate the probabilities. We chose the following multinomial likelihood function:

$$\mathcal{L}(\mathbf{B}) = \prod_s \left\{ \left[\prod_{t,i} p(t, i \mid \mathbf{B}, \tau, j)^{r_{t,i,s}} \right] \left[1 - \sum_{t,i} p(t, i \mid \mathbf{B}, \tau, j) \right]^{\left(T_{\tau,j} - \sum_{t,i} r_{t,i,s}\right)} \right\} \tag{5}$$

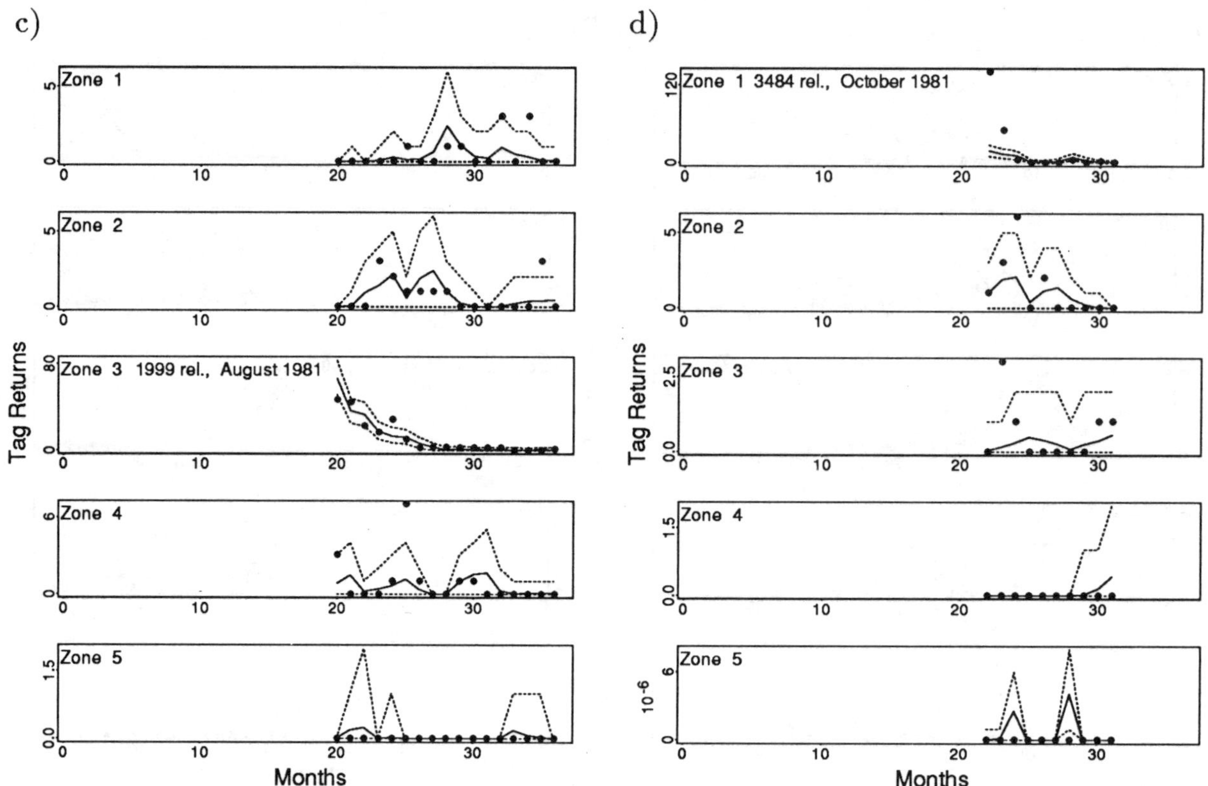

Figure 3 c–d. Predicted and actual tag return data. The "•" marks show real tag returns, and the solid lines show predicted returns from the first model with first month returns and zone 1 releases excluded from parameter estimation. The dotted lines show the 95% confidence band around the predicted values based on a poisson distribution. **c)** Returns from 1999 tagged skipjack released in zone 3 during August 1981. **d)** Returns from 3484 tagged skipjack released in zone 1 during July 1981.

where we have ignored the combinatoric factor because it does not depend on **B**, and where $r_{t,i,s}$ is the real number of tag returns in time slot t and zone i from tag set s. With more than one set of tag releases, a combined likelihood function is simply the product of individual likelihoods for each set. Fitting the model boils down to finding the set of parameter values, **B**, that gives the maximum possible likelihood. To accomplish that, we used the Nelder-Mead searching algorithm (Press *et al.*, 1988, p. 305).

2.3 <u>Results of Fitting the First Model</u>

We estimated five parameters with the first model: catchability q, natural mortality M, diffusivity D, maximum advection $V_{\max}$, and phase ϕ. We fixed the frequency parameter ω to $2\pi/12$, which produces a 12 month cycle in advective movement.

Before fitting our model to real data, we tried fitting it to tag data generated by the model. We therefore knew the parameter values that the fitting procedure should

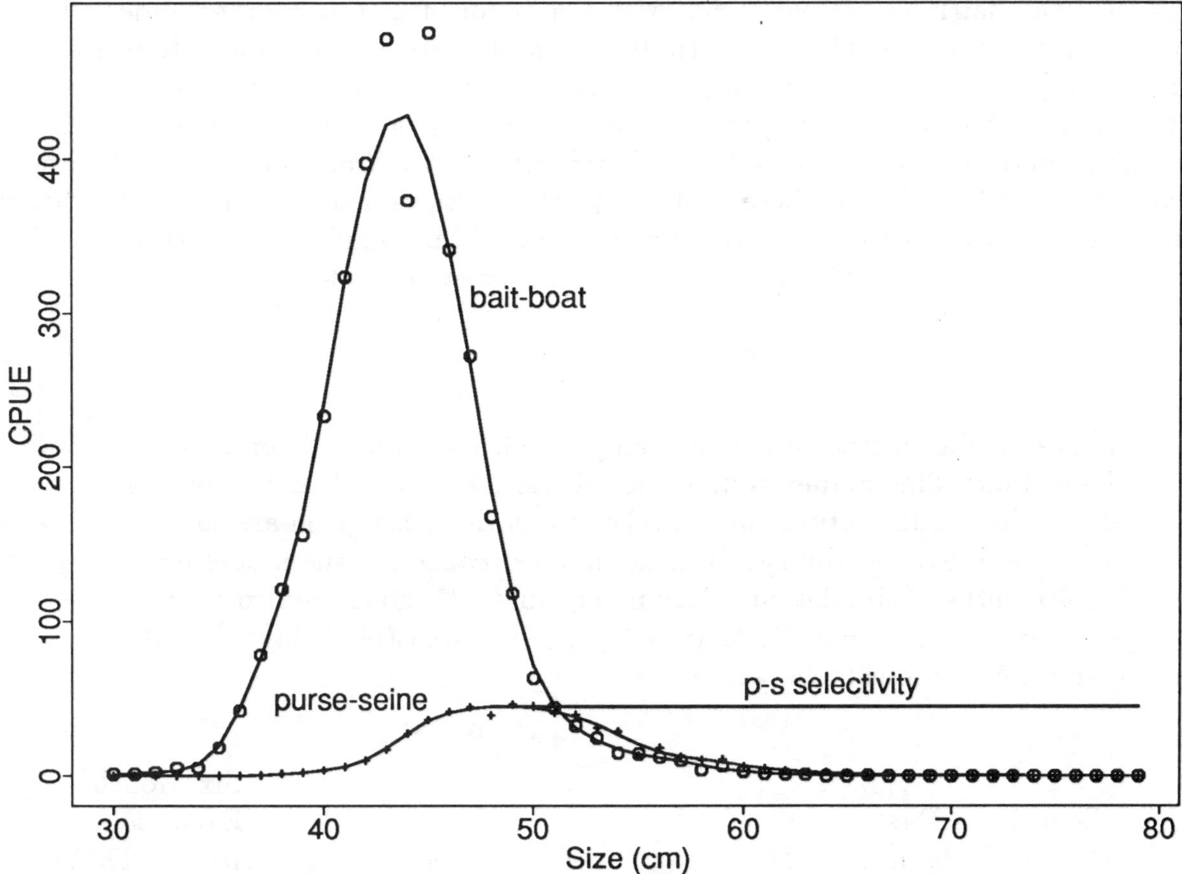

Figure 4. Average catch per unit of effort at size in zone 3 during 1980 through 1983 for pole-and-line vessels and purse seiners. The smoothed lines through the pole-and-line data and the left limb of the purse-seine data are our assumed selectivity curves.

home in on. We found that the Nelder-Mead procedure would converge readily and give reasonable estimates of the parameters for a variety of conditions of high and low values of the parameters.

In fitting the model to real data, we again found that it converged readily. We tried all combinations of including or excluding the returns in the month of release and including and excluding the only release set for zone 1. The argument for excluding data for the first month of returns is that the recovery rate within that month is expected to be anomalous because the tagged fish have not had a chance to mix with the untagged population. This could lead to anomalously high or low recovery rates depending on whether the fish were released close to or far away from concentrations of fishing effort. The argument for excluding the zone 1 tag set is that the whole recovery pattern appears anomalous. The recovery rate in that zone drops precipitously suggesting that there was bulk movement out of the zone, but the tagged fish do not appear in significant numbers in any of the other zones (Figure 3d). Perhaps the initial tag survival rate was unusually low.

Regardless of the arguments for and against excluding data, the parameter estimates are fairly consistent, being within a factor of 2 of each other under all inclusion–exclusion combinations (Table 1). In all cases exploitation rate is low, with the fishing mortality at the average prevailing effort levels being an order of magnitude smaller than the high apparent natural mortality of 5–10% per month. The apparent natural mortality could include migration away from the coast. The low exploitation rate estimated here confirms previous assessments of under-exploitation in ETA skipjack fisheries based on analyses that did not explicitly incorporate fish movement (Bard, 1986; Cayré *et al.*, 1986; Kleiber *et al.*, 1984).

Table 1. Parameter estimates from the first model for all combinations of including and excluding the release set in zone 1 and including and excluding the returns in the release month. Also given are the advective and average diffusive fish excursions realized in the model during 6 months of simulation. Parameter units: D (zone area)(mo)$^{-1}$; V_{max} (zone width)(mo)$^{-1}$; M (mo)$^{-1}$; q (zone area)(boat day)$^{-1}$(mo)$^{-1}$; ϕ (mo); excursions (zone width).

Zone 1 Releases	First Month	D	V_{max}	M	q	ϕ	Six Month Excursions Adv.	Diff.
y	y	0.041	0.31	0.10	2.1×10^{-5}	6.3	1.2	0.98
y	n	0.050	0.14	0.061	1.1×10^{-5}	6.1	0.58	0.78
n	y	0.033	0.19	0.076	1.3×10^{-5}	5.9	0.77	0.77
n	n	0.027	0.17	0.053	9.7×10^{-6}	5.6	0.69	0.71

Although the diffusion parameter estimates are smaller than the advection parameter estimates, the actual amounts of diffusive and advective movement are about the same. There is a certain amount of diffusive movement from the advection term of the model that is an artifact of the spatial discretization. Thus there is a lower limit to the amount of diffusion in the model, even with the diffusion parameter set to zero. If the actual diffusive movement of the skipjack were less than that limit, the diffusion parameter estimate would presumably have been pushed to zero. To measure the diffusive and advective movements realized in the model, we ran the model with all mortality turned off and noted the excursion of the center of gravity of the population calculated as follows:

$$\text{centre of gravity} = \bar{i}_t = \frac{\sum_i i T_{t,i}}{\sum_i T_{t,i}} \tag{6}$$

The amplitude of annual variation in $\bar{i}_t$ is the advective excursion that the population makes in six months. In Table 1, this measure of advective movement is compared with the average diffusive movement in six months, calculated as the average displacement of the population from its centre of gravity following six months of simulation starting with releases at one location:

$$\text{average displacement} = \frac{\sum_i \left|i - \bar{i}_6\right| T_{6,i}}{\sum_i T_{6,i}} \tag{7}$$

These excursions are in units of model zone widths which are of the order of 1000 km. This is only a rough approximation because of the loose correspondence between zones in Figure 1 and the linear array of zones in the model.

The predicted tag returns under the last set of parameter values in Table 1 are shown in Figures 3a–d along with the real returns. For each tag set, the best fit seems to be in the release zone for that tag set with the exception of the one set released in zone 1 (Figure 3d) with its anomalously high attrition rate in the first two months. This tag set was excluded from the parameter estimating procedure for the results shown in these figures, yet the fit of predicted returns in zones other than the release zone is about as good as is the fit in non-release zones for the other tag sets (Figures 3a–c), that is, not so bad when the high stochastic variability evidenced by the 95% confidence bands is considered. The broad bands result from the low number of predicted tags in individual zone–time strata.

3. SECOND MODEL

3.1 Description of the Model

The second model is very similar to the first. It deals with the total fish population in each zone rather than just tagged fish. As such, it needs to incorporate recruitment by growth of fish into the vulnerable size class, and there need to be non-zero initial conditions in all zones rather than a single release zone. For our second model, the change in the population of fish in any zone P_i is as follows:

$$\frac{dP_{t,i}}{dt} = D(P_{t,i-1} - 2P_{t,i} + P_{t,i-1}) - V_{\max}\left|\cos\left(\omega(t - \phi)\right)\right|(P_{t,i} - P_{t,i\pm1})$$
$$- MP_{t,i} - \hat{c}_{t,i} + g_i \tag{8}$$

where

$$\hat{c}_{t,i} = qe_{t,i}P_{t,i} \tag{9}$$

is the predicted catch in time slot t and zone i, and where recruitment and initial conditions are given by

$$g_i = G\exp(-\frac{(i - I)^2}{2\sigma^2}) \quad ; \quad P_{0,i} = P_{\max}\exp(-\frac{(i - I)^2}{2\sigma^2}) \tag{10}$$

The rationale for the last pair of equations, which give respectively recruitment and initial conditions, is that we wanted to avoid estimating 30 parameters of recruitment and starting population for all of the 15 zones. Because the population should be close to zero at the extremes of the array of zones with a maximum somewhere in the middle, we assumed that the north–south distribution could be characterized by a Gaussian curve with mean I and standard deviation σ. As further approximations we assumed that the parameters of the Gaussian curve would be the same for both initial population and recruitment except for the scale factors, G and $P_{\max}$, and we assumed that recruitment would be constant in time. We are aware that these are only rough approximations, and not even entirely self-consistent (σ would be expected to be larger for the population distribution than for the recruitment distribution), but considering the short life span of fish in the population (high apparent natural mortality) and the high degree of noise in the data, we believe that these are minor considerations.

3.2 **Fitting Second Model to Catch Data**

The parameters of this second model that were not either fixed *a priori* or estimated from the first model are the scale factors of recruitment and population, G and $P_{\max}$, and the location and spread of the Gaussian curve, I, and σ. For a given set of values for the above four parameters, the model predicts catch by time slot and zone from which we could calculate a sum of squares of differences from the real catch data. Again using the Nelder-Mead algorithm, we found the set of those parameters that minimized the sum of squares.

For all runs of the second model, we fixed D, $V_{\max}$, M, q, and ϕ, which were estimated from the first model, to the values in the last row of Table 1. As before, the Nelder-Mead algorithm converged readily to give estimates for G, $P_{\max}$, I, and σ (Table 2). The corresponding predicted population levels increase during the 36 months of simulation and approach a quasi-steady-state of annual north–south movement due to cyclical advection (Figure 2b). The north–south shifting of population abundance in the model compares well with the shifting CPUE pattern in Figure 2a, particularly in the seasonal timing of abundance peaks. This correspondence is remarkable in that the model parameters that determine fish movement are all estimated from tag data that is independent of the CPUE data in Figure 2a. The detailed historical sequence of CPUE is not so well predicted by the model, which is not surprising because the model has no mechanism for predicting the highly variable and episodic events in the CPUE data.

The reason for the growth in the predicted population is that the estimated initial population parameter is low relative to the equilibrium population level supported by the estimate of recruitment. These parameter estimates do depend on the catch and effort data used to calculate the CPUE levels shown in Figure 2a. They reflect growth in CPUE that is evident in that figure, particularly in zone 3 in which CPUE was lower than usual in the first year and extra high towards the end in month 28. Zone 3 accounts for the predominant amount of catch and therefore has a predominant influence in the estimation procedure (Table 3). We are uncertain as to what extent this growth in the model population represents real growth of the population during 1980–1982, or some other effect such as increasing catchability or availability, or simply an anomalous event in zone 3 during month 28.

Table 2. Parameter estimates from the second model. D, V_{max}, M, q, and ϕ were fixed to the values in the first row of Table 1. Parameter units: G (#)(zone area)$^{-1}$(month)$^{-1}$; P_{max} (#)(zone area)$^{-1}$; I (zone width); σ (zone width).

G	P_{max}	I	σ
0.42×10^9	0.32×10^9	8.7	0.84

Table 3. The nominal catch (10^6 fish) in each zone during 36 months of simulation under the nominal effort regime.

		Zone		
1	2	3	4	5
13	24	110	57	3.3

Table 4. Matrices showing percentage change of the simulated catch with respect to the nominal catch (Table 2) over 36 months in each zone under each of a number of altered effort regimes. The first matrix shows the effect of halving the effort of each zone, and the next shows the effect of doubling effort. Row labels indicate zones with altered effort, and column labels indicate affected zones.

Halve Effort:

	1	2	3	4	5
1	-49.	0.26	0.056	0.017	0.015
2	1.2	-49.	0.19	0.096	0.083
3	2.6	2.6	-48.	1.1	0.88
4	0.42	0.43	0.42	-49.	1.2
5	0.019	0.020	0.018	0.032	-50.

Double Effort:

	1	2	3	4	5
1	91.	-0.49	-0.11	-0.032	-0.029
2	-2.3	95.	-0.37	-0.18	-0.16
3	-4.8	-4.8	88.	-1.9	-1.6
4	-0.81	-0.82	-0.80	94.	-2.4
5	-0.038	-0.039	-0.036	-0.064	99.

4. USING SECOND MODEL TO ESTIMATE FISHERY INTERACTION

To investigate interaction, we collected the total nominal catch by zone (Table 2) from a run of the second model with best fitting parameter values and the actual (nominal) effort levels. We then made several runs of the model, two for each of the five fished zones, halving the effort in the zone for one of the runs and doubling the effort in the other run. For each run we compared the total catch by zone with the corresponding nominal catch (Table 4). By far the largest effect of such alteration of the effort is on the catch in the zone of the altered effort, the change in catch being almost proportional to the change in effort. This would be what we would expect for an under-exploited population. By contrast, in the zones other than the one with altered effort, the catch was only lightly affected, the largest effect being a 4.8% cut in the catch of zones 1 and 2 caused by a doubling of effort in zone 3.

The message of Table 4 is that interaction among the ETA skipjack fishing zones is negligible under the 1980–1982 pattern and levels of fishing effort. If anything, the interaction estimates probably err on the high side because our analysis used values of movement parameters from the first row of Table 1, that is, the highest of the range of such estimates.

5. DISCUSSION

We have confirmed previous assessments indicating that there is a low exploitation rate in ETA skipjack fisheries, and, within the limitations of our model, we have excluded the possibility that local areas with above average exploitation have a significant effect on neighbouring areas even though the overall exploitation rate is low. One of the limitations of the model is its degree of spatial resolution. We cannot exclude interactive effects on a smaller spatial scale within the model zones, for example, interaction between artisanal fisheries operating very close to the coast and commercial fisheries operating further off-shore.

It would be possible to adjust our model to finer spatial scales, but this would cause increasing problems of stochastic variability because as the data are divided into smaller and smaller strata, the number of tag returns in those strata diminishes giving rise to higher stochastic variability. Also, at finer resolution, our simplification of a linear array of zones would become untenable, and the greater complexity of two spatial dimensions would have to be introduced. Our assumptions about the distribution of recruitment and starting population would then clearly need to be revised. These are not insurmountable difficulties. The most problematic would be the need for more tagging data. A comparison of the current model with a finer scale simulation model (probably in two spatial dimensions), might be a good way to demonstrate the need for and help justify the expense of more tagging. However, justification of the modeling effort (and tagging effort) from the point of view of interaction concerns might depend on a substantial increase in fishing activity in the region so that there is a more substantial expectation of significant interaction than at present.

The current model could be refined in other ways besides increasing the spatial resolution. The timing and direction of advective movement, for example, might be linked to environmental parameters for which data exist, which could allow the cosine function in the advection term of the model to be replaced by an environmental driving function. Or the recruitment function could be relaxed so that it is not strictly symmetrical, and it could be allowed to vary, or shift north–south with season. Our guess is that as long as the skipjack population is under-exploited, our conclusions with respect to interaction would be robust to such refinements.

6. REFERENCES CITED

Bard, F.X.   1986.   Analyse des taux de décroissance numerique des listaos marqué en Atlantic Est. *In* Proceedings of the ICCAT Conference on the International Skipjack Program, edited by P.E.K. Symons, P.M. Miyake, and G.T. Sakagawa, Madrid, Spain, ICCAT, pp. 348–62.

Cayré, P., A. Fonteneau, and T. Diouf.   1986.   Elements de biologie affectant la composition en taille des listaos (*Katsuwonus pelamis*) exploités dans l'Atlantic tropical oriental et leur effet sur l'analyse de la croissance par la methode de Petersen. *In* Proceedings of the ICCAT Conference on the International Skipjack Program, edited by P.E.K. Symons, P.M. Miyake, and G.T. Sakagawa, Madrid, Spain, ICCAT, pp. 326–34.

Hilborn, R.   1990.   Determination of fish movement patterns from tag recoveries using maximum likelihood estimators. *Can.J.Fish.Aquat.Sci.*, 47(3):635–43.

Kleiber, P., S. Chivers, and E. Weber.   1984.   Analysis of ISYP skipjack tagging results using the methods of the South Pacific Commission. *Collect.Vol.Sci.Pap. ICCAT*, 20(1):212–16.

Press, W.H., B.P. Flannery, S.A. Teukolsky, and W.T. Vetterling.   1988.   Numerical recipes in C. The art of scientific computing.   Cambridge University Press, 735 p.

EVALUATION OF ADVECTION-DIFFUSION EQUATIONS FOR ESTIMATION OF MOVEMENT PATTERNS FROM TAG RECAPTURE DATA

John R. Sibert and David A. Fournier
Otter Research Ltd.
P. O. Box 265, Station A
Nanaimo, British Columbia, Canada

ABSTRACT

A version of the diffusion-advection differential equation is evaluated a as method of describing the large scale movement patterns of tunas. The equation is solved on a 50×50 grid to model a 3,000 nmi square section of the ocean. Simulated tag recaptures display features that are qualitatively similar to observed tag recaptures. An estimation procedure based on the same equation is able to recover the parameters used in the simulation from the simulated tag recaptures.

1.   INTRODUCTION

Expansion of fishing grounds and total fishery yields have broadened management issues from simple questions of sustainable yield to include questions about the effects of fisheries in one area on fisheries in another area. If these effects indeed occur, they will occur because the fish stocks on which these fisheries depend may move between areas. The ability to describe quantitatively the movements of fish populations would, therefore, be critical to a quantitative assessment of interaction between fisheries in different areas.

Models traditionally used to assess stocks, for example the general production model of Pella and Tomlinson (1969), are "closed" in that there are no explicit terms to account for immigration and emigration of individuals to and from the vulnerable population. If fisheries in different areas interact, interaction is presumably due to exchange of fish between the two vulnerable populations. (Excluding secondary effects such as habitat destruction or ghost fishing.)

There are remarkably few fisheries models described in the literature that consider the question of movement and that also include methods for estimating movement parameters. One such model was briefly discussed by Beverton and Holt (1957), but no parameter estimation method was presented. Ishii (1979a, 1979b) used tag return data to estimate exchange rates of yellowfin tuna between different regions in the eastern tropical Pacific. Sibert (1984), using a model of skipjack movement very similar to that proposed by Beverton and Holt (1957), derived a formula to describe the effects of catches in one fishery on catches in another based on exchange rates between the two fisheries estimated from tag return data. An important result of this study was to show explicitly that the potential for interaction between areas depended not only on movement between areas but also on the magnitude of natural mortality and the intensity of fishing. A fishery exploiting a population with a high natural mortality in one area is not likely to have a large impact on a fishery in a different area. Hampton (1991) applied a similar model to southern bluefin tuna with three recapture fisheries. Hilborn (1990) applied an approximation of the Sibert (1984) model to a larger number of fisheries and also to different species of fish.

All of these models parameterize movement by instantaneous bulk transfer coefficients which express the proportion of the population in one area moving to another in a unit of time. Such transfer coefficients implicitly carry the assumption that the population is uniformly distributed within an area, and depend on the sizes of the areas selected and the distances between them. Thus, it is difficult to use these models to predict movement to areas not explicitly included in the model when its parameters were estimated.

General transport models based on the so-called "advection-diffusion" equation are much less restrictive and do not have the same scale dependency as the bulk transfer models. There are other important advantages to using models based on general transport equations to describe movements of fish. These models have well known properties and have enjoyed considerable success in predicting the transport of various materials. Furthermore, there are general procedures for obtaining stable numerical solutions to these equations in several dimensions (Carnahan *et al.*, 1969; Press *et al.*, 1988).

The general transport model not only describes transport of materials. It can also be interpreted as a limiting case of certain classes of biased random walks. Various derivations of the general transport model and its application to biological systems are extensively discussed by Okubo (1980). The more recent review by Othmer *et al.* (1988) presents some alternative variations of the same general class of models.

Perhaps the simplest assumption about the movement of fish is that there are two components to movement: directed movement and random movement. Directed movement is represented by a vector of movement velocities (u, v) where u is the eastward component of directed movement and v is the northward component of directed movement. Random movement is represented by a single scalar quantity σ which is the same in all directions. The movement parameters of a general transport model applied to fish have an obvious relation to the similar parameters in models of physical transport in the sea. The vector (u, v) corresponds to advection by water currents and σ corresponds to the diffusion coefficient (hence the name diffusion-advection equation). General transport models may thus be used to model fish movement in conjunction with models in which physical transport mechanisms also modify fish distribution.

Although the general transport equation would appear to be well suited to the quantitative description of movement, estimation of model parameters from biological data has received relatively little attention. Banks *et al.* (1985 and 1988) describe procedures for estimating parameters of a one-dimensional transport model from beetle tagging data and a two-dimensional model of fly dispersion. Dalgaard and Larsen (1990) discuss parameter estimation in models involving numerical solutions of differential equations in the context of physiology.

Preliminary efforts to produce a parameter estimation procedure for a small-scale two dimensional general transport model suggested that such a model would be a useful tool for the planning and analysis of large scale tuna tagging experiments (Sibert and Fournier, 1989). The purpose of this report is to evaluate the usefulness of the this model on a more realistic scale. Three criteria will be used: Does the model move tags in a reasonable fashion? Does the model estimate the parameters accurately? Is the estimation procedure computationally practical? This report should be interpreted as an interim report of work in progress. Suggestions for further work will be included.

2.   GENERALIZED MODEL

Three steps are involved in the development of a procedure for estimating the parameters of a general transport model from tagging data. First, the appropriate partial differential equation must be defined. Second, the partial differential equation must be solved on a computer using a numerically-stable discrete approximation (usually a finite differencing scheme) in such a way that the dynamics of the equation are retained. Finally, a numerical parameter estimation procedure must be applied to estimate the model parameters from tag return data.

The derivation of the partial differential equation for the advection-diffusion model can be found in Okubo (1980) and will not be reproduced here. The general transport model for tagged fish in two dimensions can be written as

$$\frac{dA}{dt} = \sigma\left(\frac{\partial^2 A}{\partial x^2} + \frac{\partial^2 A}{\partial y^2}\right) - \frac{\partial}{\partial x}(u_s A) - \frac{\partial}{\partial y}(v_s A) - (F_s + M)A \tag{1}$$

where $A(x,y,t)$ denotes the population density of a cohort of tagged fish (or more simply tags) at the point (x,y,t). The parameters F_s and M are components of mortality attributable to fishing and "natural" causes respectively, that is, fractions of the local population which are caught or die in a unit of time. The subscript s indicates seasonal dependence of some model parameters. A "season" is defined as a period of time over which all model parameters are constant, and is discussed in more detail below. The directed components of movement and fishing mortality are assumed to have seasonal dependency. The random component of movement and natural mortality are assumed to be constant over all seasons. The subscript s is dropped in the following discussions to simplify the notation unless season is pertinent.

Several types of boundary conditions are possible. In general, boundaries may be closed or permeable, straight or regular, or these conditions may be combined. Closed or reflective boundaries simulate continental land masses and islands. Open or permeable boundaries simulate conditions when fish migrate out of the study region or through a lethal boundary (such as a low temperature isotherm for tropical tunas). For the purposes of this evaluation, only closed straight boundaries are considered:

$$\frac{\partial A}{dx} = 0; \qquad x = 0, x = X$$

$$\frac{\partial A}{dy} = 0; \qquad y = 0, y = Y$$

The obvious initial conditions for A are

$$A(x,y,0) = \begin{cases} A_r; & x = x_r \text{ and } y = y_r \\ 0; & \text{otherwise,} \end{cases}$$

where A_r is the number of tagged fished released at point (x_r, y_r) at time 0.

3. NUMERICAL SOLUTION OF THE DIFFERENTIAL EQUATION

The above differential equations are solved by finite-difference techniques using a network of regularly-spaced grid points through the region of interest. The spacing of grid points are x, y, and t which are indexed by the subscripts i, j, and k respectively where $i = 1, 2, \ldots, m; m = X/x; j = 1, 2, \ldots, n; n = Y/y$ and $k = K_1, K_1 + 1, \ldots, K_2 - 1, K_2$ where K_1 and K_2 are the first and last time steps in season s. Thus $A_{i,j,k}$ is used to approximate $A(ix, jy, kt)$ or $A(x, y, t)$.

The following finite difference approximations are used:

$$\frac{\partial A}{\partial t} \sim \frac{A_{i,j,k+1} - A_{i,j,k}}{\Delta t} \tag{2}$$

$$\frac{\partial^2 A}{\partial x^2} \sim \frac{A_{i-1,j,k+1} - 2A_{i,j,k+1} + A_{i+1,j,k+1}}{(\Delta x)^2} \tag{3}$$

$$\frac{\partial^2 A}{\partial y^2} \sim \frac{A_{i,j-1,k+1} - 2A_{i,j,k+1} + A_{i,j+1,k+1}}{(\Delta y)^2} \tag{4}$$

$$\frac{\partial}{\partial x}(uA) \sim \begin{cases} u_{i,j} \dfrac{A_{i,j,k+1} - A_{i-1,j,k+1}}{\Delta x} + A_{i-1,j,k} \dfrac{u_{i,j} - u_{i-1,j}}{\Delta x}, & u_{i,j} > 0 \\[2mm] u_{i,j} \dfrac{A_{i+1,j,k+1} - A_{i,j,k+1}}{\Delta x} + A_{i+1,j,k} \dfrac{u_{i+1,j} - u_{i,j}}{\Delta x}, & u_{i,j} < 0 \end{cases} \tag{5}$$

$$\frac{\partial}{\partial y}(vA) \sim \begin{cases} v_{i,j} \dfrac{A_{i,j,k+1} - A_{i,j-1,k+1}}{\Delta y} + A_{i,j-1,k} \dfrac{v_{i,j} - v_{i,j-1}}{\Delta y}, & v_{i,j} > 0 \\[2mm] v_{i,j} \dfrac{A_{i,j+1,k+1} - A_{i,j,k+1}}{\Delta y} + A_{i,j+1,k} \dfrac{v_{i,j+1} - u_{i,j}}{\Delta y}, & v_{i,j} < 0 \end{cases} \tag{6}$$

Dependency on time step $k + 1$ leads to "implicit" solution methods. The implicit method is used because it converges to a solution of the partial differential equations regardless of the value of the ratios $t/(x)^2$ and $t/(y)^2$ (Carnahan *et al.*, 1969). The alternating direction method is used to find solution of the equations resulting from the implicit differencing scheme (Press *et al.*, 1988). This method solves the equations in one direction after one half time step then solves the equations in the other direction in the second half time step.

Defining $H_{i,j,k} = A_{i,j,k+\frac{1}{2}}$, the complete finite difference equation for $u_{i,j} > 0$, $v_{i,j} > 0, 1 < i < M, 1 < j < N$ for the first half time step in the x direction is

$$\frac{H_{i,j,k} - A_{i,j,k}}{\Delta t/2} + (F_{i,j} + M)H_{i,j,k} -$$

$$\sigma \frac{H_{i-1,j,k} - 2H_{i,j} + H_{i+1,j,k}}{(\Delta x)^2} + u_{i,j} \frac{H_{i,j,k} - H_{i-1,j,k}}{\Delta x} + H_{i-1,j,k} \frac{u_{i,j} - u_{i-1,j}}{\Delta x}$$

$$= \sigma \frac{A_{i,j-1,k} - 2A_{i,j,k} + A_{i,j+1,k}}{(\Delta y)^2} - v_{i,j} \frac{A_{i,j,k} - A_{i,j-1,k}}{\Delta y} - A_{i,j-1,k} \frac{v_{i,j} - v_{i,j-1}}{\Delta y} \tag{7}$$

The terms on the right hand side of the equation depend only on the values of A at the end of time step k. Re-arrangement yields

$$\left(-\frac{\sigma}{(\Delta x)^2} - \frac{u_{i-1,j}}{\Delta x}\right)H_{i-1,j,k} + \left(\frac{2}{\Delta t} + \frac{2\sigma}{(\Delta x)^2} + \frac{u_{i,j}}{\Delta x} + F_{i,j} + M\right)H_{i,j,k} + \left(-\frac{\sigma}{(\Delta x)^2}\right)H_{i+1,j,k} =$$

$$- \left[\left(-\frac{\sigma}{(\Delta y)^2} - \frac{v_{i,j-1}}{\Delta y}\right)A_{i,j-1,k} + \left(-\frac{2}{\Delta t} + \frac{2\sigma}{(\Delta y)^2} + \frac{v_{i,j}}{\Delta y}\right)A_{i,j,k} + \left(-\frac{\sigma}{(\Delta y)^2}\right)A_{i,j+1,k}\right] \tag{8}$$

For the second half time step in the y directions the equation is

$$\frac{A_{i,j,k+1} - H_{i,j,k}}{\Delta t/2} -$$

$$\sigma\frac{A_{i,j-1,k+1} - 2A_{i,j,k+1} + A_{i,j+1,k+1}}{(\Delta y)^2} + v_{i,j}\frac{A_{i,j,k+1} - A_{i,j-1,k+1}}{\Delta y} + A_{i,j-1,k+1}\frac{v_{i,j} - v_{i,j-1}}{\Delta y} =$$

$$\sigma\frac{H_{i-1,j,k} - 2H_{i,j,k} + H_{i+1,j,k}}{(\Delta x)^2} - u_{i,j}\frac{H_{i,j,k} - H_{i-1,j,k}}{\Delta x} - H_{i-1,j,k}\frac{u_{i,j,k} - u_{i-1,j,k}}{\Delta x} -$$

$$(F_{i,j} + M)H_{i,j,k} \tag{9}$$

Here the terms on the left depend on the values of A at the end of time step $k + 1$, while those on the right depend on the values of H obtained by solving (8). Re-arrangement yields

$$\left(-\frac{\sigma}{(\Delta y)^2} - \frac{v_{i,j-1}}{\Delta y}\right)A_{i,j-1,k+1} + \left(\frac{2}{\Delta t} + \frac{2\sigma}{(\Delta y)^2} + \frac{v_{i,j}}{\Delta y}\right)A_{i,j,k+1} + \left(-\frac{\sigma}{(\Delta y)^2}\right)A_{i,j+1,k+1} =$$

$$-\left[\left(-\frac{\sigma}{(\Delta x)^2} - \frac{u_{i-1,j}}{\Delta x}\right)H_{i-1,j,k} + \left(-\frac{2}{\Delta t} + \frac{2\sigma}{(\Delta x)^2} + \frac{u_{i,j}}{\Delta x} + F_{i,j} + M\right)H_{i,j,k} + \right.$$

$$\left.\left(-\frac{\sigma}{(\Delta x)^2}\right)H_{i+1,j,k}\right] \tag{10}$$

Solution of (8) and then (10) yields the values of A at the end of time step $k + 1$.

Equations (8) and (10) may be more clearly expressed in general form as

$$
\begin{aligned}
b_{1,j}H_{1,j,k} + c_{1,j}H_{2,j,k} &= g_{1,j}\\
a_{2,j}H_{1,j,k} + b_{2,j}H_{2,j,k} + c_{2,j}H_{3,j,k} &= g_{2,j}\\
a_{3,j}H_{2,j,k} + b_{3,j}H_{3,j,k} + c_{3,j}H_{4,j,k} &= g_{3,j}\\
&\;\;\vdots\\
a_{i,j}H_{i-1,j,k} + b_{i,j}H_{i,j,k} + c_{i,j}H_{i+1,j,k} &= g_{i,j}\\
&\;\;\vdots\\
a_{m-1,j}H_{m-2,j,k} + b_{m-1,j}H_{m-1,j,k} + c_{m-1,j}H_{m,j,k} &= g_{m-1,j}\\
a_{m,j}H_{m-1,j,k} + b_{m,j}H_{m,j,k} &= g_{m,j}
\end{aligned}
$$

for the x direction. The corresponding equations for the second half time step in the y direction are

$$
\begin{aligned}
e_{i,1}A_{i,1,k+1} + f_{1,j}A_{i,2,k+1} &= h_{i,1}\\
d_{i,2}A_{i,1,k+1} + e_{i,2}A_{i,2,k+1} + f_{i,2}A_{i,3,k+1} &= h_{i,2}\\
d_{i,3}A_{1,2,k+1} + e_{i,3,k+1}A_{i,3} + f_{i,3}A_{i,4,k+1} &= h_{i,3}\\
&\;\;\vdots\\
d_{i,j}A_{i,j-1,k+1} + e_{i,j}A_{i,j,k+1} + f_{i,j}A_{i,j+1,k+1} &= h_{i,j}\\
&\;\;\vdots\\
d_{i,n-1}A_{i,n-2,k+1} + e_{i,n-1}A_{i,n-1,k+1} + f_{i,n-1}A_{i,n,k+1} &= h_{i,n-1}\\
d_{i,n}A_{i,n-1,k+1} + f_{i,n}A_{i,n} &= h_{i,n}
\end{aligned}
$$

which are two systems of linear equations with tridiagonal matrices of coefficients.

The coefficients $a_{i,j}, b_{i,j}, c_{i,j}$, and $d_{i,j}, e_{i,j}, f_{i,j}$ are the diagonal and off-diagonal elements of two tridiagonal matrices. The formulas for the tridiagonal coefficients and for the right hand terms $g_{i,j}$ and $h_{i,j}$ can be deduced by inspection of equations 8 and 10. Special cases are required for $i = 1, j = 1, i = m$, and $j = n$ (edges and corners) for both reflective and open boundaries and for $u_{i,j} < 0, v_{i,j} < 0$ for all values of i and j. Since the movement parameters depend on season, the tridiagonal coefficients also depend on season. Defining $z_{i,j} = F_{i,j} + M$, the formulas for the tridiagonal coefficients for all special cases are:

$$a_{i,j} = \begin{cases} -\left(\dfrac{\sigma}{(\Delta x)^2} + \dfrac{u_{i-1,j}}{\Delta x}\right) & 1 < i \le m; u_{i,j} > 0 \\[3ex] -\dfrac{\sigma}{(\Delta x)^2} & 1 < i \le m; u_{i,j} < 0 \\[3ex] 0 & i = 1 \end{cases}$$

$$d_{i,j} = \begin{cases} -\left(\dfrac{\sigma}{(\Delta y)^2} + \dfrac{v_{i,j-1}}{\Delta y}\right) & 1 < j \le n; v_{i,j} > 0 \\[3ex] -\dfrac{\sigma}{(\Delta y)^2} & 1 < j \le n; v_{i,j} < 0 \\[3ex] 0 & j = 1 \end{cases}$$

$$b_{i,j} = \begin{cases} \dfrac{2}{\Delta t} + \dfrac{2\sigma}{(\Delta x)^2} + \dfrac{u_{i,j}}{\Delta x} + z_{i,j} & 1 < i < m; u_{i,j} > 0 \\[3ex] \dfrac{2}{\Delta t} + \dfrac{2\sigma}{(\Delta x)^2} - \dfrac{u_{i,j}}{\Delta x} + z_{i,j} & 1 < i < m; u_{i,j} < 0 \\[3ex] \dfrac{2}{\Delta t} + \dfrac{\sigma}{(\Delta x)^2} + z_{i,j} & i = 1, i = m \end{cases}$$

$$e_{i,j} = \begin{cases} \dfrac{2}{\Delta t} + \dfrac{2\sigma}{(\Delta y)^2} + \dfrac{v_{i,j}}{\Delta y} & 1 < j < n; v_{i,j} > 0 \\[3ex] \dfrac{2}{\Delta t} + \dfrac{2\sigma}{(\Delta y)^2} - \dfrac{v_{i,j}}{\Delta y} & 1 < j < n; v_{i,j} < 0 \\[3ex] \dfrac{2}{\Delta t} + \dfrac{\sigma}{(\Delta y)^2} & j = 1, j = n \end{cases}$$

$$c_{i,j} = \begin{cases} -\dfrac{\sigma}{(\Delta x)^2} & 1 \le i < m; u_{i,j} > 0 \\[3ex] -\left(\dfrac{\sigma}{(\Delta x)^2} - \dfrac{u_{i,j}}{\Delta x}\right) & 1 \le i < m; u_{i,j} < 0 \\[3ex] 0 & i = m \end{cases}$$

$$f_{i,j} = \begin{cases} -\dfrac{\sigma}{(\Delta y)^2} & 1 \le j < n; v_{i,j+1} > 0 \\[3ex] -\left(\dfrac{\sigma}{(\Delta y)^2} - \dfrac{v_{i,j+1}}{\Delta y}\right) & 1 \le j < n; v_{i,j+1} < 0 \\[3ex] 0 & j = n \end{cases}$$

$$g_{i,j} = -d_{i,j}A_{i,j-1,k} + \left(\frac{2}{t} - e_{i,j}\right)A_{i,j,k} - f_{i,j}A_{i,j+1,k}$$

$$h_{i,j} = -a_{i,j}H_{i-1,j,k} + \left(\frac{2}{t} - b_{i,j}\right)H_{i,j,k} - c_{i,j}H_{i+1,j,k}$$

The tridiagonal coefficients $a_{i,j}, b_{i,j}, c_{i,j}, d_{i,j}, e_{i,j}$, and $f_{i,j}$ depend only on the movement parameters $\sigma, u_{i,j}$, and $v_{i,j}$, and on $z_{i,j}$, but not on A. They need only be computed once for each season. The right hand terms, $g_{i,j}$ and $h_{i,j}$, depend on the the current values of the state variable ($A_{i,j,k}, H_{i,j,k}$), and must be calculated at each time step. They can be easily computed from the tridiagonal coefficients. The tridiagonal systems of equations are solved by the recursive alogrithm in Press *et al.* (1988).

4. MODEL PARAMETERIZATION

4.1 Regional and Seasonal variation

For a realistic problems $m \sim n \sim 50$, which if all $u_{i,j}, v_{i,j}, \ldots$ are estimated, leads to a very large number of parameters. It is not likely that so many parameters would be estimable. In the interest of parsimony, it would seem more fruitful to be able to describe movement patterns with a much smaller number of parameters. Furthermore, with $x \sim y \sim 60$ miles, variation in the movement parameters at such a fine scale of resolution is not likely to be of much interest over a large region. Fisheries managers are more likely to be interested in movement patterns and potential fisheries interactions occurring on a larger scale.

Define a "mapping matrix" composed of elements $l_{i,j}$ such that $1 \leq l_{i,j} \leq L \ll mn$ for $i = 1, 2, \ldots, m;\ j = 1, 2, \ldots, n$. The $l_{i,j}$ define L regions of the computational grid over which the model parameters are constant with respect to (x, y). Define a vector of generalized model parameter $\tilde{p}$ indexed by $l_{i,j}$ such that $p_{i,j} = \tilde{p}_{l_{i,j}}$. Then all model parameters in the partitioned grid may be represented by the matrix

$$G_s = \left\{ \tilde{u}, \tilde{v}, \tilde{F}, \ldots, \tilde{E} \right\}$$

where the parameters under the $\tilde{\ }$ have their usual meaning, $i.e.\ u_{i,j} = \tilde{u}_{l_{i,j}}$. Typically, one would expect that $L \sim 20$. The model can be reparameterized in terms of the much smaller number of parameters $\tilde{u}, \tilde{v}, \tilde{F}, \ldots$ indexed by $l_{i,j}$.

The mapping matrix l and parameter matrix G_s completely define regions and seasons. Regions are areas over which all model parameters are assumed constant and may be as large as the entire grid or as small as a single computational cell $(x \times y)$. Seasons are periods of time over which all model parameters are assumed to be constant. A season may be as short as one month or as long as twelve months. Seasons are cyclical so that season 1 in year two repeats season 1 in year one.

4.2 Fishing Mortality

Assume that the fishing mortality F_{ijk} is related to the observed fishing effort E_{ijk} by the relationship

$$F_{i,j,k} = qE_{i,j,k}. \tag{11}$$

Fishing effort is assumed to be constant during a season.

5. FITTING THE MODEL TO DATA

Let C^{obs}_{rijk} be the observed number of tag returns from tag release cohort r in computational grid i, j during month k. Let C^{pred}_{rijk} be the predicted number of tag returns from tag release cohort r in computational grid i, j during month k. Then

$$C^{\text{pred}}_{rijk} = \frac{F_{ijk}}{M + F_{ijk}} \left(1 - \exp(-(M + F_{ijk})t) \right) A_{rijk} \tag{12}$$

where A_{rijk} is the populations of tags from release group r in cell i, j at then end of time step k obtained by numerical solution of equation (1).

At present, the model parameters are estimated by minimizing the robust χ^2 function

$$\chi^2 = \sum_{rijk} \frac{\left(C_{rijk}^{\text{pred}} - C_{rijk\cdot}^{\text{obs}}\right)^2}{1 + C_{rijk}^{\text{pred}}} \tag{13}$$

The constant 1 in the denominator of (13) is to reduce the influence of improbable events (such as the occasional aberrant tag recapture) on the parameter estimates. Equation (13) is minimized by a quasi-Newton function minimization algorithm which uses the first partial derivatives of (13) with respect to the parameters. These derivatives are computed by analytically correct methods to the limit of machine accuracy using an algorithm related to the reverse mode of automatic differentiation (Griewank and Corliss, 1991).

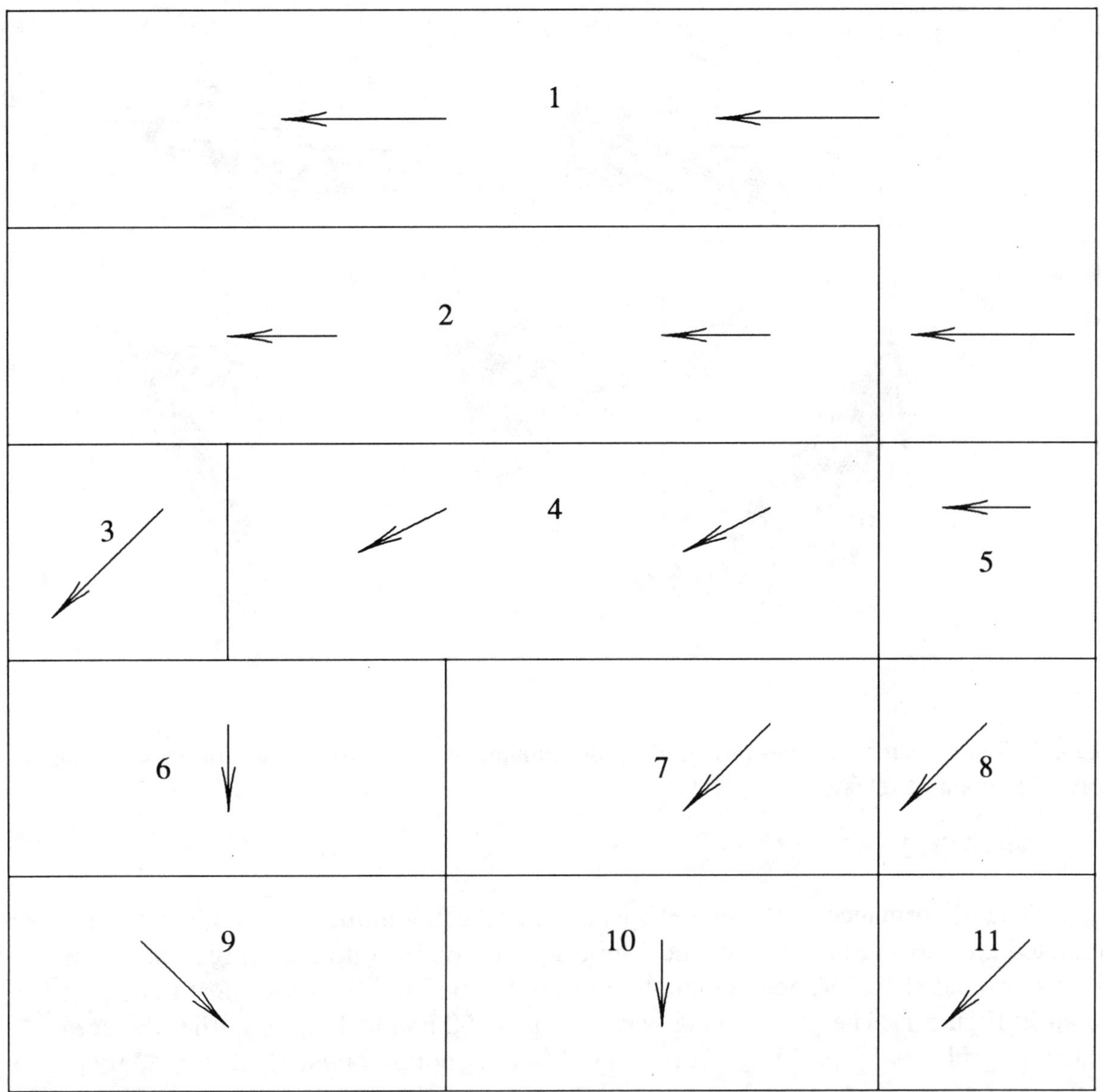

Figure 1. Regions and movement pattern used in the simulations. Numbers indicate region numbers used in Table 1 (see below). Arrows represent movement vectors (u, v) in each region.

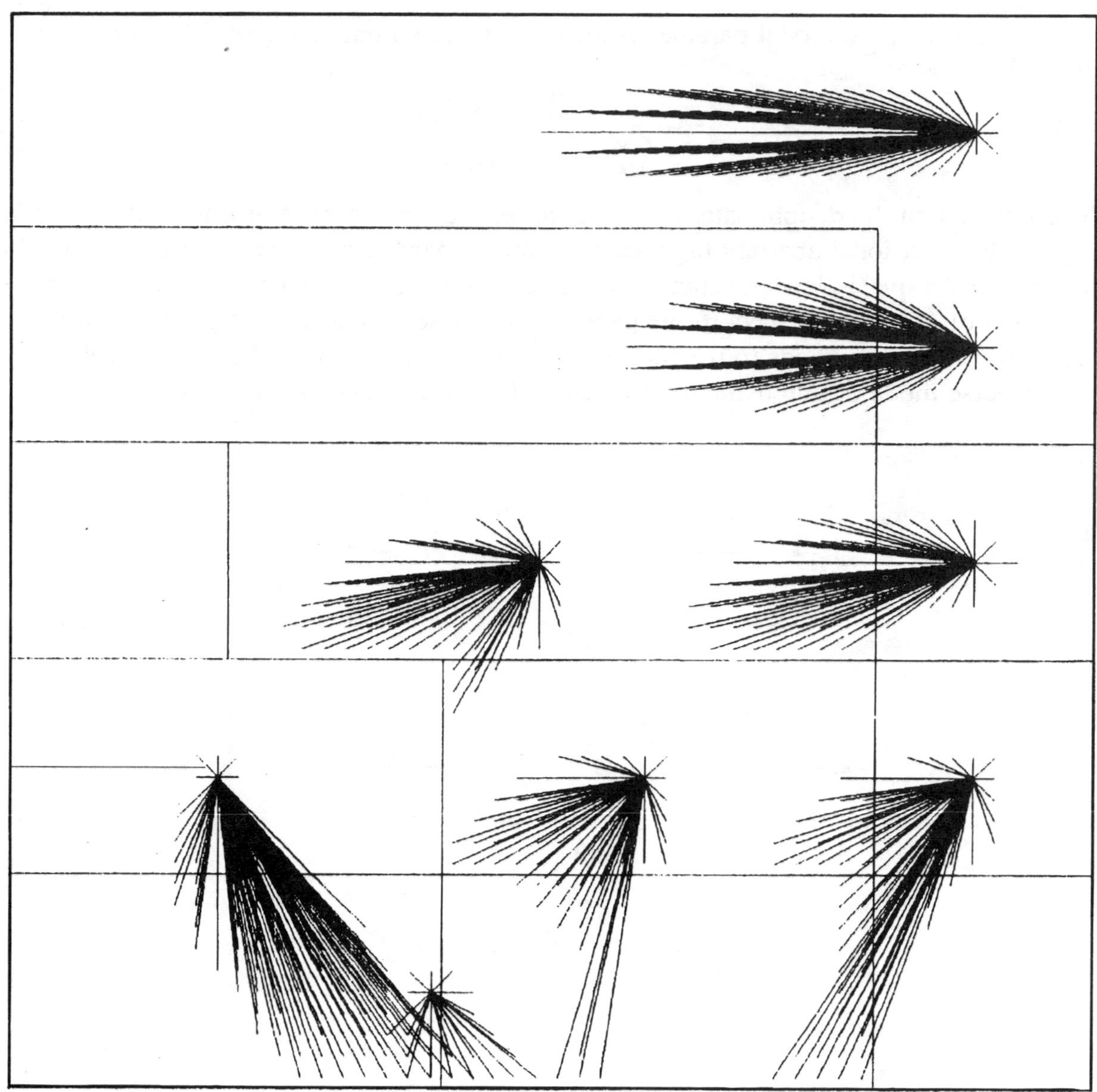

Figure 2. Tag movement arrows produced by the simulation. Lines represent movement of tags away from point of release.

6.   MODEL EVALUATION

The performance of the model was evaluated using numerical simulations. The simulated area covered a 3,000 nautical mile square divided into eleven regions of different sizes and shapes. The configuration of the regions and the movement pattern are shown in Figure 1. The grid spacing was $x = y = 60$ nautical miles so that the computational grid was 50×50. The time step, t, was kept constant at 1 month. Only a single season was used. Two thousand tags were released every 3 months in different areas from month one through month 19. The simulations were allowed to run for 24 months. Fishing effort was set to be 1000 units per month in all regions. The simulation used the finite difference approximation of equation (1) to generate populations of tags. Simulated tags recaptures were computed from

$$\eta q E_{i,j,k} A_{rijk}$$

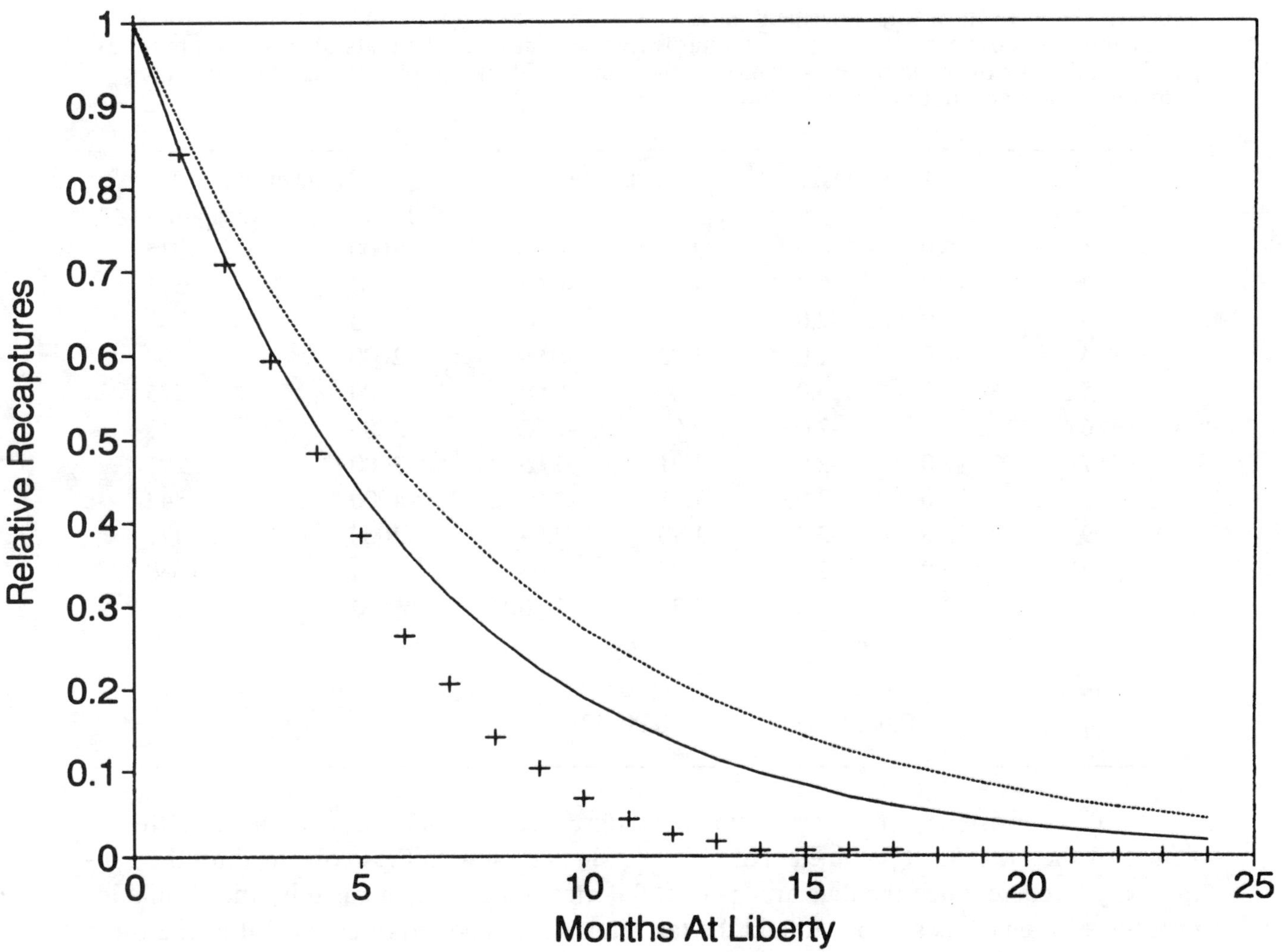

Figure 3. Tag attrition curves. The crosses represent the number of tags returned in each month relative to the number returned in month 1. The lines are plots of $(1-\exp(-(M+1000q)t))/(1-\exp(-(M+1000q)))$ with different values of M and q. The dotted line is plotted with $M = 0.03$ and $q = 0.0001$ (the values used in the simulation); the solid line is plotted with $M = 0.0662$ and $q = 0.000123$ (the values estimated from the simulated tag returns).

where $\log(\eta)$ is a normally distributed random variable with mean 0 and standard deviation σ_η. The simulation produces some fraction of a tag return in every cell in every month resulting in an impossibly large data set. Therefore, only tag returns greater than 0.1 were recorded.

No problems with lack of numerical stability were noted for values of $u_{i,j}$ and $v_{i,j}$ less than about 120 miles per month. For higher rates of directed movement, it is likely that t must be decreased. The resulting tag movement arrows are shown in Figure 2. The model would appear to move tags in a reasonably realistic way, although the value of σ used in the simulation is probably too low. The number of tags returned in each month relative to the number returned in month 1 are plotted as crosses (+) in Figure 3. The shape of the attrition curve is fairly realistic although somewhat lower than expected (dotted line in Figure 3).

Table 1. Comparison of true movement parameters with estimated movement parameters for a simulation with near perfect data and tag returns truncated at 0.1. The units of u and v are miles per day; the units of σ are miles^2 per month; the units of M are month^{-1}; the units of q are numbers of tags per unit of fishing effort.

	True Value		Estimate		Number of	
Region	u	v	u	v	Releases	Recaptures
1	-3.0	0.0	-2.97	0.04	6000	1915
2	-2.0	0.0	-1.95	-0.04	0	521
3	-2.0	-2.0	-1.36	-1.01	0	0
4	-2.0	-1.0	-1.92	-0.99	2000	1008
5	-2.0	0.0	-2.00	0.04	2000	525
6	0.0	-2.0	0.06	-2.02	2000	613
7	-2.0	-2.0	-1.91	-2.05	2000	297
8	-2.0	-2.0	-1.95	-2.04	4000	843
9	2.0	-2.0	1.90	-2.04	2000	477
10	0.0	-2.0	0.05	-2.08	0	1630
11	-2.0	-2.0	-2.34	-1.80	0	196
σ	750		580			
M	0.03		0.0662			
q	0.0001		0.000123			

The consistency of the estimation procedure can be evaluated by comparing the values of the movement parameters used in the simulation with those obtained by the estimation procedure when the data are "perfect." Perfect data are generated by the simulation procedure when σ_η is set to zero and the tag returns are reported in every cell to the limit of computational accuracy. Such precision is only possible for very small computational grids and short periods of time. When presented with "perfect" data on a small (6×5) computational grid, the estimation procedure functions very well. It recovers all parameter values used in the simulation with a very high accuracy.

A more interesting case occurs when the simulated tag returns are truncated. In this case, the simulator rounds off the tag returns to an arbitrary threshold and only reports returns greater than the threshold. Table 1 presents the results when $\sigma_\eta = 0$ and the reporting threshold is 0.1 tag.

The results show that the estimation procedure is able to accurately recover the parameters of directed movement for most regions provided there have been recaptures in that region. It appears that it is not necessary to release tags in all regions in order to estimate the directed movement parameters. This result, if sustained by further simulations, has obvious implications for planning tagging experiments. The estimation procedure seems to over-estimate natural mortality and under-estimate random movement when tag returns are truncated. This result is understandable bacause the model predicts some fraction of a tag for every computational cell but the observed tags are only reported for cells where the number of returns is greater than 0.1 tag. This discrepancy is interpreted by the estimation procedure as mortality. The magnitude of the bias is related to the truncation threshold. If the threshold is decreased from 0.1 to 0.01, the estimate for M is 0.042. Further work is currently in progress to improve the simulation and to develop an alternative to equation (13) that will be less influenced by fractional tags.

Table 2. Comparison of χ^2 from parameter estimations to different regional configurations. Only the movement parameters (u, v, σ) were estimated; M and q were held constant at their true values. σ_η was set to 0.5 and the simulated recaptures were truncated to 1.

Case 1	Case 2 "truth"	Case 3	Case 4

Case	Alteration	χ^2	Parameters
1	Split Region 2	2690	25
2	As simulated ("truth")	2694	23
3	Combine regions 3, 4, & 5	3043	19
4	Move boundary between regions 4 & 5	2757	23

The sensitivity of the estimation procedure to different movement patterns can be evaluated by comparing the values of the χ^2 calculated by the model for different configurations of the regions. The test configurations and statistical results are presented in Table 2.

Comparison of cases 1, 2, and 3 is a situation in which the number of regions and their configuration is unknown. The comparison starts with more regions than are required and combines regions until some statistical test indicates that the goodness of fit has decreased significantly. There are 12 regions in case 1, and 11 regions in case 2. If the χ^2 calculated from (13) could be rigorously interpreted, it could be concluded that the degradation in fit gained by removing a region is not significant ($\chi^2 = 4$ with 2 degrees of freedom). There are 9 regions in case 3 and there is a large degradation in fit ($\chi^2 = 349$ with 4 degrees of freedom). It could be concluded from this analysis that the hypothesis of 11 regions (case 2) gives the best agreement with the observations. Case 4 is a situation in which the number of regions is correct but the boundary between two regions is incorrect. The number of parameters remains at 23 but the χ^2 value is much higher.

These results show that the estimation procedure is sensitive to different movement patterns. The sensitivity could be exploited in a statistical procedure to discriminate between competing hypotheses. Further work is required to refine the objective function (13) so that such hypothesis testing can be rigorously supported.

This model is demanding of computer resources. To date, it has been compiled to run on 80386 or 80486 microcomputers in protected mode, and a 80860 RISC processor (the MicroWay "Number Smasher 860"). The total memory required is not known exactly, but the minimum would probably be around 4 Mb for the size of problem considered here. The times required for a single evaluation of equation (13) are given in Table 3. The total time to complete a parameter estimation from initial estimates using the 80860 is about $5\frac{1}{2}$ hours. The computer program is implemented in C++ computer language using array extensions from the AUTODIF[1] library. It has been compiled with 4 different compilers and appears to be fairly portable among platforms.

Table 3. Times required to evaluate equation (13) in different computer systems.

Platform	Seconds
80386 (33 MHz)	692
80486 (33 MHz)	308
80860 (33 Mhz)	124

7.   CONCLUSIONS AND POTENTIAL EXTENSIONS

An important advantage of the approach discussed in this paper is that it provides a uniform quantitative framework to articulate and evaluate movement hypotheses. Such a framework has been absent from some previous discussions of tuna movement and interaction.

The estimation procedure appears to have little difficulty recovering parameter estimates from the relatively simple scenarios simulated to date. In addition, it may be possible to estimate previously intractable parameters such as the tag reporting rate and the rate of initial tagging mortality. Another reasonable question is to what extent movement patterns and fisheries interactions may be estimated with this model, in the absence of fishing effort data. Further simulations are required to address these points. These simulations would explore the ability of the model to recover parameter values from increasingly pathological situations. In particular, the simulator should be revised so that the movement is less regular.

Further development is required to produce a movement pattern estimation procedure that can be applied to real situations. Irregular boundaries and islands should be accommodated. The possibilities of incorporating anisotropic and variable diffusion to more realistically capture behaviour in coastal regions could be investigated. One of the great advantages of equation (1) is its compatibility with general circulation models used in oceanography. It may eventually be possible to combine these models to distinguish between behavioral movement and passive transport of fish by physical processes. There is considerable scope for optimization of the computations to take advantage of the 80860 processor which has a maximum potential throughput of 66 megaflops. Finally, improved graphics capabilities would be helpful to set up movement hypotheses and to examine the estimates.

---

[1] AUTODIF is a proprietary C++ array language extension with automatic differentiation for use in nonlinear modeling and statistics.

8. REFERENCES CITED

Banks, H.T., P.M. Kareiva, and P.K. Lamm. 1985. Modeling insect dispersal and estimating parameters when mark-release techniques may cause initial disturbances. *J.Math.Biol.*, 22:259-77.

Banks, H.T., P.M. Kareiva, and L. Zia. 1988. Analyzing field studies of insect dispersal using two-dimensional transport equations. *Environmental Entomology*, 17:815-20.

Beverton, R. J. H. and S. J. Holt. 1957. On the dynamics of exploited fish populations. U. K. Min. Agric. Fish., Fish. Invest. (Ser. 2)19:533 p.

Carnahan, B., H.A. Luther, and J.O. Wilkes. 1969. Applied numerical methods. New York, John Wiley & Sons, 604 p.

Dalgaard, P. and M. Larsen. 1990. Fitting numerical solutions of differential equations to experimental data:a case study and some general remarks. *Biometrics* 46:1097-1109.

Griewank, A. and G.F. Corliss. 1991. Automatic differentiation of algorithms: Theory, practice and application. Philadelphia, SIAM.

Hilborn, R. 1990. Determination of fish movement patterns from tag recoveries using maximum likelihood estimators. *Can.J.Fish.Aquat.Sci.*, 47:635-43.

Hampton, J. 1991. Estimation of southern bluefin tuna *Thunnus maccoyii* mortality and movement rates from tagging experiments. *Fish.Bull.NOAA-NMFS* 89:591-610.

Ishii, T. 1979a. Estimating parameters of a fish population supplied by sequential recruitment. *Inv.Pesq.*, 43:123-37.

Ishii, T. 1979b. Attempt to estimate migration of fish population with survival parameters from tagging experiment data by the simulation method. *Inv.Pesq.*, 43:301-17.

Okubo, A. 1980. Diffusion and ecological problems:mathematical models. New York, Springer, 254 p.

Othmer, H.G., S.R. Dunbar, and W. Alt. 1988. Models of dispersal in biological systems. *J.Math.Biol,.* 26:263-98.

Pella, J.J. and P.K. Tomlinson. 1969. A generalized stock production model. *Bull. I-ATTC* 13(3):421-96.

Press, W.H., B.P. Flannery, S.A. Teukolsky, and W.T. Vetterling. 1988. Numerical recipes in C. New York, Cambridge Univ. Press, 735 p.

Sibert, J. 1984. A two-fishery tag attrition model for the analysis of mortality, recruitment, and fishery interaction. *Tech.Rep.Tuna Billfish Assess.Programme, S.Pac.Comm.* 13:27 p.

Sibert, J. and D. Fournier. 1989. Statistical Modeling of Tuna Movement. Proceedings of the 40th Annual Tuna Conference, May 1989, edited by M. Hinton.

DISCRETE POPULATION FIELD THEORY
FOR TAG ANALYSIS AND FISHERY MODELING

Carlos A. M. Salvadó[1]
Southwest Fisheries Science Center
National Marine Fisheries Service, NOAA
Box 271
La Jolla, CA 92038-0271

ABSTRACT

I consider discrete fields representing population processes: population density and population density rate. Using the fields for a tagged fish population I construct the probability density of movement from cell to cell over an interval of time of length equal to the time the tagged fish survive in their domain. I formulate rules for the construction of expressions of these population processes using the principle of linear superposition, where each population process at a particular cell and at a particular interval of time is weighted by the probability density of an individual surviving movement into another cell over that interval of time. These expressions can be used to answer questions of fisheries interaction such as how increased effort in a portion of the population domain will affect the catch in another portion. They will also be found to be of the same form as the discrete version of the integral equation solutions of partial differential equations that model the dynamics in space-time of tagged and untagged fish populations if we identify the probability density of movement from point to point in space-time with the point-source solution, or Green function, of the differential equations.

1.   INTRODUCTION

To begin, for the purposes of this paper I define "field" to mean a quantity that has a definite value at any point in space-time. As in a field in the country, with depressions and promontories, deep valleys and high mountains, every point in a field has an elevation, a certain value to its height relative to some level, which in the case of the country field is sometimes chosen to be sea level. A population distributed in space-time constitutes a field because there is a definite value, relative to the zero population level, for example to the number of individuals in any region at any time I choose. In this work applied to fisheries, some of the fields I will be considering are the population density (*e.g.*, numbers of fish per unit area), the effort density (*e.g.*, number of hooks or boat-time spent fishing per unit area), the catch density rate (*e.g.*, number of fish caught per unit area per unit time) and the recruitment density rate (*e.g.*, number of fish recruited per unit area per unit time).

In formulating a continuous population field theory based on tagging data Salvadó

---

[1]Graduate Program, CICESE, Ensenada, Baja California.

(1993) derived field equations (*i.e.*, partial differential equations) for the population densities of tagged and untagged fish. These were solved in terms of integral equations by invoking the point-source solutions (or Green functions) that are associated with the differential operators of the field equations (Courant and Hilbert 1953; Bjorken and Drell 1964; Salvadó 1993) . The Green function, or propagator (Bjorken and Drell 1964), for the fishery can be interpreted as the probability density of surviving the movement from point to point in space-time (Salvadó 1993). This formulation of fishery problems based on tagging data has several features that makes it quite simple to use. First of all, the whole fisheries interaction problem can be solved by matrix multiplication. These matrix equations are the result of the discretization of the integral equations mentioned above, which for a single interval of time coincide with the Markovian formulation of the interaction between the discrete areas of the fishery. The advantage of the method by Salvadó (1993) over the strictly Markovian approach is that the probability density of fish movement is derivable from a field equation with an assumed parametrization. The parameters can then be evaluated by simply taking the moments of the probability density. However, the approach by Salvadó (1993) has the problem that the mathematics necessary to derive it are abstract, requiring methods that are not usually taught to fishery biologists. Therefore the work by Salvadó (1993) will not have a broad appeal to the fishery biology community unless it can be derived in a simpler way using mathematics with which most fishery biologists are familiar.

This work is a different approach to the same problem. The linear superposition of fields representing population processes is considered. The probability density of fish movement from point to point in discrete space-time is defined as the tagged fish population density divided by the number of tagged fish released. Based on the definition of probability density of fish movement, rules are formulated for the linear superposition of fields representing population processes. Expressions for the population densities of tagged and untagged fish are constructed given that there is an untagged fish population source due to recruitment, and tagged and untagged fish population sinks due to catch. The mathematical expressions that will result from this superposition for the population density and the catch density rate are matrix equations which correspond to the discretized integral equations solutions of the continuum field equations (Salvadó 1993). The probability density of causal movement from cell to cell will be identified with the causal Green function or propagator associated with the field equations (Salvadó 1993). This will be done strictly algebraically making no use of differential equations. The price that I will pay for this simplification is an inability to derive discretely the deeper results attainable with the continuum approach. However, because the linkage to the continuum field theory of Salvadó (1993) is made, all the results derived in the continuum can also be used discretely. So I warn the reader that the approach in this work is not mathematically rigourous. For the precise derivation of the results the reader must go to Salvadó (1993). The approach in this work is intuitive.

The symbols, and their meaning, used in this document are as follows:

$\mathcal{A}$ = area inhabited by a fish population, called the domain of the population
$\mathcal{A}_i$ = area of the the ith cell of $\mathcal{A}$

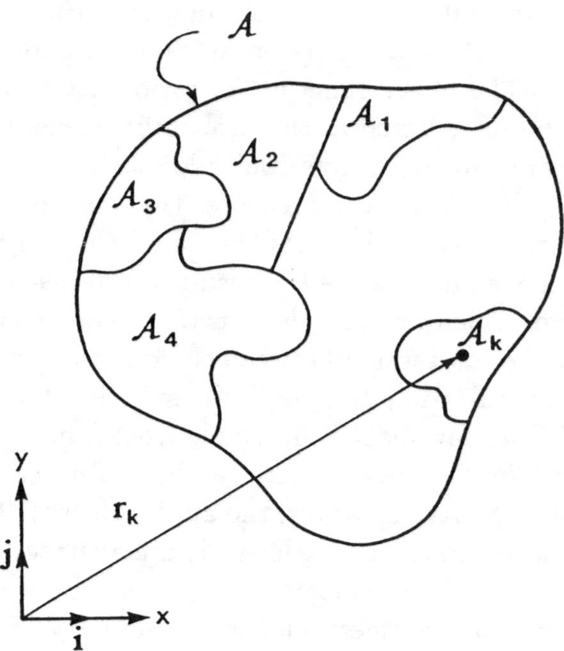

Figure 1. The area $\mathcal{A}$ inhabited by the fish population is divided into n nonoverlapping cells. The ith cell is defined as the neighborhood of area $\mathcal{A}_i$ about point $\mathbf{r}_i$.

n = total number of cells in $\mathcal{A}$

$\sum_{i=1}^{M} \mathcal{W}_i \phi(w_i) = \mathcal{W}_1 \phi(w_1) + \mathcal{W}_2 \phi(w_2) + \ldots + \mathcal{W}_M \phi(w_M)$

$\mathbf{i}$ and $\mathbf{j}$ = unit vectors, respectively, in the positive x and y directions

$\in$ = member of

$\mathbf{r}_k = x_k \mathbf{i} + y_k \mathbf{j}$ = a vector $\in \mathcal{A}_k$

t_j = jth time point

m = total number of intervals of time

$\{O_i : i = 1, 2, \ldots, M\}$ = the set $O_1, O_2, \ldots, O_M$

$\subset$ = subset of

$\subseteq$ = subset of or equal to

$<$ = less than

$>$ = greater than

$\leq$ = less than or equal to

$\geq$ = greater than or equal to

$\ll$ = much less than

$\approx$ = approximately equal to

$\sim$ = similar to, of the same order of magnitude

$[t_l, t_j] = t_j - t_l$

$\mathcal{T}_I = [t_1, t_{m+1}]$

$\mathcal{T}_T = [t_l, t_{m+1}]$ for $1 \leq l \leq m$ and therefore $\subseteq \mathcal{T}_I$

$\mathcal{T}_j = [t_j, t_{j+1}]$

$\mathcal{T}_l$ = interval of time in which tagging experiment is initiated

$[t_l, t_{l+\delta l}] \subset \mathcal{T}_l$

$e(\mathbf{r}_i, t_j)$ = effort density in cell $\mathcal{A}_i$ in the interval of time $\mathcal{T}_j$

$q(\mathbf{r}_i, t_j)$ = catchability in cell $\mathcal{A}_i$ in the interval of time $\mathcal{T}_j$, with units of $area \times time^{-1}$

$f(\mathbf{r}_i, t_j) = q(\mathbf{r}_i, t_j)e(\mathbf{r}_i, t_j)$ = death rate due to fishing (*i.e.*, fishing mortality) in the interval of time $\mathcal{T}_j$

$c_T(\mathbf{r}_i, t_j)$ = tagged fish catch density rate in cell $\mathcal{A}_i$ in the interval of time $\mathcal{T}_j$

$N_T(\mathbf{r}_k, t_l)$ = number of tagged fish released in cell $\mathcal{A}_k$ during interval of time $[t_l, t_{l+\delta l}]$

$N_D(\mathbf{r}_k, t_l)$ = number of tagged fish released in cell $\mathcal{A}_k$ during interval of time $[t_l, t_{l+\delta l}]$ that do not survive the trauma of tagging

$N_R(t_{j+1}) = \sum_{s=l}^{j \leq m} \mathcal{T}_s \sum_{v=1}^{n} \mathcal{A}_v c_T(\mathbf{r}_v, t_s)$ = the number of tags recovered in the interval of time $[t_l, t_{j+1}]$

$p_T(\mathbf{r}_i, t_j)$ = population density of tagged fish in cell $\mathcal{A}_i$ in the interval of time $\mathcal{T}_j$

$g_0(\mathbf{r}_i, t_j | \mathbf{r}_k, t_l)$ = probability density of a fish surviving the movement to cell $\mathcal{A}_i$ in the interval of time $[t_l, t_j]$ if at time t_l it was located in cell $\mathcal{A}_k$ when fishing occurs in that time interval, therefore called the effort-dependent probability density.

$r_T(\mathbf{r}_k, t_j)$ = recruitment density rate of tagged fish in cell $\mathcal{A}_k$ in the interval of time $\mathcal{T}_j$

$p(\mathbf{r}_i, t_j)$ = population density of untagged fish in cell $\mathcal{A}_i$ in the interval of time $\mathcal{T}_j$

$c(\mathbf{r}_i, t_j)$ = untagged fish catch density rate in cell $\mathcal{A}_i$ in the interval of time $\mathcal{T}_j$

$r(\mathbf{r}_k, t_j)$ = recruitment density rate of untagged fish in cell $\mathcal{A}_k$ in the interval of time $\mathcal{T}_j$

$u_0(\mathbf{r}_i, t_j)$ = population density due to recruitment in cell $\mathcal{A}_i$ in the interval of time $\mathcal{T}_j$ propagated in space-time with the effort-dependent probability density g_0

$g(\mathbf{r}_i, t_j | \mathbf{r}_k, t_l)$ = probability density of a fish surviving the movement to cell $\mathcal{A}_i$ in the interval of time $[t_l, t_j]$ if at time t_l it was located in cell $\mathcal{A}_k$ when fishing is absent in that time interval, therefore called the effort-independent probability density

$u(\mathbf{r}_i, t_j)$ = population density due to recruitment in cell $\mathcal{A}_i$ in the interval of time $\mathcal{T}_j$ propagated in space-time with the effort-independent probability density g.

2.   PROBLEM STATEMENT

Consider an area $\mathcal{A}$ inhabited by a fish population, and which is divided into n non-overlapping cells (Fig. 1):

$$\mathcal{A} = \sum_{i=1}^{n} \mathcal{A}_i .$$

The cell of area $\mathcal{A}_i$ is defined as the neighbourhood of point $\mathbf{r}_i$. An economic zone can be composed of several cells. The length of time $\mathcal{T}_I = [t_1, t_{m+1}]$ is divided into m intervals $\{\mathcal{T}_j = [t_j, t_{j+1}] : j = 1, 2, ..., m\}$ such that

$$\mathcal{T}_I = \sum_{j=1}^{m} \mathcal{T}_j = [t_1, t_{m+1}] .$$

Let $\mathbf{i}$ and $\mathbf{j}$ respectively be the orthogonal unit vectors in the positive x and y directions in a right-handed cartesian coordinate system. Let $\mathbf{r}_i = x_i\mathbf{i} + y_i\mathbf{j} \in \mathcal{A}_i$ (*i.e.*, the vector $\mathbf{r}_i$ indexes the position of a point within cell $\mathcal{A}_i$). The value of a field ϕ within cell $\mathcal{A}_i$ during the interval of time $\mathcal{T}_j$ will be written as $\phi(\mathbf{r}_i, t_j)$.

During the interval of time $\mathcal{T}_I$ the fish population in $\mathcal{A}$ is exploited with effort density given by $\{e(\mathbf{r}_i, t_j) : i = 1, 2, ..., n; j = 1, 2, ..., m+1\}$, which results in a catch density rate $\{c(\mathbf{r}_i, t_j) : i = 1, 2, ..., n; j = 1, 2, ..., m+1\}$. The fishery interaction issue I want to address is the question of how much the catch in a zone of $\mathcal{A}$ is affected by the effort applied in other zones. The linkage between fisheries in different zones depends on the pattern of fish movement. To determine this pattern, a tagging experiment is initiated at time t_l for $t_l \geq t_1$ which in the interval of time $\mathcal{T}_T = [t_l, t_{m+1}] \subseteq \mathcal{T}_I$ results in a tagged fish catch density rate $\{c_T(\mathbf{r}_i, t_j) : i = 1, 2, ..., n; j = l, l+1, ..., m+1\}$.

Using the data from the tagging experiment it is possible to define the probability of surviving the movement from cell to cell over an interval of time. This will be done by developing the rules of linear superposition of population processes.

3. CONSTRUCTION OF THE POPULATION DENSITY BY LINEAR SUPERPOSITION

3.1 Analysis of Tagged Fish: Part I

I will assume that the movement, and natural and fishing death density rates obey linear processes of population density. However, the recruitment density rate is free to be a nonlinear process of population density (Salvadó 1993). The reader should not be put off by the impracticality of some of the cases that will be presented here. They are being considered simply because of pedagogical reasons as examples of sources or sinks of population.

Let $p_T(\mathbf{r}_i, t_j)$ and $e(\mathbf{r}_i, t_j)$ respectively be the population density of tagged fish and the effort density in area $\mathcal{A}_i$ in the interval of time $\mathcal{T}_j$. If $q(\mathbf{r}_i, t_j)$ is the catchability in cell $\mathcal{A}_i$ in the interval of time $\mathcal{T}_j$, the relationship between catch density rate, and effort and population densities for the tagged portion of the population is given by

$$c_T(\mathbf{r}_i, t_j) = q(\mathbf{r}_i, t_j)\, e(\mathbf{r}_i, t_j)\, p_T(\mathbf{r}_i, t_j)\,, \tag{1}$$

and a similiar expression holds for the untagged portion:

$$c(\mathbf{r}_i, t_j) = q(\mathbf{r}_i, t_j)\, e(\mathbf{r}_i, t_j)\, p(\mathbf{r}_i, t_j)\,.$$

I consider a tagging experiment in which N_T tagged fish are released in cell $\mathcal{A}_k$ in the interval of time $[t_l, t_{l+\delta l}] \ll \mathcal{T}_l = [t_l, t_{l+1}]$. By this I mean that if a release of tagged fish is initiated at time t_l it must be completed in a time short compared to $\mathcal{T}_l$. I shall index the number of tagged fish, their location and initial time of release by writing $N_T(\mathbf{r}_k, t_l)$. The "initial condition" of the tagging experiment for the population density of tagged fish is then $p_T(\mathbf{r}_k, t_l) = N_T(\mathbf{r}_k, t_l)/\mathcal{A}_k$. Therefore, for the intervals of time $\mathcal{T}_j$ for $j \geq l$, the population density of tagged fish $\{p_T(\mathbf{r}_i, t_j) : i = 1, 2, ..., n; j = l, l+1, ..., m+1\}$ must always be such that $p_T(\mathbf{r}_i, t_j)\, \mathcal{A}_i \leq N_T(\mathbf{r}_k, t_l)$ because of possible losses due to natural death and recapture by fishing. It then follows that if I define a field

$$g_0(\mathbf{r}_i, t_j | \mathbf{r}_k, t_l) = \frac{p_T(\mathbf{r}_i, t_j)}{N_T(\mathbf{r}_k, t_l)} = \frac{c_T(\mathbf{r}_i, t_j)}{q(\mathbf{r}_i, t_j)\, e(\mathbf{r}_i, t_j)\, N_T(\mathbf{r}_k, t_l)}\,, \tag{2}$$

this four-indexed quantity g_0 has the following properties:

A. Dimensional

A dimensional analysis of (2) reveals that g_0 is a density. That is, g_0 has the dimensions of area^{-1}.

B. Causal

Because tagged fish can exist in their domain at time t_j only if they are present at an equal or earlier time t_l, g_0 must satisfy

$$g_0(\mathbf{r}_i, t_j | \mathbf{r}_k, t_l) = 0 \text{ for } t_j < t_l.$$

C. Initial

If the tagged fish are released in cell A_k at time t_l, assuming they disperse at a finite rate, initially they must be present in the cell of release only. Therefore, g_0 must satisfy

$$g_0(\mathbf{r}_i, t_l | \mathbf{r}_k, t_l) = \frac{\delta_{ik}}{A_i},$$

where the Kroenecker delta function δ_{ik} is defined by

$$\delta_{ik} = \begin{cases} 1 & \text{if } i = k \\ 0 & \text{otherwise}. \end{cases}$$

However, it is understoood that the time t_l here refers to the time $t_{l+\delta l}$ when all the tagged fish have been released.

D. Normalization

As a consequence of the initial property

$$\sum_{i=1}^{n} A_i g_0(\mathbf{r}_i, t_l | \mathbf{r}_k, t_l) = 1,$$

where I have weighted the sum with the corresponding area because of the dimensional property of g_0. Here again, the time t_l refers to time $t_{l+\delta l}$ when all the tagged fish have been released.

E. Probabilistic

Because the number of tagged fish recaptured cannot be greater than those released, for $t_j \geq t_l$

$$0 \leq g_0(\mathbf{r}_i, t_j | \mathbf{r}_k, t_l) \leq \frac{\delta_{ik}}{A_i},$$

and therefore

$$0 \leq \sum_{i=1}^{n} A_i g_0(\mathbf{r}_i, t_j | \mathbf{r}_k, t_l) \leq 1.$$

Hence $g_0(\mathbf{r}_i, t_j | \mathbf{r}_k, t_l)$ can be regarded as the probability density of a tagged fish surviving the movement to cell $\mathcal{A}_i$ in the interval of time $[t_l, t_j]$ if at time t_l it was in cell $\mathcal{A}_k$ when fishing takes place in the same interval of time. I will therefore call g_0 the effort-dependent probability density.

F. Nonconservation of probability

Because tagged fish are being removed from their domain by natural and fishing deaths, in general it must be true that for $t_j > t_l$

$$\sum_{i=1}^{n} \mathcal{A}_i g_0(\mathbf{r}_i, t_j | \mathbf{r}_k, t_l) < 1.$$

For $g_0(\mathbf{r}_i, t_j | \mathbf{r}_k, t_l)$ I shall refer to $(\mathbf{r}_i, t_j)$ as the receiver space-time coordinates, while $(\mathbf{r}_k, t_l)$ will be referred to as the source space-time coordinates.

The initial property of g_0 makes it possible to estimate $q(\mathbf{r}_k, t_l)$. Multiply (2) by $q(\mathbf{r}_i, t_j)$, sum over the index of the receiver spatial coordinate, and let $t_j \to t_l$. Using the initial property of g_0 the result is

$$q(\mathbf{r}_k, t_l) = \frac{1}{N_T(\mathbf{r}_k, t_l)} \sum_{i=1}^{n} \mathcal{A}_i \frac{c_T(\mathbf{r}_i, t_l)}{e(\mathbf{r}_i, t_l)}. \tag{3}$$

Dimensional analysis of this equation or Eq. (1) reveals that q has the dimensions of $area \times time^{-1}$. The dependence of its value on spatio-temporal location will ultimately dictate in the number of cells and intervals of time that tagged fish must be released in order to correctly characterize the pattern of fish movement within $\mathcal{A}$. A discussion of this will follow after I present some necessary material. The accurate estimation of q using (3) requires that the loss of tagged fish due to natural mortality during the time of release be negligible. This condition is fullfilled if natural mortality is much less than $1/[t_l, t_{l+\delta l}]$. Doing otherwise will result in underestimating q. It is, therefore, important to release the tagged fish in as short a time as possible.

A simple rearrangement of equation (2) gives the population density of tagged fish in cell $\mathcal{A}_i$ in the interval of time $\mathcal{T}_j$ if N_T tagged fish are released in area $\mathcal{A}_k$ at an earlier time t_l:

$$p_T(\mathbf{r}_i, t_j) = g_0(\mathbf{r}_i, t_j | \mathbf{r}_k, t_l)\, N_T(\mathbf{r}_k, t_l) = \mathcal{A}_k\, g_0(\mathbf{r}_i, t_j | \mathbf{r}_k, t_l) \frac{N_T(\mathbf{r}_k, t_l)}{\mathcal{A}_k}$$

$$\tag{4}$$

$$= \mathcal{A}_k\, g_0(\mathbf{r}_i, t_j | \mathbf{r}_k, t_l) p_T(\mathbf{r}_k, t_l).$$

This is an example of how a single source of population density is mathematically propagated from point to point in space-time. It is for this reason that g_0 can be called the effort-dependent probability density or propagator.

Had it been specified that of the N_T tagged fish released in area $\mathcal{A}_k$ at time t_l, N_D tagged fish do not survive the trauma of tagging, the initial condition for expression (4) would have to be modified to $p_T(\mathbf{r}_i, t_l) = [\, N_T(\mathbf{r}_k, t_l) - N_D(\mathbf{r}_k, t_l)\,]/\mathcal{A}_k$. Although in priciple it is possible to know the number of tagged fish that do not survive the trauma of tagging once they are released, it is very difficult to make such a determination. This is one of the pedagogical examples that I referred to at the beginning of this section. The case is being considered here as an example of a sink of population density which will enable me to formulate the rules of superposition for the construction of expressions for the population density involving both sources and sinks of population.

Suppose now that tagged fish are released simultaneously in several cells over an interval of time $[t_l, t_{l+\delta l}] \ll T_l$. Letting $N_T(\mathbf{r}_v, t_l)$ be the number of tagged fish released correspondingly in cells $\{\mathcal{A}_v : v = 1, 2, \ldots, n\}$ during interval of time $[t_l, t_{l+\delta l}]$, then the "initial condition" for the tagged fish population density is given by $p_T(\mathbf{r}_v, t_l) = N_T(\mathbf{r}_v, t_l)/\mathcal{A}_v$ for $v = 1, 2, \ldots, n$. Following the prescription of Eq. (4), the population density of tagged fish in cell $\mathcal{A}_i$ in the interval of time T_j is then given by the linear superposition

$$p_T(\mathbf{r}_i, t_j) = g_0(\mathbf{r}_i, t_j|\mathbf{r}_1, t_l)N_T(\mathbf{r}_1, t_l) + g_0(\mathbf{r}_i, t_j|\mathbf{r}_2, t_l)N_T(\mathbf{r}_2, t_l) + \ldots$$

$$\ldots + g_0(\mathbf{r}_i, t_j|\mathbf{r}_n, t_l)N_T(\mathbf{r}_n, t_l) = \mathcal{A}_1 g_0(\mathbf{r}_i, t_j|\mathbf{r}_1, t_l)p_T(\mathbf{r}_1, t_l)$$

$$+ \mathcal{A}_2 g_0(\mathbf{r}_i, t_j|\mathbf{r}_2, t_l)p_T(\mathbf{r}_2, t_l) + \ldots + \mathcal{A}_n g_0(\mathbf{r}_i, t_j|\mathbf{r}_n, t_l)p_T(\mathbf{r}_n, t_l) \qquad (5)$$

$$= \sum_{v=1}^{n} \mathcal{A}_v g_0(\mathbf{r}_i, t_j|\mathbf{r}_v, t_l)p_T(\mathbf{r}_v, t_l) \, .$$

This is an example of how multiple simultaneous sources of population density are mathematically propagated from point to point in space-time.

As was considered for the tagging experiment in a single cell, suppose that of the $N_T(\mathbf{r}_v, t_l)$ tagged fish that are released in cell $\mathcal{A}_v$ during interval of time $[t_l, t_{l+\delta l}] \ll T_l$, $N_D(\mathbf{r}_v, t_l)$ tagged fish do not survive the trauma of tagging. Then the initial condition for the population density in expression (5) would have to be modified to

$$p_T(\mathbf{r}_v, t_l) = [N_T(\mathbf{r}_v, t_l) - N_D(\mathbf{r}_v, t_l)]/\mathcal{A}_v \, .$$

With the result given in (5) I can deduce a property for the summation of probability densities. I can use $p_T(\mathbf{r}_i, t_j)$, the final condition of (5), as the initial condition for propagation to a later time, say $t_s \geq t_j$:

$$p_T(\mathbf{r}_k, t_s) = \sum_{i=1}^{n} \mathcal{A}_i g_0(\mathbf{r}_k, t_s|\mathbf{r}_i, t_j)p_T(\mathbf{r}_i, t_j) \, . \qquad (6)$$

Substituting (5) into (6) and interchanging summations, I get

$$p_T(\mathbf{r}_k, t_s) = \sum_{v=1}^{n} \mathcal{A}_v \left[\sum_{i=1}^{n} \mathcal{A}_i g_0(\mathbf{r}_k, t_s|\mathbf{r}_i, t_j)g_0(\mathbf{r}_i, t_j|\mathbf{r}_v, t_l) \right] p_T(\mathbf{r}_v, t_l) \, . \qquad (7)$$

Because (7) must be of the form of (5) and (6) I can deduce the following property of g_0:

G. Summation of probability densities

The probability density g_0 satisfies

$$g_0(\mathbf{r}_k, t_s | \mathbf{r}_v, t_l) = \sum_{i=1}^{n} \mathcal{A}_i g_0(\mathbf{r}_k, t_s | \mathbf{r}_i, t_j) g_0(\mathbf{r}_i, t_j | \mathbf{r}_v, t_l) \,.$$

In particular, for consecutive intervals of time and $l \leq m - 1$

$$g_0(\mathbf{r}_k, t_{l+2} | \mathbf{r}_v, t_l) = \sum_{i=1}^{n} \mathcal{A}_i g_0(\mathbf{r}_k, t_{l+2} | \mathbf{r}_i, t_{l+1}) g_0(\mathbf{r}_i, t_{l+1} | \mathbf{r}_v, t_l) \,.$$

The expression for the summation of probability densities for consecutive intervals of time is useful for the construction of the transition probability density elements $\{g_0(\mathbf{r}_i, t_{s+1} | \mathbf{r}_j, t_s) : i, j = 1, 2, \ldots, m; s = l, l+1, \ldots, m\}$. However, in order to define a completely determined system of equations with these over an interval of time it is necessary to have constructed as many probability densities as there are cells, one per cell. If q varies significantly with time, these statements imply that to get a complete description of the movement, it is necessary to conduct a tagging experiment in every cell at every interval of time, a total of $n \times m$ experiments. However if q does not vary significantly with time, it is sufficient to conduct n experiments, one per cell (Salvadó 1993).

Suppose now that tagged fish are released simultaneously in several cells during the intervals of time $\{[t_s, t_{s+\delta s}] \ll T_s : s = l, l+1, l+2, \ldots, m\}$. Let $N_T(\mathbf{r}_v, t_s)$ be the number of tagged fish released in cell $\mathcal{A}_v$, for $v = 1, 2, \ldots, n$, during interval of time $[t_s, t_{s+\delta s}]$, for $s = l, l+1, \ldots, m$. The population density rate of released tagged fish is given by the recruitment density rate of tagged fish $r_T(\mathbf{r}_v, t_s) = N_T(\mathbf{r}_v, t_s)/(\mathcal{A}_v T_s)$. The population density of tagged fish at $(\mathcal{A}_i, T_{j+1})$ is given by the linear superposition

$$p_T(\mathbf{r}_i, t_{j+1}) = g_0(\mathbf{r}_i, t_{j+1} | \mathbf{r}_1, t_l) N_T(\mathbf{r}_1, t_l) + g_0(\mathbf{r}_i, t_{j+1} | \mathbf{r}_2, t_l) N_T(\mathbf{r}_2, t_l) + \cdots$$

$$\cdots + g_0(\mathbf{r}_i, t_{j+1} | \mathbf{r}_n, t_l) N_T(\mathbf{r}_n, t_l) + g_0(\mathbf{r}_i, t_{j+1} | \mathbf{r}_1, t_{l+1}) N_T(\mathbf{r}_1, t_{l+1})$$

$$+ g_0(\mathbf{r}_i, t_{j+1} | \mathbf{r}_2, t_{l+1}) N_T(\mathbf{r}_2, t_{l+1}) + \cdots + g_0(\mathbf{r}_i, t_{j+1} | \mathbf{r}_n, t_{l+1}) N_T(\mathbf{r}_n, t_{l+1}) + \cdots$$

$$\cdots + g_0(\mathbf{r}_i, t_{j+1} | \mathbf{r}_1, t_j) N_T(\mathbf{r}_1, t_j) + g_0(\mathbf{r}_i, t_{j+1} | \mathbf{r}_2, t_j) N_T(\mathbf{r}_2, t_j) + \cdots$$

$$\cdots + g_0(\mathbf{r}_i, t_{j+1} | \mathbf{r}_n, t_j) N_T(\mathbf{r}_n, t_j) = T_l \, \mathcal{A}_1 g_0(\mathbf{r}_i, t_{j+1} | \mathbf{r}_1, t_l) \, r_T(\mathbf{r}_1, t_l) \tag{8}$$

$$+ T_l \, \mathcal{A}_2 g_0(\mathbf{r}_i, t_{j+1} | \mathbf{r}_2, t_l) \, r_T(\mathbf{r}_2, t_l) + \cdots + T_l \, \mathcal{A}_n g_0(\mathbf{r}_i, t_{j+1} | \mathbf{r}_n, t_l) \, r_T(\mathbf{r}_n, t_l)$$

$$+ T_{l+1} \, \mathcal{A}_1 g_0(\mathbf{r}_i, t_{j+1} | \mathbf{r}_1, t_{l+1}) r_T(\mathbf{r}_1, t_{l+1}) + T_{l+1} \, \mathcal{A}_2 g_0(\mathbf{r}_i, t_{j+1} | \mathbf{r}_2, t_{l+1}) r_T(\mathbf{r}_2, t_{l+1}) + \cdots$$

$$\cdots + T_{l+1} \, \mathcal{A}_n g_0(\mathbf{r}_i, t_{j+1} | \mathbf{r}_n, t_{l+1}) r_T(\mathbf{r}_n, t_{l+1}) + \cdots + T_j \, \mathcal{A}_1 g_0(\mathbf{r}_i, t_{j+1} | \mathbf{r}_1, t_j) r_T(\mathbf{r}_1, t_j)$$

$$+ T_j \, \mathcal{A}_2 g_0(\mathbf{r}_i, t_{j+1} | \mathbf{r}_2, t_j) r_T(\mathbf{r}_2, t_j) + \cdots + T_j \, \mathcal{A}_n g_0(\mathbf{r}_i, t_{j+1} | \mathbf{r}_n, t_j) r_T(\mathbf{r}_n, t_j)$$

$$= \sum_{s=l}^{j \leq m} T_s \sum_{v=1}^{n} \mathcal{A}_v g_0(\mathbf{r}_i, t_{j+1} | \mathbf{r}_v, t_s) r_T(\mathbf{r}_v, t_s) \,.$$

This is an example of how multiple simultaneous sources of population density rate at different intervals of time are propagated from point to point in space-time.

131

As before, if in addition to $N_T(\mathbf{r}_v, t_s)$, the number of tagged fish which are released in cell $\mathcal{A}_v$ during the intervals of time $\{[t_s, t_{s+\delta s}] : s = l, l+1, \ldots, m\}$, there is a number of tagged fish $N_D(\mathbf{r}_v, t_s)$ that do not survive the tagging process, the recruitment density rate of tagged fish in (8) must be modified to

$$r_T(\mathbf{r}_v, t_s) = [N_T(\mathbf{r}_v, t_s) - N_D(\mathbf{r}_v, t_s)]/(\mathcal{A}_v \mathcal{T}_s).$$

With these examples the rule for linearly superposing the fields representing population processes can be encapsulated:

H. Rule of superposition of population processes

Let $\{\phi(\mathbf{r}_v, t_s) : v = 1, 2, \ldots, n; s = l, l+1, l+2, \ldots, m+1\}$ be a field representing a population process with units of $(quantity\ of\ fish) \times (area)^{-1}$ ($i.e.$, a population density) and $\{\psi(\mathbf{r}_v, t_s) : v = 1, 2, \ldots, n; s = l, l+1, l+2, \ldots, m+1\}$ a field representing a population process with units of $(quantity\ of\ fish) \times (area \times time)^{-1}$ ($i.e.$, a population density rate). Let $\{\pi(\mathbf{r}_i, t_j | \mathbf{r}_v, t_s) : i, v = 1, 2, \ldots, n; j \geq s = l, l+1, \ldots, m\}$ be the probability density of a fish surviving the movement to cell $\mathcal{A}_i$ in the interval of time $[t_s, t_j]$ if at time t_s it was in cell $\mathcal{A}_v$. The contribution of $\phi(\mathbf{r}_v, t_s)$ and $\psi(\mathbf{r}_v, t_s)$ to the population density $p(\mathbf{r}_i, t_{j+1})$ in the interval of time $\mathcal{T}_T$ obeys the following rule of linear superposition:

$$p(\mathbf{r}_i, t_{j+1}) = \lambda_\phi \sum_{v=1}^{n} \mathcal{A}_v \pi(\mathbf{r}_i, t_{j+1}|\mathbf{r}_v, t_l)\phi(\mathbf{r}_v, t_l) + \lambda_\psi \sum_{s=l}^{j<m} \mathcal{T}_s \sum_{v=1}^{n} \mathcal{A}_v \pi(\mathbf{r}_i, t_{j+1}|\mathbf{r}_v, t_s)\psi(\mathbf{r}_v, t_s)$$

where

$$\lambda_\kappa = \begin{cases} +1 & \text{for sources} \\ \\ -1 & \text{for sinks}. \end{cases} \qquad \text{for } \kappa = \phi \text{ or } \psi$$

such that $p(\mathbf{r}_i, t_{j+1}) \geq 0$ for $i = 1, 2, \ldots, n$ and $j = l, l+1, \ldots, m$.

The expressions for the tagged fish catch density rate corresponding to the tagged fish population densities in this subsection are computed by multiplying these by the death rate due to fishing, $f(\mathbf{r}_i, t_{j+1}) = q(\mathbf{r}_i, t_{j+1})e(\mathbf{r}_i, t_{j+1})$.

3.2 Analysis of Untagged Fish: Part I

The population density of untagged fish has an "initial condition" at time t_l given by

$$p_{in}(\mathbf{r}_k, t_l) = \frac{c(\mathbf{r}_k, t_l)}{q(\mathbf{r}_k, t_l)e(\mathbf{r}_k, t_l)} \quad \text{for } k = 1, 2, \ldots, n. \tag{9}$$

In addition, there is a recruitment density rate given by $\{r(\mathbf{r}_k, t_s) : k = 1, 2, \ldots, n; s = l, l+1, \ldots, m+1\}$. By applying the rule of superposition of population processes in subsection (3.1), the population density of untagged fish at $(\mathcal{A}_i, \mathcal{T}_{j+1})$ is given by

$$p(\mathbf{r}_i, t_{j+1}) = \sum_{v=1}^{n} \mathcal{A}_v g_0(\mathbf{r}_i, t_{j+1}|\mathbf{r}_v, t_l)p_{in}(\mathbf{r}_v, t_l) + \sum_{s=l}^{j\leq m} \mathcal{T}_s \sum_{v=1}^{n} \mathcal{A}_v g_0(\mathbf{r}_i, t_{j+1}|\mathbf{r}_v, t_s)r(\mathbf{r}_v, t_s),$$

and it has been assumed that the untagged fish have the same movement patterns as those of tagged ones. This assumption is correct if many tagged fish are released so that g_0, constructed as (2) indicates, approximately represents the average movement of the fish population.

The catch density rate of untagged fish is computed from this last equation by multiplication by $f(\mathbf{r}_i, t_{j+1}) = q(\mathbf{r}_i, t_{j+1})e(\mathbf{r}_i, t_{j+1})$:

$$c(\mathbf{r}_i, t_{j+1}) = f(\mathbf{r}_i, t_{j+1}) \left[\sum_{v=1}^{n} \mathcal{A}_v g_0(\mathbf{r}_i, t_{j+1}|\mathbf{r}_v, t_l)p_{in}(\mathbf{r}_v, t_l) + u_0(\mathbf{r}_i, t_{j+1}) \right] , \qquad (10)$$

where $u_0(\mathbf{r}_i, t_{j+1})$ is the contribution to the population density due to recruitment and is given by

$$u_0(\mathbf{r}_i, t_{j+1}) = \sum_{s=l}^{j\leq m} \mathcal{T}_s \sum_{v=1}^{n} \mathcal{A}_v \, g_0(\mathbf{r}_i, t_{j+1}|\mathbf{r}_v, t_s)r(\mathbf{r}_v, t_s) .$$

Because I have knowledge of the fields g_0, f and c in the interval of time $\mathcal{T}_T$, u_0 can be computed in that interval of time by use of

$$u_0(\mathbf{r}_i, t_{j+1}) = \frac{c(\mathbf{r}_i, t_{j+1})}{f(\mathbf{r}_i, t_{j+1})} - \sum_{v=1}^{n} \mathcal{A}_v g_0(\mathbf{r}_i, t_{j+1}|\mathbf{r}_v, t_l)p_{in}(\mathbf{r}_v, t_l)$$

for $j \leq m$. Once this is done, it is possible to use (10) to model catch density rate of untagged fish in the interval of time $\mathcal{T}_T$, but only at the level of effort density for which g_0 was constructed. No other case may be considered rigorously. To consider other levels of effort, other tagging experiments at the new levels of effort must be performed, because g_0 is the effort-dependent probability density of fish movement as can be appreciated in Eq. (2). The history of recapture of tagged fish will affect the value of g_0. Although in principle I could conduct a tagging experiment at any level of effort, it is a very impractical and expensive method. It would be more convenient to be able to construct an effort-independent propagator, that is, the probability density of surviving movement from point to point in space from one time to a later one when no fishing takes place in the interval of time under consideration. This will be accomplished by extracting from the effort-dependent propagator the fishing death rate. What will be left in the propagator will be a measure of movement and natural death rate of the fish.

3.3   Analysis of Tagged Fish: Part II

The goal here is to derive expressions for the population density of tagged fish using a probability density of movement from point to point in space-time when fishing effort is not being applied. I will call this probability density g.

As in the first example of subsection (3.1), N_T tagged fish are released in cell $\mathcal{A}_k$ in a interval of time $[t_l, t_{l+\delta l}] \subset \mathcal{T}_l$. During the interval of time $\mathcal{T}_T$ the fishing effort applied resulted in a tagged-fish catch density rate $\{c_T(\mathbf{r}_i, t_j) : i = 1, 2, \ldots, n; j = l, l+1, \ldots, m+1\}$. Let $g(\mathbf{r}_i, t_j|\mathbf{r}_k, t_l)$ be the effort-independent probability density of a fish surviving the movement to cell $\mathcal{A}_i$ in the interval of time $[t_l, t_j]$ if at an earlier time t_l it was in cell $\mathcal{A}_k$. By effort-independent I mean that it is constructed as if there is no fishing effort in the interval of time $[t_l, t_j]$. According to the rule of superposition of population processes in subsection (3.1), the population density of tagged fish in this case is given by

$$p_T(\mathbf{r}_i, t_{j+1}) = g(\mathbf{r}_i, t_{j+1}|\mathbf{r}_k, t_l)N_T(\mathbf{r}_k, t_l) - \sum_{s=l}^{j \leq m} \mathcal{T}_s \sum_{v=1}^{n} \mathcal{A}_v g(\mathbf{r}_i, t_{j+1}|\mathbf{r}_v, t_s) c_T(\mathbf{r}_v, t_s), \quad (11)$$

where for this case the propagated catch density rate has been included because this information is not contained in g as it is in g_0. The sign of the contribution must be negative because catch is a sink of population. Dividing through this equation by $N_T(\mathbf{r}_k, t_l)$ and the use of Eq. (2) results in

$$g_0(\mathbf{r}_i, t_{j+1}|\mathbf{r}_k, t_l) = g(\mathbf{r}_i, t_{j+1}|\mathbf{r}_k, t_l)$$

$$\tag{12}$$

$$-\frac{1}{N_T(\mathbf{r}_k, t_l)} \sum_{s=l}^{j \leq m} \mathcal{T}_s \sum_{v=1}^{n} \mathcal{A}_v g(\mathbf{r}_i, t_{j+1}|\mathbf{r}_v, t_s) c_T(\mathbf{r}_v, t_s),$$

which makes it possible to compute g from knowledge of g_0. Because (12) is a system of linear algebraic equations, they can be inverted to resolve g in terms of g_0 (Salvadó 1993). However, there are approximate solutions that can be invoked whose range of validity are based on the magnitude of the ratio of recovered to released tagged fish: the exact solution of (12) for g is given by

$$g(\mathbf{r}_i, t_{j+1}|\mathbf{r}_k, t_l) = \sum_{r=0}^{\infty} g_r(\mathbf{r}_i, t_{j+1}|\mathbf{r}_k, t_l) \tag{13}$$

where the series arises from the Neumann expansion (*i.e.*, successive approximations) of (12):

$$g_r(\mathbf{r}_i, t_{j+1}|\mathbf{r}_k, t_l) = \frac{1}{N_T(\mathbf{r}_k, t_l)} \sum_{s=l}^{j \leq m} \mathcal{T}_s \sum_{v=1}^{n} \mathcal{A}_v g_{r-1}(\mathbf{r}_i, t_{j+1}|\mathbf{r}_v, t_s) c_T(\mathbf{r}_v, t_s)$$

for $r = 1, 2, \ldots$. The total number of tags recovered in the interval of time $\mathcal{T}_T$ is given by

$$N_R(t_{j+1}) = \sum_{s=l}^{j \leq m} \mathcal{T}_s \sum_{v=1}^{n} \mathcal{A}_v c_T(\mathbf{r}_v, t_s).$$

The Neumann expansion of (12) converges uniformly to the solution in the interval of time $[t_l, t_{j+1}]$ if $N_R(t_{j+1}) < N_T(\mathbf{r}_k, t_l)$ (Salvadó 1993). The Born approximation (Bjorken and Drell 1964) is given by the first order term of the Neumann series:

$$g(\mathbf{r}_i, t_{j+1}|\mathbf{r}_k, t_l) \approx g_0(\mathbf{r}_i, t_{j+1}|\mathbf{r}_k, t_l) + \frac{1}{N_T(\mathbf{r}_k, t_l)} \sum_{s=l}^{j \leq m} \mathcal{T}_s \sum_{v=1}^{n} \mathcal{A}_v g_0(\mathbf{r}_i, t_{j+1}|\mathbf{r}_v, t_s) c_T(\mathbf{r}_v, t_s)$$

and is a good approximation to (12) unless $N_R(t_{j+1}) \sim N_T(\mathbf{r}_k, t_l)$. However, if $N_R(t_{j+1}) \ll N_T(\mathbf{r}_k, t_l)$ (*i.e.*, the natural death rate is much greater than that due to fishing) the approximation

$$g(\mathbf{r}_i, t_{j+1}|\mathbf{r}_k, t_l) \approx g_0(\mathbf{r}_i, t_{j+1}|\mathbf{r}_k, t_l)$$

is adequate.

The properties of g can be deduced with the help of (13). It satisfies all the properties that g_0 satisfies. In addition it has the property

$$g(\mathbf{r}_i, t_j | \mathbf{r}_v, t_l) \geq g_0(\mathbf{r}_i, t_j | \mathbf{r}_v, t_l),$$

where necessarily the equality holds at $t_j = t_l$.

Multiplying (11) by the death rate due to fishing in cell $\mathcal{A}_i$ at time t_{j+1} results in

$$c_T(\mathbf{r}_i, t_{j+1}) = f(\mathbf{r}_i, t_{j+1}) g(\mathbf{r}_i, t_{j+1} | \mathbf{r}_k, t_l) N_T(\mathbf{r}_k, t_l)$$

$$-f(\mathbf{r}_i, t_{j+1}) \sum_{s=l}^{j \leq m} \mathcal{T}_s \sum_{v=1}^{n} \mathcal{A}_v g(\mathbf{r}_i, t_{j+1} | \mathbf{r}_v, t_s) \, c_T(\mathbf{r}_v, t_s),$$

which is system of algebraic equations for the catch density rate of tagged fish, and is a nonlinear function of effort density.

3.4   Analysis of Untagged Fish: Part II

The goal here is to derive expressions for the catch density rate for the untagged portion of the fish population using the effort-independent propagator. For an initial condition given by (9), a catch density rate of untagged fish $\{c(\mathbf{r}_i, t_j) : i = 1, 2, \ldots, n; j = l, l+1, \ldots, m+1\}$ and a recruitment density rate $\{r(\mathbf{r}_i, t_j) : i = 1, 2, \ldots, n; j = l, l+1, \ldots, m+1\}$, using the rules of superposition developed in subsection (3.1), I can write for the population density in $(\mathcal{A}_i, \mathcal{T}_{j+1})$ as

$$p(\mathbf{r}_i, t_{j+1}) = \sum_{v=1}^{n} \mathcal{A}_v g(\mathbf{r}_i, t_{j+1} | \mathbf{r}_v, t_l) p_{in}(\mathbf{r}_v, t_l) + u(\mathbf{r}_i, t_{j+1})$$

$$- \sum_{s=l}^{j \leq m} \mathcal{T}_s \sum_{v=1}^{n} \mathcal{A}_v g(\mathbf{r}_i, t_{j+1} | \mathbf{r}_v, t_s) \, c(\mathbf{r}_v, t_s),$$

where

$$u(\mathbf{r}_i, t_{j+1}) = \sum_{s=l}^{j \leq m} \mathcal{T}_s \sum_{v=1}^{n} \mathcal{A}_v g(\mathbf{r}_i, t_{j+1} | \mathbf{r}_v, t_s) \, r(\mathbf{r}_v, t_s).$$

The dynamic catch density rate is given by

$$c(\mathbf{r}_i, t_{j+1}) = f(\mathbf{r}_i, t_{j+1}) [\sum_{v=1}^{n} \mathcal{A}_v g(\mathbf{r}_i, t_{j+1} | \mathbf{r}_v, t_l) p_{in}(\mathbf{r}_v, t_l)$$

$$+ u(\mathbf{r}_i, t_{j+1}) - \sum_{s=l}^{j \leq m} \mathcal{T}_s \sum_{v=1}^{n} \mathcal{A}_v g(\mathbf{r}_i, t_{j+1} | \mathbf{r}_v, t_s) \, c(\mathbf{r}_v, t_s)].$$

$$(14)$$

As can be appreciated in the equation above, the dynamic catch density rate is in general a nonlinear function of effort density.

Solving for the contribution to the population density due to recruitment u using (14), I have

$$u(\mathbf{r}_i, t_{j+1}) = \frac{c(\mathbf{r}_i, t_{j+1})}{f(\mathbf{r}_i, t_{j+1})} - \sum_{v=1}^{n} \mathcal{A}_v g_0(\mathbf{r}_i, t_{j+1} | \mathbf{r}_v, t_l) p_{in}(\mathbf{r}_v, t_l)$$

$$+ \sum_{s=l}^{j \leq m} \mathcal{T}_s \sum_{v=1}^{n} \mathcal{A}_v g(\mathbf{r}_i, t_{j+1} | \mathbf{r}_v, t_s) \, c(\mathbf{r}_v, t_s).$$

$$(15)$$

Using these equations the process for studying the fishery interaction problem posed in section 2 is the following: 1) compute q as indicated in Eq. (3); 2) estimate g_0 as indicated in Eq. (2), 3) compute g using Eq. (13) and the formula for the summation of probability densities; 4) compute the recruitment using (15); 5) compute catch density rate at new effort level using Eq. (14). The final step requires the assumption that recruitment will not change significantly at the new level of effort.

Finally, if in the expressions developed in this work $\mathcal{A}_i \to 0$ for $i = 1, 2, ..., n$, and $\mathcal{T}_j = t_{j+1} - t_j \to 0$ for $j = l, l+1, ..., m$ as $n, m \to \infty$, the continous expressions that result are identical to the integral equation solutions of the continuous population field equations in Salvadó (1993). In addition, the properties of the Green functions of the integral equations are identical to those of the probability densities g and g_0. Therefore the probability densities of movement of this paper are identical to the discretized causal Green function (Salvadó 1993) associated with the continuum population field equations. Thus, all of the formalism developed in the continuum (Salvadó 1993) can be used in the discrete case. In particular, as is shown in (Salvadó 1993), the moments of the Green function yield the parameters of the fishery such as natural death rate, advective velocity, diffusivity, etc.

4.   SUMMARY

I have presented a rule for constructing discrete expressions for the population density by the linear superposition of population processes distributed in space-time. Each population process at a particular cell is weighted by the probability of surviving the movement from that cell into another over an interval of time, and was applied to fish population processes. The probability density of movement was constructed empirically using the population density of tagged fish (as related to the catch density rate of these) at the end of the interval of time divided by the number of tagged fish at the begining of the interval of time. Because the probability that the tagged fish are in the cell of release at the time of release is unity, this allows me to compute the catchability. However, because tagged fish are being removed from their domain by natural and fishing deaths, probability is not conserved for times after their release.

The probability density constructed with the catch density rate and effort density was identified as the effort-dependent probability density. I constructed an effort-independent probability density which is useful to answer questions of fisheries interaction. By expressing the catch density rate in terms of the effort-independent propagator it is not necessary to perform tagging experiments at all the levels of effort that are to be considered. Only n tagging experiments, one per cell, at one level of effort is necessary if the recruitment and movement behaviour does not change significantly at the different levels of effort.

The expressions constructed for the population density are identical to the discretized integral equations which are solutions of partial differential equations if the probability density of movement is identified with the Green function associated with the differential operators.

5.   REFERENCES CITED

Bjorken, J.D. and S.D. Drell.   1964. Relativistic quantum mechanics. McGraw-Hill Book Co. New York, NY. 300 p.

Courant, R. and D. Hilbert.   1953. Methods of mathematical physics, v. I. Interscience. New York, NY. 560 p.

Salvadó, C.A.M.   1993. Population field theory with applications to tag analysis and fisheries modeling: the empirical Green function. *Can. J. Fish. Aquat. Sci.* (In Press).

A MARKOV MOVEMENT MODEL OF YELLOWFIN TUNA IN THE EASTERN PACIFIC OCEAN AND SOME ANALYSES FOR INTERNATIONAL MANAGEMENT[1]

Richard B. Deriso, Richard G. Punsly, and William H. Bayliff
Inter-American Tropical Tuna Commission
Scripps Institution of Oceanography
La Jolla, CA 92037, USA

ABSTRACT

A movement model is developed to quantify the probabilities of movement for yellowfin tuna in the eastern Pacific Ocean. The model specifies a probability transition matrix for the odds of a given-aged individual moving from one spacial stratum to an adjacent spatial stratum during each month. A novel feature of the model is the description of discrete movement with three parameters (velocity, diffusion, and direction) in a manner analogous to continuous movement models. A maximum likelihood approach is taken to estimate movement and catchability parameters for data obtained from tagging experiments. Yield-per-recruit analysis is a critical, though often neglected, tool required to evaluate the need for international management of tunas. A modified mean residence time calculation is shown which addresses some consequences of eliminating the harvest of smaller sub-optimal sized yellowfin tuna.

---

[1] Published: Deriso, R.B., R.G. Punsly, and W.H. Bayliff. 1991. A Markov movement model of yellowfin tuna in the eastern Pacific Ocean and some analyses for international management. *Fisheries Research*, Elsevier Scence Publishers B.V., Amsterdam, 11(1991):375-95.

A REVIEW OF TUNA FISHERY-INTERACTION ISSUES IN THE WESTERN AND CENTRAL PACIFIC OCEAN

John Hampton
South Pacific Commission
Noumea, New Caledonia

ABSTRACT

The rapid expansion of tuna fisheries in the western and central Pacific Ocean in the past ten years has brought into prominence a number of fisheries-interaction issues. Most of these issues fall into three general categories: A) competition for fish at the same stage in their life cycle in the same general area by two or more fisheries; B) the effect of fishing a stock at an early stage of its life cycle upon a fishery that exploits the stock at a later stage, typically with a different gear; and C) the effect of fishing a stock in one area upon a fishery that exploits the stock elsewhere. Some of the specific issues include interaction between large-scale fisheries operated by distant-water fishing nations (DWFNs) using different gear types; the effect of DWFN fisheries on locally-based, commercial fisheries; the effect of industrial fisheries on artisanal/subsistence fisheries; the effect of fisheries in different EEZs on one another; and the effect of different fishing companies operating in the same EEZ on one another. The issues cover the whole range of major commercial tuna fishing methods in the Pacific Islands and Southeast Asia (purse seine, pole and line, longline, troll, handline and drift gillnet) and all of the tuna species that support major fisheries in these areas (skipjack, yellowfin, bigeye and albacore). Much of the data required to assess many of these interaction issues (detailed catch and effort, tagging and biological data) exist, although their restricted availability, in many cases, has hindered thorough analytical treatment. Some of these problems have now been overcome through cooperative research programmes and the increasing quality of the South Pacific Commission's fisheries and tagging databases. Future work recommended includes the compilation of fisheries data for interaction studies; the continued development of models of tagged-tuna dynamics explicitly incorporating movement; the development of abundance indices; analyses and experiments to investigate the question of possible heterogeneity of the yellowfin population with respect to surface and longline gears; estimation of the geographical distribution of skipjack and yellowfin recruitment relative to the surface fisheries and the effect this might have on interaction; and further tagging and modelling work on South Pacific albacore.

1.   INTRODUCTION

The tuna fisheries of the western and central Pacific Ocean (WCPO), in this paper defined as the South Pacific Commission (SPC) statistical area (Figure 1), are extremely diverse, ranging from artisanal/subsistence fishing in Pacific Island and Southeast Asian countries, through small-scale commercial tuna fishing in several of those countries, to the large, distant-water purse-seine, pole-and-line and longline fisheries active on the high seas and, by way of licensing agreements, in the exclusive economic zones (EEZs) of many countries.

AS	American Samoa
AU	Australia
CK	Cook Islands
FJ	Fiji
FM	Federated States of Micronesia
GU	Guam
HB	Howland and Baker
HW	Hawaii
JT	Johnston
JV	Jarvis
KI	Kiribati
LN	Line Islands
MH	Matthew and Hunter
MI	Marshall Islands
MR	Northern Mariana Islands
NC	New Caledonia
NF	Norfolk
NR	Nauru
NU	Niue
NZ	New Zealand
PF	French Polynesia
PG	Papua New Guinea
PN	Pitcairn
PU	Palau
PX	Phoenix
PY	Palmyra
SB	Solomon Islands
TK	Tokelau
TO	Tonga
TV	Tuvalu
VU	Vanuatu
WF	Wallis and Futuna
WK	Wake
WS	Western Samoa

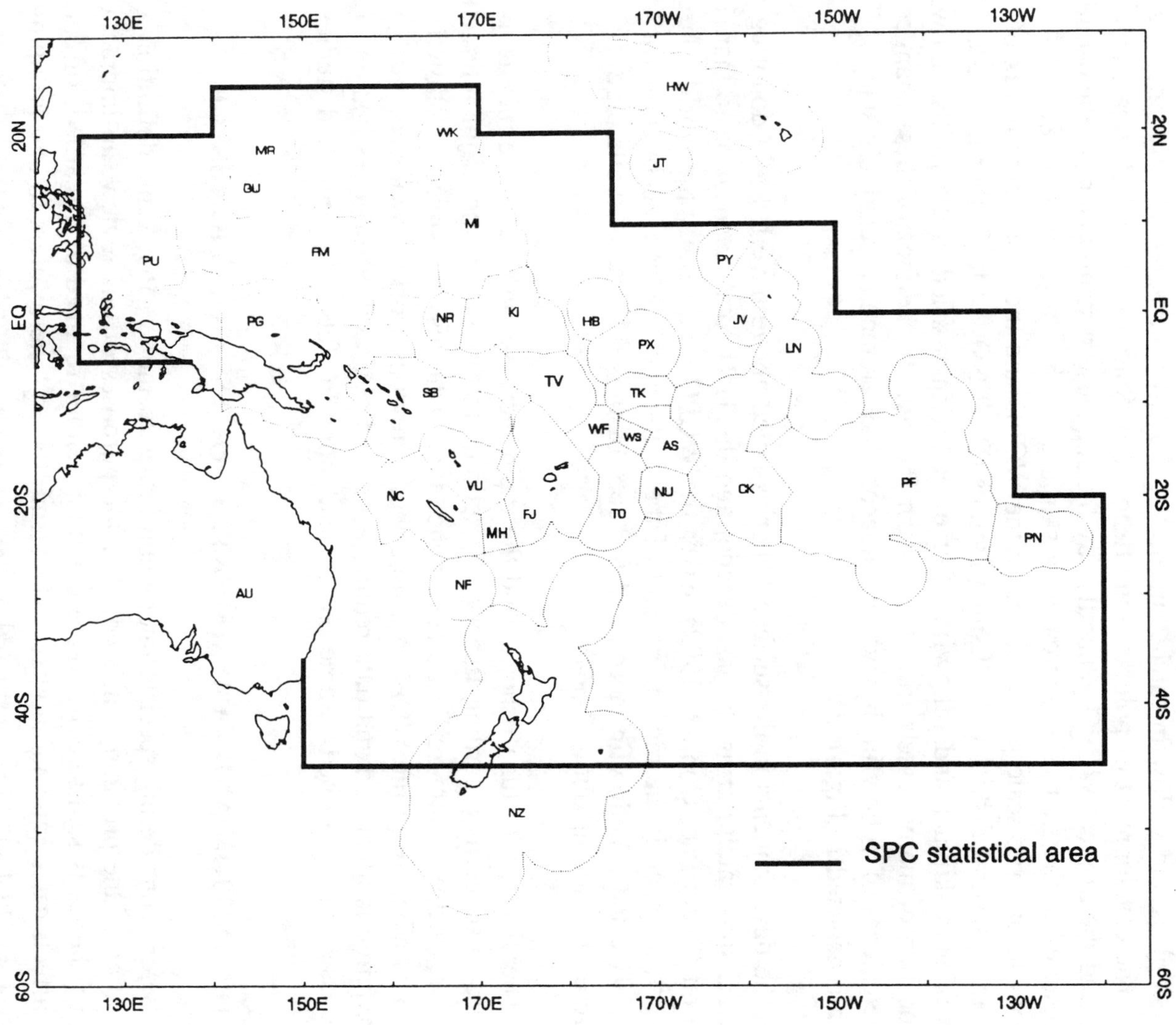

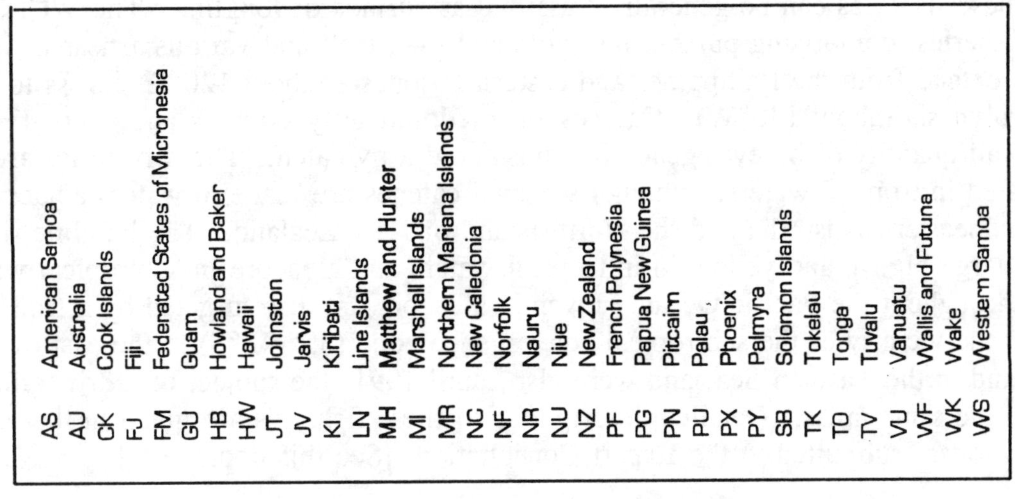

Figure 1. The South Pacific Commission statistical area.

These fisheries can be generally classified as surface or longline. The WCPO surface fisheries, comprising purse-seine, pole-and-line, troll and various artisanal fishing methods, extend from the Philippines and eastern Indonesia (about 120°E) across to French Polynesia (about 130°W). Catches are predominantly skipjack and yellowfin, with a small quantity of bigeye, generally considered a by-catch. These fisheries are concentrated in tropical waters, although seasonal catches are made in waters adjacent to Japan, southeastern Australia and the North Island of New Zealand. The longline fishery, targeting large bigeye and yellowfin in tropical waters and albacore in subtropical waters, extends throughout the Pacific Ocean. Juvenile albacore are also targeted by a troll fishery in the vicinity of the subtropical convergence zone (35°-45°S) to the east of New Zealand and in the Tasman Sea, and were also, until 1991, the subject of a drift gillnet fishery in the same areas. Detailed descriptions of these fisheries are given in the various species synopses submitted to the Expert Consultation. [See this document.]

Skipjack and yellowfin catches in the WCPO have increased rapidly since the early 1970s. The development of pole-and-line fisheries in Solomon Islands, Papua New Guinea and the tropical WCPO generally (by the Japanese distant-water pole-and-line fleet) resulted in the first large increases in skipjack catch. In the early 1980s, development of large-scale purse seining in the WCPO and the subsequent influx of vessels from several distant-water fishing nations (DWFNs) led to further increases in the catches of both skipjack and yellowfin. Purse seining intensified in the 1990s following the relocation of many vessels from the eastern Pacific. In the face of these changes, longline catches of yellowfin, bigeye and albacore have remained relatively stable. These trends are depicted in Figure 2.

The developments in the surface fisheries noted above have led to a doubling of the WCPO tuna catch during the last decade, and the 1991 total catch of more than 1.4 million mt (South Pac. Comm., 1992) makes the WCPO tuna fishery the world's largest. By weight, skipjack is the most important of the four major species, accounting for 69% of the 1991 catch. Yellowfin accounted for 26% of the 1991 catch, while bigeye and albacore each made up about 2-3%.

The diversity of tuna fisheries and the rapid increases in catch have inevitably given rise to concerns regarding fishery interaction. In this paper, the major interaction issues in respect of commercial and subsistence tuna fisheries for skipjack, yellowfin, bigeye, and albacore in the WCPO are described, previous work is reviewed, further work required is identified and data requirements and availability for investigation of the specific issues are described. Some recommendations for future or continued studies are made.

2.   DEFINITION AND CLASSIFICATION OF INTERACTION ISSUES

Before considering specific interaction issues in the WCPO, some definition and classification of the problem is needed. For the purposes of this review, an interaction between fisheries is said to exist where the catch, either retained or discarded, of one fishery affects catch rates in another. The term "fishery" here refers to any grouping of resource users that may be considered to have common interest; "fisheries" may be defined by gear types, nationalities, areas fished, fishing companies or social/cultural groupings.

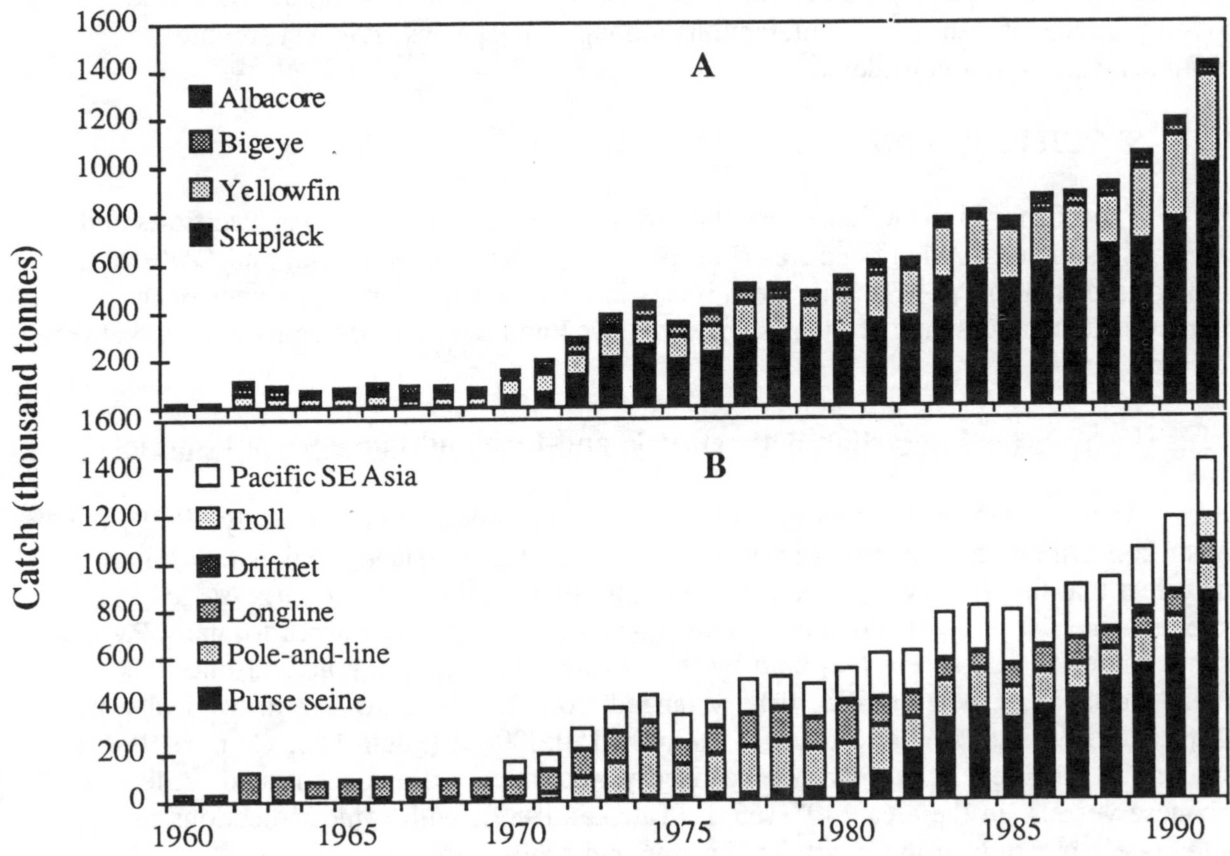

Figure 2.   Tuna catch trends (A) in the Pacific Islands area (excluding Philippines and Indonesia) by gear type and (B) in the WCPO (including Philippines and Indonesia) by species. Source: SPC (1992).

Some classification of the types of fishery-interaction problems that occur or could occur in the WCPO is useful for a number of reasons, *e.g.* deciding which analytical and/or experimental methods might be appropriate to investigate the problem. One such classification that might be useful is as follows:

A.   Competition for fish at the same stage in their life cycle in the same general area by two or more fisheries;

B.   The effect of fishing a stock at an early stage in its life cycle upon a fishery that exploits the stock at a later stage, typically with a different gear; and

C.   The effect of fishing a stock in one area upon a fishery that exploits the stock elsewhere.

Most of the tuna fishery-interaction problems that occur or that potentially could occur in the WCPO fall into one of these categories, or are a combination of types B and C. This classification is restricted to interaction between fisheries exploiting the same generation of fish, *i.e.* the possibility (which is probably minimal for WCPO tuna fisheries at current levels of exploitation) of interaction resulting from the effects of heavy exploitation of the spawning stock on subsequent recruitment is ignored. Also, the possible effects of fisheries on interactions among tuna species, *e.g.* via trophic relationships, are not considered.

3.   SPECIFIC ISSUES

Tuna fishery-interaction issues that are of current concern among Pacific Island countries, Southeast Asian countries (Philippines and eastern Indonesia) and DWFNs are summarised in Table 1. Each of these issues is reviewed with an assessment of the current state of interaction, further studies required and data requirements and availability for such studies.

3.1   Large-Scale Interaction Between Pole-and-Line and Purse-Seine Fisheries

As the distant-water purse-seine fishery began to develop in the early 1980s, the major concern among fisheries administrators in the Pacific Islands region was initially the effect that this fishery may have on catch rates of skipjack by the large-scale pole-and-line fishery. (At this time, a major benefit from tuna resources to many Pacific Island countries was access fees paid by the Japanese long-range, pole-and-line fleet.) Then, the Japanese pole-and-line fishery ranged from 130°E to 160°W in tropical waters (Tanaka, 1989) with annual catches of 100,000-150,000 mt (South Pac. Comm., 1992). By contrast, the developing purse-seine fishery was much more concentrated, at least for Japanese vessels, in the area 140°-165°E (Tanaka, 1989), with some indications of deliberate separation of the two fleets in time and space.

At the same time as the purse-seine fishery was developing, catches by the Japanese long-range, pole-and-line fishery began to decline because of effort reduction. That, along with the emergence of other important interaction issues, has resulted in some lessening in the concern by Pacific Island countries regarding interaction between the large-scale, purse-seine and pole-and-line fisheries.

3.1.1 Current assessment

No specific analyses of this interaction problem have been undertaken, but observations of the time series of Japanese pole-and-line catch per unit effort (CPUE) would suggest that there has been little, if any, effect of catches by purse seiners on pole-and-line CPUE (Figure 3). However, the increasing trend evident in CPUE may be due to increased efficacy of the Japanese pole-and-line fleet, which could also mask any negative impact the purse-seine fishery may be having.

3.1.2 Further work and data required

Because of the differences in spatial distribution of the fisheries alluded to earlier, a more thorough investigation of this problem would ideally incorporate the movement dynamics of skipjack. The development of such a model is currently underway and shows considerable promise (Sibert and Fournier, 1993). The model describes the dynamics of tagged tuna, explicitly incorporating a generalised movement sub-model,

Table 1. Tuna fishery-interaction issues in the western and central Pacific Ocean.

Description	Species	Affecting fishery		Affected fishery		Class
		Area	Gear	Area	Gear	
Large-scale interaction between PS and PL fisheries	SJ	WCPO	PS	WCPO	PL	A or C
Interaction between DWFN surface fisheries and locally-based commercial surface fisheries	SJ, YF	WCPO	PS, PL	ID	PS, PL	C
	SJ, YF	WCPO	PS, PL	PH	Various	C
	SJ	WCPO/PU	PS, PL	PU	PL	A, C
	SJ, YF	WCPO/SB	PS, PL	SB	PS, PL	A, C
	SJ	WCPO/KI	PS, PL	KI	PL	A, C
	SJ	WCPO	PS, PL	FJ	PL	C
	SJ	WCPO	PS, PL	PF	PL	C
Interaction between industrial fisheries and artisanal/subsistence fisheries	SJ, YF, BE	WCPO	PS, PL, LL	All Pac. Is. & SE Asian countries	Various	A, C
	SJ, YF	SB	PS, PL	SB	TR, HL	A, C
	SJ	KI	PL	KI	PL, TR, HL	A
	SJ, YF	ID	PL, PS	ID	TR, HL	A, C
	SJ, YF	PH	PS	PH	Various	A, C
Interaction between DWFN fisheries in different EEZs	SJ, YF, BE, AL	EEZs in WCPO	PS, PL, LL	EEZs in WCPO	PS, PL, LL	C
Interaction between different fleets in Solomon Islands	SJ, YF	SB (non-archipelagic)	PS (Co. #1)	SB (non-archipelagic)	PS (Co. #2)	A
	SJ	SB (archi-pelagic)	PL (Co. #1)	SB (archi-pelagic)	PL (Co. #2)	A
	SJ	SB (non-archipelagic)	PS	SB (archi-pelagic)	PL	A or C
Interaction between DWFN PS fishery and locally based "fresh-sashimi" LL fisheries	YF, BE	WCPO/PU	PS	PU	LL	B, C
		WCPO/FM	PS	FM	LL	B, C
		WCPO/MI	PS	MI	LL	B, C
		WCPO	PS	GU	LL	B, C
		WCPO/NC	PS	NC	LL	B, C
		WCPO	PS	FJ	LL	B, C
		WCPO/AU	PS	AU	LL	B, C
Interaction between DWFN PS fishery and "frozen-sashimi" LL fishery	YF, BE	WCPO	PS	WCPO	LL	B, C
Interaction between DWFN "frozen-sashimi" LL and locally-based "fresh-sashimi" fisheries	YF, BE	WCPO/PU	LL	PU	LL	A,C
		WCPO/FM	LL	FM	LL	A,C
		WCPO/MI	LL	MI	LL	A,C
		WCPO	LL	GU	LL	C
		WCPO	LL	NC	LL	C
		WCPO/FJ	LL	FJ	LL	A,C
		WCPO/AU	LL	AU	LL	A,C
Interaction between albacore driftnet and troll fisheries	AL	STCZ, Tasman Sea	DN	STCZ, NZ	TR	A, C
Interaction between surface and longline fisheries for albacore	AL	STCZ, Tasman Sea	DN, TR	South Pacific	LL	B, C

Gear codes:	PS: purse seine, PL: pole-and-line, LL: longline, DN: driftnet, TR: troll, HL: handline
Country/area codes:	AU: Australia, FM: Federated States of Micronesia, FJ: Fiji, PF: French Polynesia, GU: Guam, ID: Indonesia, KI: Kiribati, MI: Marshall Islands, NC: New Caledonia, NZ: New Zealand, PU: Palau, PH: Philippines, SB: Solomon Islands, WCPO: western Pacific Ocean, STCZ: sub-tropical convergence zone
Species codes:	AL: albacore, BE: bigeye, SJ: skipjack, YF: yellowfin

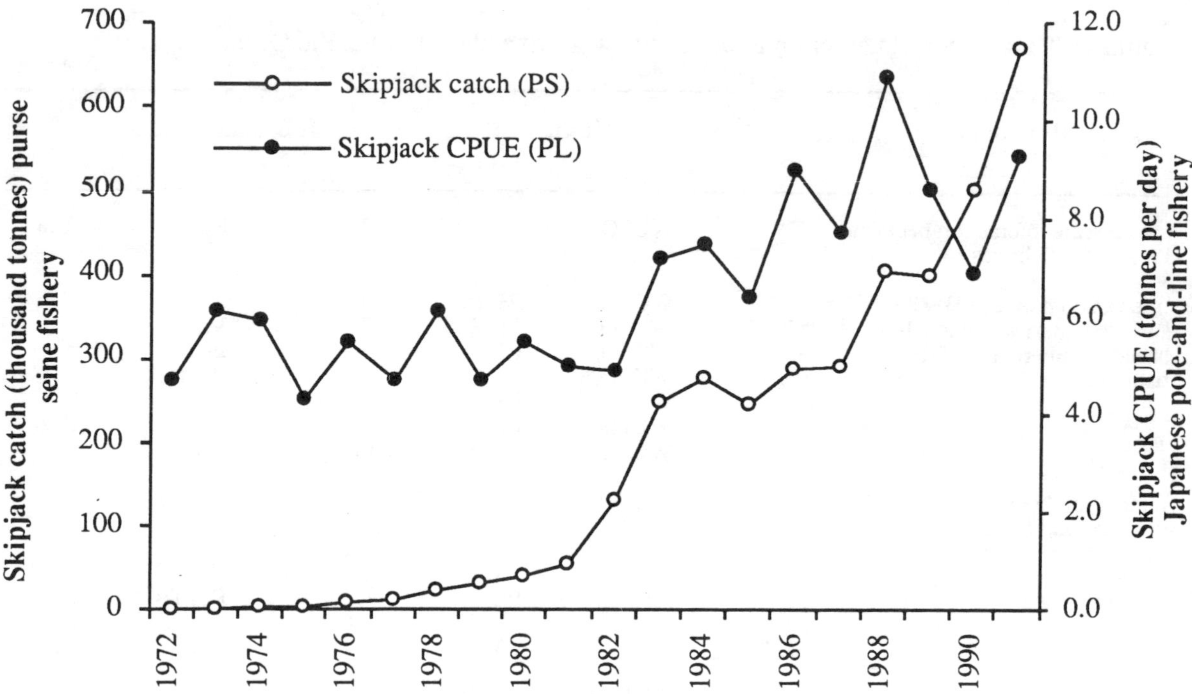

Figure 3.   Trends in skipjack CPUE by the Japanese pole-and-line (PL) fishery and skipjack catch by the purse-seine (PS) fishery. Sources: Skipjack CPUE data from SPC Regional Tuna Database; skipjack catch data from SPC (1992).

population size, natural mortality and fishing mortality, and is directly applicable to this and a number of other interaction problems reviewed in this paper.

The SPC has two skipjack tagging data sets that could be used to estimate the parameters of a generalised skipjack model useful for investigating this and other interaction problems. The first is that of the Skipjack Survey and Assessment Programme (SSAP), for which 140,000 skipjack were tagged during the period 1977-1980. From these releases, approximately 7,000 tags were returned to the SPC, the majority with reliable recapture data. Most recoveries were from pole-and-line fisheries, which accounted for most of the skipjack catch at that time. The second data set is that of the Regional Tuna Tagging Project (RTTP), which tagged a further 98,000 skipjack from mid-1989 to late 1992. As at 21 December 1992, 10,543 skipjack recaptures had been reported. As expected, most recaptures have been made by purse seiners. These two contrasting data sets provide an excellent opportunity to estimate skipjack movement dynamics, and so investigate a variety of interaction questions. However, some critical catch-and-effort data sets that are also required for such analyses have yet to be made available. These unavailable data are mainly from the earlier period when the coverage of the SPC Regional Tuna Database was poor. The minimum data required are estimates of total effort and catch by species stratified by vessel nationality, gear type, 5° quadrangle and month.

As noted above, the time-series CPUE trend for the Japanese pole-and-line fishery gives no indication of any negative effect of increased purse-seine catches. However, changes in the pole-and-line fishery make conclusions based on unadjusted CPUE rather tenuous. A more detailed analysis of pole-and-line CPUE, by way of a general linear model (GLM), may provide more information. Such a model would require detailed information on pole-and-line CPUE, characteristics of vessels and crew, environmental variables likely to affect pole-and-line fishing success and purse-seine catch.

3.2   Interaction Between DWFN Surface Fisheries and Locally-Based Commercial Surface Fisheries

With the development of the large-scale DWFN fisheries for skipjack and, to a lesser extent, yellowfin, there was concern at the national level regarding the impact that this development might have on locally-based, primarily pole-and-line, commercial tuna fisheries. This issue has been, or could be, important in Palau, Papua New Guinea, Solomon Islands, Kiribati, Fiji and French Polynesia in the Pacific Islands as well as Indonesia and Philippines in Southeast Asia. In some countries (*e.g.* Palau, Solomon Islands and Kiribati), there may be specific concern that DWFN surface fisheries in the immediate vicinity of the locally-based fisheries may be resulting in a type-A interaction. In countries where little DWFN surface fishing takes place (*e.g.* Indonesia, Philippines, Fiji and French Polynesia), the concern is more generally directed at the activities of the DWFN fleets throughout the WCPO, which would constitute a type-C interaction.

3.2.1   Current assessment

All other factors being equal, we would expect a type-A interaction to be stronger than a type-C interaction. Other important factors are the sizes of the affecting and affected fisheries, the natural mortality rate and the distance and rate of movement between the two fisheries (for type-C interactions). For example, we might predict that the effect of the DWFN surface fisheries concentrated in equatorial waters on the relatively-small skipjack fisheries in Fiji (3,000-5,000 mt per year) and French Polynesia (500-900 mt per year) would be small in relation to other factors, such as various environmental influences, that may affect local tuna abundance. Somewhat greater potential for interaction would exist where DWFN fisheries operate near to the locally-based fisheries (*e.g.* Palau, Federated States of Micronesia and sometimes Kiribati) and where the catch of the locally-based fisheries is large (*e.g.* Solomon Islands, Indonesia, Philippines), although there is little direct evidence for such interaction currently being significant.

As noted above, the rates of fish movement between fisheries in different areas are of key importance for type-C interactions. Closely related to this is how spawning and recruitment are distributed in relation to the fisheries in question. For example, the effect of skipjack fisheries outside the Solomon Islands EEZ on the surface fisheries for skipjack in the Solomon Islands EEZ will depend to a large extent on whether the Solomon Islands fisheries are sustained primarily from local recruitment or from immigration of skipjack from outside the Solomon Islands EEZ. There is evidence that larvae of skipjack and other tuna are much more abundant in the vicinity of reefs and Islands than in oceanic waters (Leis *et al.*, 1991). If this phenomenon also extends to tuna of a size recruiting to surface fisheries, local recruitment may represent a significant source of the tuna biomass available to some locally-based surface fisheries. This would be a double-edged sword for the larger locally-based fisheries. While large local recruitment would reduce the potential for negative interaction with outside fisheries, there would be a greater chance that the activities of the locally-based fisheries themselves may affect the population; the need for local regulation of catches may be greater under these circumstances.

3.2.2 Further work and data required

No detailed analyses of these types of interaction questions have yet been undertaken. The assessment of type-C interaction problems involving relatively large locally-based fisheries such as those of Solomon Islands could be undertaken by estimating explicit movement rates and other parameters of a two-fishery tag-attrition model (Sibert, 1984) fitted to the RTTP tagging data. For a complete picture, the question of local recruitment would also need to be addressed. For similar problems involving small locally-based fisheries, sufficient numbers of tag recoveries in the locally-based fishery are unlikely to be generated from a regional tagging project and obtaining of answers would therefore require some extrapolation from general movement characteristics estimated as described in section 3.2.1. If type-A interactions are identified (*i.e.* if the assumption of a homogeneous "stock" relative to the two fisheries is reasonable), simpler methods, such as a multi-gear, yield-per-recruit model, can be employed. The data required to undertake these types of analyses will be available on completion of the RTTP.

3.3 <u>Interaction Between Industrial Fisheries and Artisanal/Subsistence Fisheries</u>

In most Pacific Island and Southeast Asian countries, small-scale artisanal or subsistence fishing is important for the economic, social and nutritional well-being of a large percentage of the population. Therefore, questions regarding the possible effects of industrial fisheries, both DWFN and locally-based, on artisanal/subsistence fisheries are frequently raised.

3.3.1 Current assessment

Specific analyses have not been undertaken, due in part to the absence of reliable catch statistics for artisanal/subsistence fisheries in many countries. Because these fisheries in the Pacific Islands are relatively small and of restricted range, it is likely that any interaction with large, industrial fisheries would be minor in comparison with catch rate variability associated with environmentally-driven variation in local tuna availability and catchability. The exception to this might be where an industrial fishery operates in the same area as the artisanal/subsistence fishery, *e.g.* Solomon Islands and Kiribati. In Kiribati, artisanal skiff fishers may, in fact, frequently benefit from fishing near to the locally-based, pole-and-line vessels by fishing schools located and chummed to the surface by the larger vessels; in this case, the interaction may be positive. However, preliminary data from the RTTP would suggest that the effect on artisanal/subsistence fisheries is slight. As of December 1992, only 34 of approximately 10,000 skipjack tag returns had been recovered by artisanal/subsistence fishers in the Pacific Islands. If the overall exploitation rate of skipjack is moderate, as suggested by the interim skipjack recovery rate of approximately 11%, the low number of recoveries by artisanal/subsistence fishers is indicative of the small effect of the industrial fisheries on their catch rates. This situation is in contrast with that in the Philippines, where the artisanal/subsistence (municipal) fishery is much larger. Here, skipjack recoveries recorded by the artisanal/ subsistence fishery account for more than 25% of the total skipjack recoveries in the Philippines (1,330 recoveries as of 21 December 1992). The potential for this type of interaction is clearly much greater in the Philippines.

3.3.2 Further work and data required

Estimates of the impact of industrial tuna fisheries on artisanal/subsistence fisheries in the Pacific Islands will also rely on the application of a model of skipjack dynamics incorporating movement, as described above. This assumes that some estimates of artisanal/subsistence catches would be available. More direct estimates of interaction, using a multi-fishery, tag-attrition model or similar model based on actual tag recoveries, may be derived for the Philippines.

3.4   Interaction Between Fisheries in Different EEZs

The effect of fisheries in one EEZ on catch rates in other EEZs is particularly important for countries that license large numbers of foreign vessels. Examples can be found that involve all major tuna species (skipjack, yellowfin, bigeye, and albacore) and gear types (purse seine, pole and line, and longline) in the WCPO. The best examples, however, probably relate to purse seining skipjack and yellowfin in the main fishing area of 10°N-10°S, 130°E-170°W. The principal countries involved are Palau, Federated States of Micronesia, Papua New Guinea, Solomon Islands, Nauru, Marshall Islands, Kiribati and Tuvalu.

3.4.1 Current assessment

Some of the information in this section is taken directly from section 13 of Wild and Hampton (1993).

The degree of interaction among EEZs such as those listed above will be determined by the size of the areas, the distances between them, skipjack and yellowfin movement rates, natural mortality rates and the intensity of the fisheries (South Pac. Comm., 1988). Some specific analyses with respect to skipjack have been carried out. Kleiber *et al.* (1984), using SSAP tagging data, calculated a series of interaction coefficients based on the proportions of total throughput in receiver EEZs derived from immigration from donor EEZs. These results are summarised in Table 2. Most of the coefficients are low, indicating that under conditions prevailing when the SSAP data were gathered, there was generally little potential for fishery interaction. Not surprisingly, most cases of significant exchange occurred between adjacent EEZs. In particular, the results suggested that 37% of "throughput" in the Marshall Islands EEZ at the time of the tagging resulted from immigration from Federated States of Micronesia. Relatively high interaction coefficients were also observed for Northern Mariana Islands ↔ Federated States of Micronesia and to a lesser extent for Palau → Federated States of Micronesia and Papua New Guinea ↔ Solomon Islands, indicating some potential for fishery interaction between those countries. The only case of a relatively high interaction coefficient for widely separated areas was New Zealand → Fiji; however this may have been the result of the timing of tag releases into the highly seasonal New Zealand fishery (Argue and Kearney, 1983).

Table 2. Coefficients of interaction between fisheries operating in various countries and territories of the WCPO. Receiver countries are listed across the top of the table and donor countries down the left margin. Source: Kleiber et al. 1984.

To →	Papua New Guinea (PG)	Solomon Islands (SB)	Palau (PU)	Federated States of Micro-nesia (FM)	Marshall Islands (MI)	Northern Mariana Islands (MR)	Fiji (FJ)	New Zealand (NZ)	Western Samoa (WS)	French Polynesia (PF)
PG		2.6	0.8	1.4	0.5					
SB 1977	1.1									
SB 1980	3.7									
PU 1978				8.6	2.2					
PU 1980	1.6	0.4		3.5	1.3	0.7				
FM	0.7	0.9			37.0	10.8				
MI										
MR				17.4						
FJ 1978								0.6		
NZ							6.5		2.1	3.6
KI RIBATI				<0.1	0.1					

This relatively-simple representation of interaction does not explicitly specify the controlling factors noted above. Sibert (1984) derived a more rigorous method to estimate interaction between two countries and applied the method to Papua New Guinea and Solomon Islands, both of which had substantial pole-and-line fisheries for skipjack at the time of the SSAP. Exchange rates between the two EEZs, losses from natural mortality and movement to other areas, the proportions that remained resident and lived and the proportions that were caught locally on a monthly basis were estimated. The Solomon Islands stock was found to be relatively stable with low rates of natural mortality and emigration (resulting in high survival) and low rate of movement to Papua New Guinea. The Papua New Guinea stock was found to be more dynamic with higher rates of natural mortality and emigration (lower survival), but with a low rate of movement to Solomon Islands. This situation is shown is Figure 4. Sibert (1984) estimated from these results that an increase in the catch in either EEZ of 1,000 mt would result in a decrease in the steady-state catch of the other of only 1-3 mt.

3.4.2 Further work and data required

The tuna-movement model currently under development as an activity of the Pacific Skipjack Working Group would be ideally suited to a more thorough analysis of this type of spatial (type C) interaction problem. It is intended that this model be applied to both SSAP and RTTP tagging data for this purpose. Complementary catch and effort data as detailed earlier would also be required.

3.5 Interaction Among Different Fleets in Solomon Islands

Solomon Islands has the largest, locally-based skipjack fishery in the Pacific Islands region. Two companies currently operate; Solomon Taiyo Ltd (STL), a

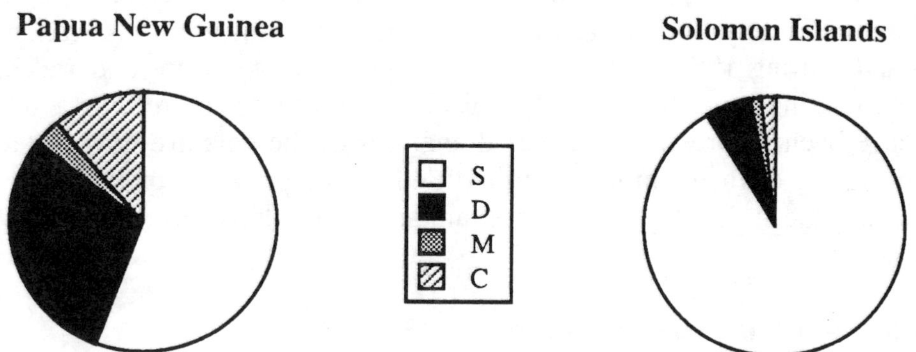

Figure 4.   Relative proportion of tagged skipjack released in Papua New Guinea and Solomon Islands that survive without migrating (S), disappear for unknown reasons (D), migrate to the other country (M), and are caught in the fishery of release (C) each month.   Source: Sibert (1984).

joint-venture company between Taiyo Fisheries (Japan) and the Solomon Islands Government, and National Fisheries Development (NFD), formerly a Government-owned enterprise but now owned by British Colombia Packers and operated by its subsidiary, Mar Fishing Company.   A third company, Makirabelle, a joint-venture between Makira Province and Frabelle Fishing Corporation of the Philippines will soon begin operations. STL operates a fleet of 20 pole-and-line vessels, predominantly 59-GRT, as well as one group-seine operation.   The NFD consists of a fleet of ten pole-and-line vessels (59-GRT) and one single seiner.   Makirabelle will operate two group seiners.   In general, pole-and-line vessels have access to all waters in the EEZ out side of three miles, whereas purse seiners, setting mainly on fish aggregation devices (FADs), are excluded from waters inside the main group archipelagic (MGA) baseline.   The Solomon Islands Government has allocated total tuna quotas of 47,500 mt, 37,500 mt and 35,000 mt, to STL, NFD and Makirabelle, respectively.   For STL and NFD, catches inside the MGA by their pole-and-line fleets are restricted by quotas of 30,000 mt and 20,000 mt, respectively.   In 1990, approximately 30,000 mt of tuna (80% skipjack) were caught by locally-based fleets in Solomon Islands.   Of the skipjack catch, about 80% was caught by the pole-and-line fleet and 20% by the purse-seine fleet.   Additionally, Japanese distant-water, pole-and-line vessels caught about 6,000 mt of skipjack in non-archipelagic waters of the Solomon Islands EEZ in 1990, while Japanese longliners caught about 3,500 mt, mostly yellowfin.

3.5.1 Current assessment

This diversity of interests in the Solomon Islands fishery has provided considerable scope for interaction issues to develop.   The recent allocation of quota to the third company (Makirabelle) caused concerns in the other two companies (STL and NFD) that their catch rates would be affected.   Also, STL's proposals to introduce an additional group-seine operation have to date been refused by the Government because of fears that pole-and-line catch rates might be reduced as a result.

In response to these types of interaction question and to questions regarding the overall potential productivity of the skipjack resource in Solomon Islands, the SPC and Solomon Islands Ministry of Natural Resources undertook a tagging programme in Solomon Islands during 1989-1990. In all, six tagging cruises were undertaken over a 12 month period, tagging more than 8,000 skipjack. From these, more than 1,000 recoveries have been recorded, and detailed analyses of the data are now in progress. A model incorporating diffusive movement modified by the presence of FADs in various areas has recently been developed, and preliminary results have been reported at the Expert Consultation.

3.5.2   Further work and data required

The need for further work on interaction problems in the Solomon Islands will be determined when the results of the above study have been fully analysed. Quality tagging data are available, as outlined above, as well as complete catch-and-effort data and size-composition samples from the fisheries concerned. The case of Solomon Islands provides an excellent example of the analysis of specific interaction problems through a directed research project and a well documented fishery.

3.6   Interaction Between the DWFN Purse-Seine Fishery and Longline Fishery

The impact of the DWFN purse-seine fishery, which catches, in addition to skipjack, large quantities of yellowfin over a broad size range, on the various longline fisheries of the region has probably been the foremost interaction issue during the past five years. Concerns have been expressed that the large catches of mainly 1 year old and 2 year old yellowfin by purse seiners may ultimately have a "downstream" negative effect on longline catch rates of predominantly 2, 3, and 4 year old yellowfin. Longline fisheries provide substantial income for many Pacific Island countries, both through licensed fishing by DWFNs and, recently in several countries, the development of transhipment bases for fleets of small, "fresh-sashimi" longliners that air-freight their product to sashimi markets in Japan and Honolulu. The consequences of any interaction of this type are greater than for the skipjack-interaction issues reviewed elsewhere in this paper because the market value of longline-caught yellowfin is typically several times greater than that of purse-seine-caught skipjack and yellowfin.

In the WCPO, the purse-seine fishery is confined largely to the area 10°N-10°S, 130°E-170°W. The longline fishery includes this area, but also extends further to the east, north and south. Interaction between the two fisheries would therefore be primarily a type-B interaction, with some spatial (type-C) characteristics. The concerns regarding the possibility of a purse-seine-longline interaction were largely responsible for the initiation of the RTTP, which has gathered information on yellowfin dynamics though a targeted tagging programme. It is worth noting that a similar, if lesser, problem could potentially exist with bigeye, also a target species of the longline fishery and caught incidentally in generally unknown numbers at small size by purse seiners.

3.6.1   Current assessment

Various analyses of catch-and-effort data, on both large and local scales, have been undertaken (Suzuki 1988, 1993). For various reasons, none of these studies has been conclusive in demonstrating an effect of the increase in purse-seine catches of yellowfin on longline-yellowfin CPUE. At a local scale (*e.g.* 5° quadrangle and month), studies such as that of Polacheck (1988) could not detect any interaction between purse seiners and longliners, presumably because of restricted vertical exchange and/or rapid horizontal mixing on those area-time scales.

For the WCPO as a whole, virtually no purse seining occurred during the 1952-1975 period. During this same period, yellowfin CPUE by Japanese longliners declined steadily (Figure 5). Given the absence of other significant yellowfin fisheries at the time, it is reasonable to assume that this decline in CPUE was either self-inflicted, *i.e.* the result of reduced yellowfin abundance brought about by longline fishing, or was the result of natural variation. The sudden increase in longline CPUE just prior to the advent of the purse-seine fishery might have resulted from a series of strong year classes recruiting to the longline fishery. The subsequent decline in longline CPUE is roughly contemporaneous with the increase in purse-seine catch; however, the current longline CPUE is approximately at the same level as just prior to the advent of purse seining (Figure 5). It is therefore unclear whether the recent decline in longline CPUE has resulted from an interaction with the purse-seine fishery or from other causes similar to those that brought about the earlier decline.

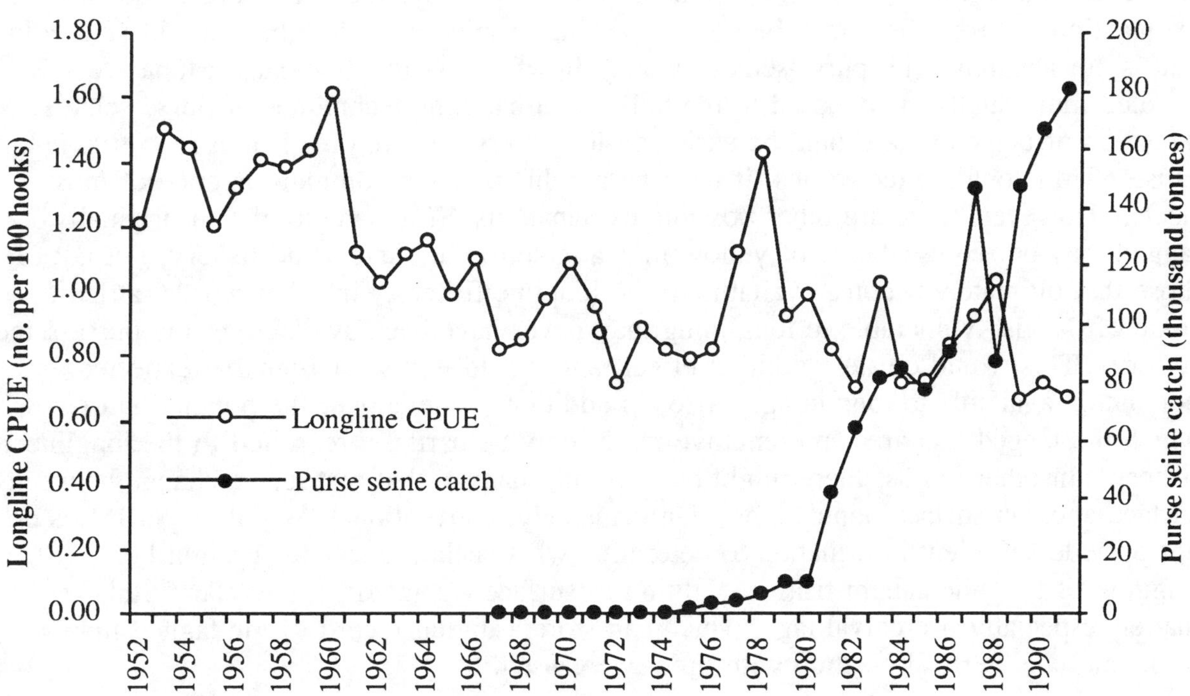

Figure 5.   Time series of yellowfin CPUE by Japanese longliners and yellowfin catch by purse seiners in the tropical WCPO. Sources: longline CPUE data for 1952-89 from Suzuki (1991); longline CPUE data for 1990-91 from SPC Regional Tuna Database; purse-seine catch data for 1967-90 from SPC (1992).

3.6.2 Further work and data required

Quantification of the purse-seine-longline interaction will require the development of a yellowfin population model that incorporates age structure, movement and other population dynamics. The yellowfin-tagging results of the RTTP will contribute significantly to the development of such a model; however detailed catch, effort and size-composition statistics, stratified by area (5° quadrangle), time (month), gear type and fishing method would be required. Some of these data are available to the SPC, although the assistance of DWFNs will be required to fill gaps in data coverage.

At this point it is worth noting that most tuna tagging experiments, where tagged fish have been released into a surface fishery with the expectation of recaptures firstly in the surface fishery and, after a period of time, in the longline fishery, have almost invariably resulted in fewer longline recoveries than expected. This has occurred with yellowfin-tagging experiments in the eastern Pacific (Lenarz and Zweifel, 1979) and Atlantic Oceans (Fonteneau, 1986), with southern bluefin tagging in the Pacific, Southern and Indian Oceans (Hampton, 1989) and may well occur in the present RTTP in the WCPO. As at 21 December 1992, only ten longline recoveries of RTTP-tagged yellowfin had been reported to the SPC, but there had been more than 100 recoveries by other gear types, mostly purse-seine, of a size (>100 cm) normally considered vulnerable to longlining. We have yet to determine if the number of longline recoveries is indeed less than expected, taking into account the size and geographical distribution of yellowfin catches in relation to tag releases. If there is a discrepancy, there are several possible explanations. First, there may be a tag-reporting problem with longline-caught yellowfin that is not apparent with purse-seine-caught yellowfin. As longline-caught tuna are handled individually, as opposed to the bulk-catch-handling techniques of purse seiners, detection of tagged tuna should be easier on longliners. Therefore, if non-reporting is the cause of low longline recoveries, it must be a deliberate and common practice of most fleets. However, there are other possible explanations. The sample of fish originally tagged may be representative of yellowfin available to the purse-seine fishery, but not of those that ultimately become available to the longline fishery. In other words, at least some of the fish vulnerable to longlining may never have been available to the purse-seine fishery. This would create a dilution effect, and would imply a larger that expected population available to longlining. Also, in addition to the above, the population of which the tagged fish are representative might only be partially recruited to the longline fishery. In other words, there might be some mechanism that restricts exchange between surface and sub-surface populations. Unfortunately, conventional tagging experiments do not provide sufficient information to determine what such a mechanism might be. Tagging of longline-caught fish, in addition to surface-caught fish, may clarify this matter, especially if archival tags, which can store a digital record of the tagged tuna's horizontal and vertical position with time, were used.

3.7 Interaction Between DWFN Longline Fishery and Locally-Based "Fresh-Sashimi" Longline Fisheries

Interaction between the large-scale DWFN longline fleet of freezer vessels and various locally-based, "fresh-sashimi" longline fleets targeting yellowfin and bigeye may become an important issue for countries in which the latter fleets are based. In some countries, the locally-based fleets are comprised of foreign-owned vessels, *e.g.* Palau,

Guam, Federated States of Micronesia and Marshall Islands, whereas in others, *e.g.* New Caledonia and Australia, the locally-based fleets are locally owned. In Fiji, there are locally-based fleets of locally-owned and Taiwanese-owned longliners. Several of these countries, *e.g.* Palau, Federated States of Micronesia, Marshall Islands and Australia, also license foreign-based longliners to fish in their EEZs. A variety of interaction questions, falling into the local, type-A category or the spatial, type-C category, can thus arise. These are analogous to the purse-seine issues discussed above.

3.7.1 Current assessment

No analyses of interaction between longline fleets have been carried out to date. As in other cases, the type-A interactions are likely to emerge as the most important, and in some cases, steps may be necessary to avoid physical conflicts of deployed gear.

3.7.2 Further work and data required

The rates of movement of larger yellowfin and bigeye are unknown, so assessment of the potential for type-C interactions are not yet possible. The difficulty in tagging longline-caught tunas in large numbers will limit the practicality of this method of assessment, although the RTTP has had recent success in tagging large yellowfin and bigeye caught by handline. A tagging experiment using archival tags could provide detailed information on the movements of large yellowfin and bigeye and should be pursued if possible. Analyses of fishery statistics using GLM techniques may also produce useful results.

3.8 Interaction Among the South Pacific Albacore Drift Gillnet, Troll, and Longline Fisheries

Interaction among drift gillnet, troll and longline fleets in the South Pacific albacore fishery emerged as an important issue in 1988 with the rapid buildup of drift gillnet vessels in the Tasman Sea and sub-tropical convergence zone, areas also fished by troll vessels. The issue of interaction between the troll and drift gillnet fisheries was largely one of direct competition (type A), demonstrable by the occurrence of drift gillnet marks on a considerable proportion (14% overall for the 1988-1989 season) of troll-caught albacore during the height of the drift gillnet fishery. (The drift gillnet fishery ceased operations in the South Pacific in 1991.) Interaction between these surface fisheries and the longline fishery, which generally operates further north on somewhat larger albacore, is essentially a type-B/C interaction similar in many respects to the yellowfin purse-seine-longline interaction.

3.8.1 Current assessment

The interaction among South Pacific albacore fisheries is currently not well understood because of a lack of information on basic population parameters, distribution of the stock and, until recently, catch, effort, and size composition from the fisheries. There have been reports of falling catch rates in the longline fishery, with the popular interpretation being that this is a direct result of the large surface fishery catches of 1988-1989 and 1989-1990. Confirmation or otherwise of this hypothesis awaits scientific evaluation of the data currently being collected.

3.8.2 Further work and data required

Efforts to mount an effective tagging programme are currently being made, and if successful, will provide much of the basic biological information required. Also, the South Pacific Albacore Research group is assembling catch-and-effort and size-composition databases to enable specific analyses to be carried out. These efforts should be continued and encouraged. Some important methodological development is also required, namely the development of an age-structured model based on size data from which estimates of interaction can be derived. This work is currently being undertaken through SPC's Albacore Research Project.

4. RECOMMENDATIONS FOR CONTINUED OR FUTURE STUDIES

4.1 <u>Fishery Data Compilation</u>

Analysis of any of the interaction issues described above requires access to some level of fisheries statistics. The detail and scope of the statistics required depends on the type of interaction problem and the method of analysis employed. For simple, type-A interaction problems, time series of total catch-and-effort data for the competing fisheries may be sufficient. For type-B problems, a time series of size composition data would also be required. For the more complex, spatial, type-C problems, catch, effort and size-composition data stratified by area for the interacting fisheries are required. Certain analytical techniques may require much more detailed data. For example, the analysis of type-A and B interaction problems using GLM techniques requires access to detailed daily catch and effort logbook data, descriptions of vessel and crew characteristics and environmental data. SPC has begun to compile an aggregated database for the WCPO that it hopes will be useful for the analysis of the majority of interaction and stock assessment problems. This database will have catch/effort and size composition components, with these variables being aggregated by vessel nationality, gear type, fishing method, 5° quadrangle and month. The database will attempt to represent the entire industrial tuna fishing activity in the WCPO, *i.e.* all estimates will be raised. In the first instance, the database will be compiled from daily catch-and-effort logbook data provided to SPC by its member countries, statistics published by DWFNs or otherwise made available to the SPC for this purpose, estimates of total catch and effort for various fleets and data currently available from various port sampling programmes. The SPC would welcome the opportunity to collaborate with other organisations on this initiative.

While the proposed database would satisfy many of the fishery-data requirements for interaction studies, some more specific data may be required in some instances. For example, the thorough analysis of interaction problems involving artisanal/subsistence fisheries would require reliable catch, effort and ideally size-composition data on those fisheries. In most cases, such data have not been collected from these fisheries, and the establishment of data-collection systems would be a time-consuming and costly exercise. Some preliminary work with simulation models incorporating movement may be useful in indicating the desirability of collecting detailed data on artisanal/subsistence fisheries.

4.2   Spatial Models of Tagged-Tuna Dynamics

It is clear that most of the type-C interaction problems can only be adequately addressed by spatially-disaggregated models of tagged-tuna dynamics that explicitly include a generalised movement component. Efforts to develop such models, for application to the SSAP, RTTP, and other tagging data sets, are currently being made and should be continued as a matter of priority. Ideally, such models should be structured to allow extrapolation to real fisheries situations, and in particular allow estimates of the effect that an existing fishery has on another, both at its current level of exploitation and for alternative levels and distributions of effort. Detailed catch/effort statistics from the fisheries involved are required for such studies.

4.3   Abundance Indices

The interpretation of various CPUE trends is confounded by improvements in searching and catching technology, variation in the spatial and temporal distributions of the fisheries and the effects of environmental variation. Without more detailed analysis, it is difficult to attribute variation in CPUE specifically to any of these factors or to the activities of another fishery. The GLM techniques may provide useful information on the relative effects of different factors on CPUE. In particular, various pole-and-line/purse-seine interactions could be investigated with models of pole-and-line CPUE incorporating the effects of spatio-temporal distribution of fishing, pole-and-line vessel characteristics and fishing techniques, environmental variables affecting pole-and-line fishing success and purse-seine catch. Detailed, daily catch-and-effort logbook data and information on affecting factors would be required for such analyses.

4.4   Availability of Yellowfin to Surface and Longline Gears

Investigation of the question of possible heterogeneity of the yellowfin population with respect to surface and longline gears is required. In the first instance, a thorough analysis of existing tagging data is needed to compare "expected" longline recaptures of tagged yellowfin, under a null hypothesis of equal availability, with reported recaptures. If differences in recapture rates are significant and cannot be attributed to non-reporting, field experiments may be required to determine the extent and the cause of the heterogeneity. One useful experiment might be to tag yellowfin in both surface and longline fisheries on a small-area scale where both gear types are present. If sufficient releases were made from both gear types, a direct comparison of the vulnerability of surface- and longline-caught releases might be possible. The use of archival tags in such an experiment would add enormously to its value, and continued efforts should be made to develop a cost-effective tag of this type.

4.5   Distribution of Recruitment

Investigation of the distribution of skipjack and yellowfin recruitment in relation to the distribution of surface fisheries, both locally-based and DWFN, operating in the WCPO is required for a more thorough understanding of type-C interactions, particularly in the case of oceanic versus island-based fisheries. Tagging experiments may provide some insight into these matters in an indirect way, *e.g.* it might be possible to estimate the proportion of recruitment in a particular area from particular sources outside that area, as was done in Kleiber *et al.* (1984). Survey approaches and analyses of size-composition and catch-and-effort data stratified by area may provide the necessary information.

4.6   South Pacific Albacore Research

The continuation of efforts to mount an effective tagging programme on South Pacific albacore should be continued, and in particular experiments carried out to determine the reasons for the very low recovery rates of tagging programmes to date. The development of an age/size-structured model from which estimates of surface-longline fishery interaction can be derived should also be undertaken.

5.   REFERENCES CITED

Argue, A.W., and R.E. Kearney. 1983. An assessment of the skipjack and baitfish resources of New Zealand. *Final Ctry.Rep.Skipjack Surv.Assess.Programme, S.Pac.Comm.*, (8):47 p.

Fonteneau, A. 1986. Surface versus longline tropical fisheries in the Atlantic Ocean - what interaction? *Coll.Vol.Indo-Pac.Tuna Dev.Mgt.Programme, Colombo, Sri Lanka,* Working Doc.:191-216.

Hampton, J. 1989. Population dynamics, stock assessment and fishery management of the southern bluefin tuna (*Thunnus maccoyii*). Ph.D. Thesis, University of New South Wales, Australia, 273 p.

Kleiber, P., A.W. Argue, J.R. Sibert, and L.S. Hammond. 1984. A parameter for estimating potential interaction between fisheries for skipjack tuna (*Katsuwonus pelamis*) in the western Pacific. *Tech.Rep.Tuna Billfish Assess.Programme, S.Pac.Comm.*, (12): 11 p.

Leis, J.M., T. Trnski, M. Harmelin-Vivien, J.-P. Renon, V. Dufour, M.K. El Moudni, and R. Gazlin. 1991. High concentrations of tuna larvae (Pisces: Scombridae) in near-reef waters of French Polynesia (Society and Tuamotu Islands). *Bull.Mar.Sci.*, 48:150-8.

Lenarz, W.H., and J.R. Zweifel. 1979. A theoretical examination of some aspects of the interaction between longline and surface fisheries for yellowfin tuna, *Thunnus albacares*. *Fish.Bull.NOAA-NMFS*, 76:807-25.

Polacheck, T. 1988. Yellowfin tuna, *Thunnus albacares*, catch rates in the western Pacific. *Fish.Bull.NOAA-NMFS*, 87:123-44.

Sibert, J.R. 1984. A two-fishery tag attrition model for the analysis of mortality, recruitment and fishery interaction. *Tech.Rep.Tuna Billfish Assess.Programme, S.Pac.Comm.*, (13):27 p.

Sibert, J. and D. Fournier. 1993. Evaluation of diffusion-advection equations for the estimation of movement patterns from tag recapture data. *In* Proceedings of the First

FAO Expert Consultation on Interactions of Pacific Tuna Fisheries, edited by R.S. Shomura, J. Majkowski, and S. Langi, 3-11 December 1991, Noumea, New Caledonia. [See this document.]

South Pacific Commission. 1988. Interaction among South Pacific skipjack fisheries. Paper presented at the South Pacific Commission Twentieth Regional Technical Meeting on Fisheries, 1-5 August 1988, Noumea, New Caledonia, Info. Pap. 4.

South Pacific Commission. 1992. Status of tuna fisheries in the SPC area during 1991, with revised annual catches since 1952. *Tech.Rep.Tuna Billfish Assess.Programme, S.Pac.Comm.*, 29:73 p.

Suzuki, Z. 1988. Study of interaction between longline and purse seine fisheries on yellowfin tuna, *Thunnus albacares* (Bonnaterre). *Bull.Far Seas Fish.Res.Lab.*, (25):73-144.

Suzuki, Z. 1993. A review of interaction between purse seine and longline on yellowfin tuna, Thunnus albacares, in the western and central Pacific. *In* Proceedings of the First FAO Expert Consultation on Interactions of Pacific Tuna Fisheries, edited by R.S. Shomura, J. Majkowski, and S. Langi, 3-11 December 1991, Noumea, New Caledonia. [See this document.]

Tanaka, T. 1989. Shift of the fishing ground and features of shoals caught by purse seine fishery in the tropical seas of the western Pacific Ocean. *Bull.Tohoku Reg.Fish.Res.Lab.*, (51):75-88.

Wild, A., and J. Hampton. 1993. A review of the biology and fisheries for skipjack tuna, *Katsuwonus pelamis*, in the Pacific Ocean. *In* Proceedings of the First FAO Expert Consultation on Interactions of Pacific Tuna Fisheries, edited by R.S. Shomura, J. Majkowski, and S. Langi, 3-11 December 1991, Noumea, New Caledonia. [See this document.]

A REVIEW OF INTERACTION BETWEEN PURSE SEINE AND LONGLINE ON YELLOWFIN (*THUNNUS ALBACARES*) IN THE WESTERN AND CENTRAL PACIFIC OCEAN

Ziro Suzuki
National Research Institute of Far Seas Fisheries
Shimizu, Japan

ABSTRACT

Interaction between purse-seine and longline fisheries for yellowfin tuna in the western and central Pacific Ocean was reviewed mainly through fisheries information such as trends in catch, fishing effort, CPUE, and size of fish captured. In spite of a significant increase of the purse-seine fishery, there has been little sign of a clear interaction between the two fisheries except a decline in the longline CPUE within the purse-seine fishing ground. In addition, it was inferred from relatively small rates of decline in the historical CPUE series of the longline fishery that the yellowfin stock in the western and central Pacific is significantly larger than other yellowfin stocks. Recommendations for improvement of the interaction studies were made.

1.   INTRODUCTION

Since the rapid development of a purse-seine fishery for skipjack and yellowfin tuna in the western and central Pacific in the early 1980s, the estimation of maximum sustainable yield (MSY) as well as possible interaction between the purse-seine and the longline fisheries for yellowfin tuna have been major subjects of concern in this region. However, inadequate basic fisheries statistics and biological information on yellowfin tuna in this area hinder the progress of the studies on these problems.

Theoretical studies suggest that the total yield from the yellowfin stock is larger in the case of coexistence of the two fisheries than in the case of exploitation by either one of the two fisheries alone (Lenarz and Zweifel, 1979), especially if the stock is somewhat discrete for the longline and the surface fisheries (Hilborn, 1989). In this regard, knowledge about mixing and migration of yellowfin tuna, especially for medium- and large-sized fish on which little information is available so far, is critical for understanding of the interaction studies and therefore for international management of this species (Deriso *et al.*, 1991). An interesting finding on transatlantic migration of the large yellowfin tuna has been reported recently, including the first recoveries of tagged yellowfin tuna in the Atlantic by the longline boats (ICCAT, 1992).

As for small fish, it is anticipated that the ongoing large scale tagging by the South Pacific Commission (SPC) will provide good material for various quantitative studies including the interactions among the fisheries. As later mentioned in the Section 3, the interactions among the fisheries on yellowfin tuna are much more complicated than previously assumed.

2. OBSERVED TRENDS IN BASIC STATISTICS ON YELLOWFIN IN THE WESTERN AND CENTRAL PACIFIC

In this section, factual aspects of fisheries information relevant to the interactions between purse-seine and longline fisheries are described.

2.1 Total Catch and Catch by Fishing Gear

As frequently mentioned elsewhere, it is difficult to estimate the total catch of yellowfin tuna from the western and central Pacific due mainly to the lack of a coordinated data acquisition system with binding power for data submission from the countries in the area. Two sets of statistics, however, are available which approximate the total catch of yellowfin tuna from the western and central Pacific. One is the Food and Agriculture Organization (FAO) yearbook for fisheries statistics of catch and landings (*e.g.*, FAO, 1991), the other is statistical publications from the SPC (*e.g.*, SPC, 1991).

There is fairly good accordance, at least in trend and magnitude, on the total catch between the two sets of statistics (Tables 1 and 2) despite difference in their data collecting areas (Figure 1) and data collecting system. The total catch given in the SPC statistics tends to be far smaller than those recorded by the FAO before 1979, due partly to exclusion of the Philippine and Indonesian catches. Other than that period, the SPC statistics show consistently larger catches than those from the FAO area 71, probably because the former statistics cover larger areas than the latter. The catch prior to 1962 is unavailable in the SPC statistics but sizable longline catches, mainly by the Japanese fleet, were recorded since 1952 (Table 3). It should be noted that in addition to the total estimate of the SPC statistical area, the catches from the contiguous area to the north, *i.e.*, FAO area 61, produces about 30 to 40 thousand tons of yellowfin, caught predominantly by Japanese and Taiwanese longline fisheries. Therefore, the overall total catch from the western and central Pacific including the temperate areas may be on the order of 370-380 thousand tons in 1990.

Table 2 shows gear breakdown of the total catch for the SPC statistical area. It is clear that the increase in the total catches is due to increase in purse-seine catch and catches by the Philippine and Indonesian fisheries. The longline catch fluctuates during the past three decades with a slight decrease in the last decade while the baitboat catch shows a clear decreasing trend after 1980. These trends in the catch by major fishing gear category are largely driven by the corresponding trends of fishing effort in each gear category.

2.2 Catch, Fishing Effort, and Catch Per Unit Effort (CPUE) by Major Fisheries

As shown in Table 2, the baitboat catch of yellowfin tuna is very small compared to the purse-seine and longline fisheries. Therefore, only purse-seine and longline fisheries as well as the fisheries of the Philippines and Indonesia are described in this study. The major distant-water fishing countries with purse-seine and longline fleets in the region are the Japanese with purse-seine and longline fleets and the United States of America (USA) with a purse-seine fleet.

160

Table 1. Catch of yellowfin tuna by countries from the FAO area
71 (Pacific, Central west).

Unit:ton

Country (Gear)/Year	1971	1972	1973	1974	1975	1976	1977	1978
FIJI				12	11	74	151	540
Indonesia					11062	8037	10859	10601
Japan Total	23531	24315	29822	31670	33070	37813	50822	73080
Longline	22807	24021	29286		29528		41987	59044
Purse seine	379		481		2176		7159	7036
Baitboat	345	294	55		55		1676	769
Other								6231
Kiribati				25	25	25	2771	2930
Korea Rep.					259	3664	5462	5088
Papua New Guinea			1420	1420	1743	8563	3695	3115
Philippines			14900	51732	52793	44478	63059	47629
Solomon Is.								
U.S.A.								
Other				4229	4319	2510	2805	1870
Total	23531	24315	46142	89088	92209	105164	138904	144853
Japan/Total	1.00	1.00	0.65	0.36	0.36	0.36	0.37	0.50

Country (Gear)/Year	1979	1980	1981	1982	1983	1984	1985	1986
FIJI	361	240	846	1157	1586	1771	1128	995
Indonesia	14663	17550	21869	24340	20200	26450	31022	34140
Japan, Total	54789	75990	77145	72362	70682	61767	76513	73108
Longline	43488	55888	49003	38162	40193	28433	30766	24872
Purse seine	10528	9918	21827	28054	25567	32057	37523	42388
Baitboat	773	6143	2706	1531	1030	1275	3229	1827
Other		4041	3609	4014	3892	2	4995	4021
Kiribati	3000	3148	3000	3000	2135	4036	4844	1065
Korea Rep.	6881	7424	2712	2528	1156	1373	1893	3251
Papua New Guinea	2881	3019	3516	0	0	372	370	400
Philippines	49224	48023	56176	51922	62036	58927	64293	59510
Solomon Is.	192	314	1167	2165	3328	2816	3698	2769
U.S.A.		772	12867	14345	51066	41455	28798	36520
Other	2084	2363	1444	1036	433	1031	1062	1009
Total	134075	158843	180762	172855	212622	199998	213621	212767
Japan/Total	0.41	0.48	0.43	0.42	0.33	0.31	0.36	0.34

Country (Gear)/Year	1987	1988	1989
Fiji	1199	628	617
Indonesia	33080	34920	46422
Japan Total	75535	48139	62818
Longline	23568	16432	21679
Purse seine	46442	26283	36078
Baitboat	1590	1112	1074
Other	3935	4312	3987
Kiribati	350	395	385
Kores Rep	18279	15710	16283
Philippines	51809	57060	62146
Solomon Is.	6565	11581	9541
Tuvalu	74	31	30
U.S.A.	69636	25272	41274
Vanuatu	60	60	60
Other	4376	7566	5573
Total	260963	201362	245149
Japan/Total	0.29	0.24	0.26

After Suzuki et al. (1989) and FAO (1991)
The Japanese catches shown in the Table
do not always accord with those by the
FAO statistics due to revision made
afterward by Statistics and Information
Division of Japanese Government.

Table 2. Yellowfin catch from the SPC statistical area and the waters of Indonesia and the Philippines

After SPC (1991)   Unit:ton

| | SPC statistical area | | | | | | GRAND TOTAL |
YEAR	LONGLINE	BAITBOAT	PURSE SEINE	SUB TOTAL	PHILIPPINES	INDONESIA	
1952	0			0			0
1953	0			0			0
1954	0			0			0
1955	0			0			0
1956	0			0			0
1957	0			0			0
1958	0			0			0
1959	0			0			0
1960	0	0		0			0
1961	0	0		0			0
1962	53,327	0		53,327			53,327
1963	49,715	0		49,715			49,715
1964	41,270	141		41,411			41,411
1965	43,563	173		43,736			43,736
1966	49,966	71		50,037			50,037
1967	28,168	52		28,220			28,220
1968	38,401	17		38,418			38,418
1969	37,598	133		37,731			37,731
1970	33,253	75	0	33,328			33,328
1971	40,677	263	0	40,940			40,940
1972	48,649	2,796	0	51,445			51,445
1973	49,494	2,688	412	52,594			52,594
1974	49,087	3,180	728	52,995			52,995
1975	28,937	4,177	1,664	34,778			34,778
1976	36,310	11,944	3,504	51,758			51,758
1977	52,657	9,759	5,189	67,605			67,605
1978	72,573	5,885	7,854	86,312			86,312
1979	57,673	5,440	11,271	74,384	49,224	17,899	141,507
1980	81,384	11,048	12,015	104,447	48,023	20,898	173,368
1981	50,735	10,204	45,320	106,259	56,176	25,239	187,674
1982	39,313	3,286	66,840	109,439	51,922	28,080	189,441
1983	46,162	2,499	86,990	135,651	62,036	26,088	223,775
1984	34,706	3,074	87,920	125,700	58,924	30,697	215,321
1985	38,761	5,808	80,625	125,194	64,293	34,130	223,617
1986	32,939	3,428	99,752	136,119	59,510	37,508	233,137
1987	40,414	3,531	148,101	192,046	51,810	35,706	279,562
1988	36,066	4,662	93,968	134,696	57,060	37,491	229,247
1989	36,480	4,050	156,494	197,024	62,146	57,995	317,165
1990	38,057	3,967	176,703	218,727	62,146	57,995	338,868

Catches of yellowfin tuna by purse seiners may include as much as 10 percent bigeye.

Statistics for Philippines and Indonesia are taken from IPTP (1991). Catch estimates for 1989 were used as preliminary estimates for 1990.

Taiwanese and Korean purse-seine catches have been increasing rapidly and may catch up with their Japanese and USA counterparts (SPC, 1991). However, the time series of those countries is rather short. Solomon Islands catch is significant among the South Pacific countries, but the catch is predominantly composed of skipjack. Therefore, the trend of catch, fishing effort, and CPUE as well as average weight are described for the following major fisheries only.

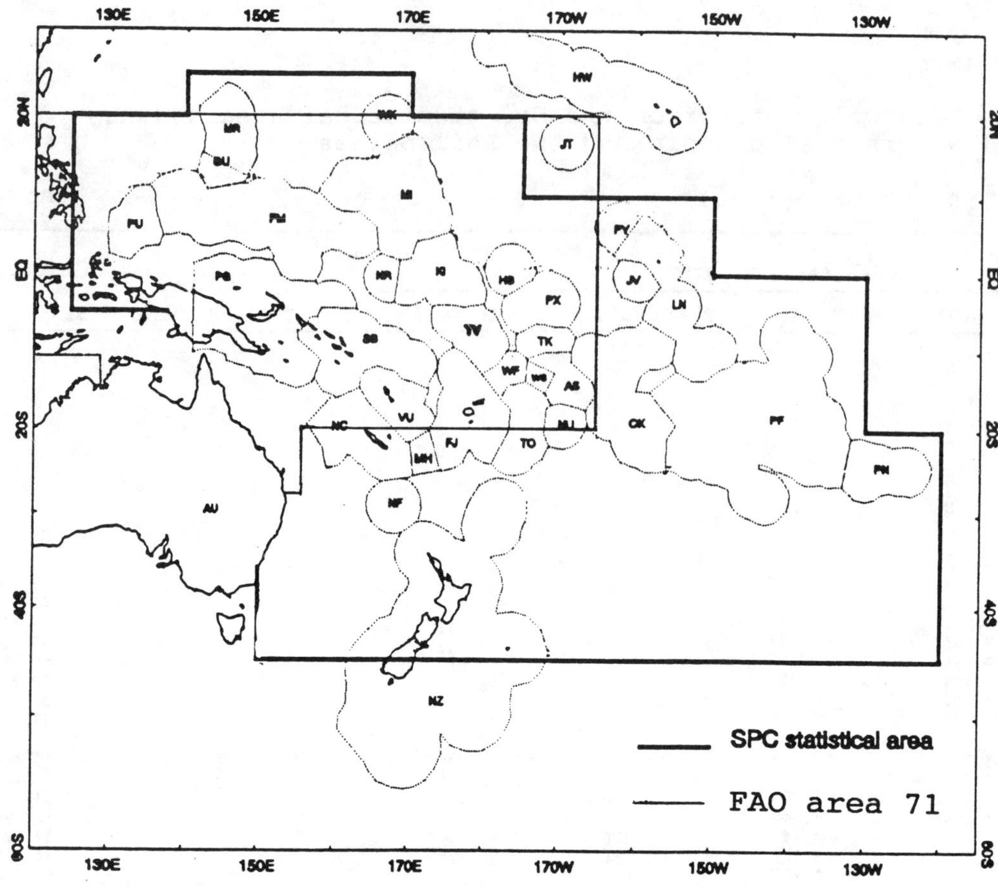

Fig. 1. SPC statistical area and FAO area 71

2.2.1 Japanese longline fishery

The Japanese longline fishery has the longest time series in the western and central Pacific distant water tuna fisheries. A pre-World War II longline fishery by the Japanese existed in the northwestern Pacific, but the exact amount of the catch is not well known.

The fishery has developed rapidly from 1952, aiming at almost all available kinds of tuna and billfishes, including albacore, yellowfin, bigeye, blue marlin, and swordfish. The fishery reached its first peak in yellowfin catch during the early half of the 1960s and then decreased up to the early 1970s, reflecting a change in target species from tropical tunas to temperate tunas (Table 3). The second peak was attained in the period from the late 1970s to the early 1980s due to the development of deep longline fishing for bigeye in the tropical areas where yellowfin is also abundant. The recent catch shows a decreasing trend due to reduction of fishing effort, including the voluntary domestic effort reduction policy and reduction in numbers of longline boats.

The CPUE trend of the Japanese longline fishery in the western and central Pacific, derived from effective fishing effort estimated by the Honma method (Honma, 1974), shows a decline of about one-half from the start of fishing in 1952 (1.29) to the record low 1989 value (0.67), as shown in Table 3 and Figure 2. Incidentally, both general linear model analysis and the Honma method gave an almost identical CPUE

Table 3. Basic statistics of Japanese longline fisheries for yellowfin tuna in the western and central Pacific (west of 180, 40N-40S)

YEAR	CATCH(ton)	CATCH(No)	AV.W(KG)	EFFORT(G)	EFFORT(E)	E/G	CPUE
1952	18481	568634	32.5	923	441	0.477	1.290
1953	27578	761368	36.2	1048	509	0.485	1.497
1954	31114	756512	41.1	1263	525	0.415	1.441
1955	21725	586207	37.1	1222	489	0.400	1.198
1956	24269	788620	30.8	1264	598	0.473	1.318
1957	42172	1312749	32.1	1445	933	0.646	1.407
1958	37481	1166603	32.1	1464	842	0.575	1.386
1959	34110	1185096	28.8	1601	826	0.516	1.434
1960	46297	1622004	28.5	1833	1010	0.551	1.607
1961	50881	1500979	33.9	1936	1346	0.695	1.116
1962	52500	1739317	30.2	1860	1701	0.914	1.023
1963	45195	1465535	30.8	1674	1317	0.787	1.113
1964	39478	1314789	30.0	1401	1139	0.813	1.155
1965	35369	1213945	29.1	1666	1235	0.741	0.983
1966	57775	1858307	31.1	1949	1694	0.869	1.097
1967	29915	922591	32.4	1950	1131	0.580	0.816
1968	29904	992523	30.1	1713	1171	0.683	0.848
1969	31016	1039432	29.8	1631	1068	0.655	0.973
1970	23631	813891	29.0	1291	753	0.583	1.081
1971	26632	927742	28.7	1390	978	0.703	0.949
1972	26262	806933	32.5	1282	1124	0.876	0.718
1973	33314	1132922	29.4	1285	1283	0.998	0.883
1974	29318	1091142	26.9	1545	1329	0.860	0.821
1975	33747	1030052	32.8	1419	1304	0.919	0.790
1976	36635	1099240	33.3	1517	1346	0.887	0.817
1977	39980	1449040	27.6	1445	1298	0.898	1.117
1978	56719	2139592	26.5	1546	1502	0.972	1.425
1979	45478	1604668	28.3	1766	1738	0.984	0.923
1980	53974	2145910	25.2	1870	2174	1.162	0.987
1981	47751	1875750	25.5	2114	2289	1.082	0.820
1982	40794	1422147	28.7	1905	2021	1.061	0.704
1983	40443	1417417	28.5	1642	1387	0.845	1.022
1984	29623	1039670	28.5	1580	1453	0.920	0.716
1985	36189	1149270	31.5	1667	1597	0.958	0.720
1986	26132	871714	30.0	1571	1047	0.666	0.833
1987	23493	802730	29.3	1501	875	0.583	0.917
1988	32257	988549	32.6	1612	963	0.597	1.027
1989	25128	779836	32.3	1472	1170	0.795	0.666

Effort(G), Effort(E) and CPUE denote nominal number of hooks (10^5), effective hooks (10^5) and catch in numbers of fish per effective hooks. E/G is concentration index.

trend for yellowfin tuna in the areas (Suzuki *et al.*, 1989). Average weight of the fish caught by the longline fishery showed a consistent decreasing trend from above 30 kg in the early phase of the fishing to 25 kg in the early 1980s, increasing in recent years up to almost the same level as the 1950s.

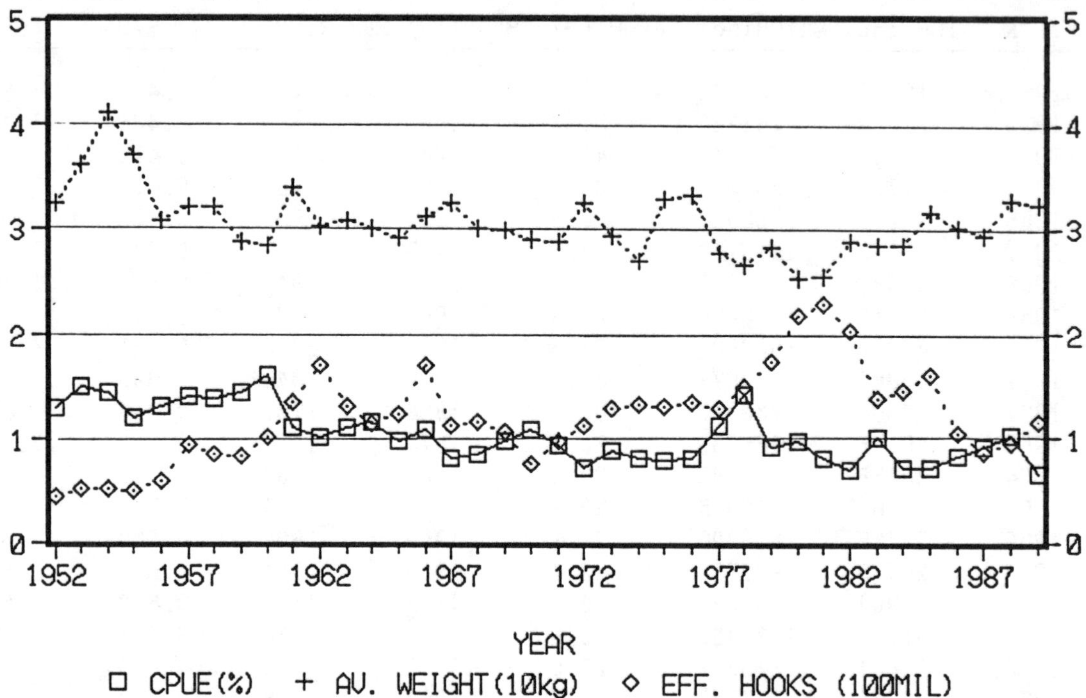

Fig. 2.   Trends of fishing effort, CPUE and average weight of yellowfin tuna caught by the Japanese longline fishery

See Table 3 for exact figures of each valuables.

2.2.2   Japanese purse-seine fishery

In the western tropical Pacific, the Japanese purse-seine boats that commenced operations in the middle of the 1970s developed into the present international fishery (Honma and Suzuki, 1978). The Japanese purse-seine fleets in the area consisted of two distinctive components: one is single boat seiners, the other so-called group seiners. The single boat seiners dominate in the purse seining in terms of catch and number of boats, operating throughout the year in the area, while the group seiners, which consist of several support boats such as reefers and scouting boats, operate only a few months.

Table 4 shows trends of catch, fishing effort, and nominal CPUE for the Japanese single purse seine boats operating in the western and central Pacific. Since skipjack dominates the purse seine catch in this area, the relevant information for skipjack is also shown in the table.

The catch of yellowfin tuna by the single purse seiners increased remarkably up to 1983 then somewhat leveled off with slight increase. Fishing effort statistics are incomplete but show a similar trend. The number of boats has been limited since 1982. Nominal CPUE showed an increasing trend through 1987 but dropped sharply in 1988 then showed a higher value. Average weight of the catch was rather stable from 1976 to 1986, around about 5 kg (Suzuki *et al.*, 1989).

Table 4 . Catch statistics for single purse seiners of Japan

YEAR	VESSELS ACTIVE	DAYS FISHED	SKIPJACK MT	SKIPJACK CPUE	SKIPJACK %	YELLOWFIN MT	YELLOWFIN CPUE	YELLOWFIN %	OTHER MT	TOTAL MT	TOTAL CPUE
1973[4]	6	...	1,245	...	71	412	...	24	95	1,752	...
1974[4]	7	...	2,437	...	72	728	...	21	227	3,392	...
1975	7[4]	...	4,566[1]	...	73	1,664[1]	...	27	...	6,229[1]	...
1976	10[4]	...	10,353[1]	...	76	3,304[1]	...	24	...	13,658[1]	...
1977	13[4]	...	13,566[1]	...	73	4,989[1]	...	27	...	18,556[1]	...
1978	16[4]	...	23,249[1]	...	75	7,654[1]	...	25	...	30,903[1]	...
1979	16[4]	...	24,875[1]	10.9	70	10,671[1]	4.7	30	...	35,546[1]	15.8
1980	14[4]	...	31,391[1]	13.7	77	9,607[1]	3.4	23	...	40,999[1]	17.3
1981	24[4]	...	37,188[1]	10.6	63	21,730[1]	5.0	37	...	58,918[1]	15.7
1982	33[4]	...	70,000[1]	11.5	71	28,774[1]	4.7	29	...	98,774[1]	16.3
1983	34[5]	6,581[1]	109,830[1]	16.7[1]	81	26,191[1]	4.0[1]	19	...	136,021[1]	20.7[1]
1984	41[5]	7,262[1]	110,052[1]	15.2[1]	78	30,836[1]	4.2[1]	12	...	140,889[1]	19.4[1]
1985	33[5]	7,209[1]	103,647[1]	14.4[1]	75	34,730[1]	4.8[1]	25	...	138,377[1]	19.2[1]
1986	34[5]	6,302[1]	108,486[1]	17.2[1]	75	39,724[1]	6.3[1]	25	...	148,210[1]	23.5[1]
1987	32[5]	6,450[1]	88,442[1]	13.7[1]	69	40,392[1]	6.3[1]	31	...	128,834[1]	20.0[1]
1988	33[5]	6,898[1]	137,965[1]	20.0[1]	85	24,928[1]	3.6[1]	15	...	162,894[1]	23.6[1]
1989	33[5]	...	115,300[2]	14.7	77	33,500[2]	5.0	22	1,035[2]	149,835[2]	19.9
1990	32[5]	...	141,952[3]	19.9	77	41,244[3]	4.4	22	1,274[3]	184,470[3]	24.5

Units: CPUE, metric tonnes per day

SOURCES

1. The number of days fished and CPUE for 1983-1988, and catches of skipjack and yellowfin for 1975-1988, were estimated during joint research conducted in 1989 by the SPC Tuna and Billfish Assessment Programme and the National Research Institute of Far Seas Fisheries. The area covered is bordered by 20°N-20°S and 120°E-180°.

2. Catches for 1989 were provided by the National Research Institute of Far Seas Fisheries (Tsuji, personal communication, August 1990). The estimates are for an area bordered by 25°N-25°S and 130°E-180°. The catch of other species includes 950 mt of bigeye.

3. Preliminary catch estimates for 1990 were determined by raising the catches for 1989 by the ratio of the catch rate in 1990 to the catch rate in 1989.

4. Catch statistics for 1973-1974 and the number of vessels during 1973-1982 are from the Fisheries Agency of Japan, quoted in Habib(1984). The number of vessels include one survey vessel in 1974-1975, two survey vessels in 1976, and three survey vessels in 1977-1982.

5. The number of vessels active for 1983-1990 were determined from data held in the Regional Tuna Fisheries Database. Purse seiners licensed in Japan for exploratory fishing may be included.

2.2.3 USA purse-seine fishery

The advent of the USA purse seiners in the western and central Pacific was due to decline of the catch rate in the eastern Pacific and the devastating effect of the 1983-1984 El Niño which drove a significant portion of the fleet from the eastern to the western Pacific. In the 1980s the USA fleet appears to have been operating over much wider areas, especially to the east and south of the other purse-seine fleets (Coan, 1993).

Table 5 shows the trend of the USA purse-seine catch and fishing effort in terms of the number of boats. Yellowfin catch increased drastically from the early 1980s and peaked in 1987; it dropped sharply in 1988 and again increased. The number of boats peaked in 1983 and has stabilized at about 35 boats in recent years reflecting the South Pacific Regional Tuna Treaty between the USA and the Forum Fisheries Agency (FFA). Only fragmental information is available for CPUE. The sharp drop of the 1988 value is the same with the Japanese single purse-seine fishery. However, since data coverage of the USA fleet is very small before July 1988, the 1987 CPUE value may be unreliable.

Table 5. Landings(ton), number of boats and CPUE (catch per days fished) for U.S. purse seiner operated in the western and central Pacific

After Coan (in press) and SPC (1991) for 1990 figures

Year	Yellowfin	Skipjack	Bigeye	Total	No. of boats	Yellowfin CPUE
1976	200	500	-	700	3	
1977	200	700	-	900	1	
1978	200	800	-	1000	2	
1979	600	8000	20	8620	8	
1980	1100	9900	0	11000	14	
1981	13000	17400	170	30570	14	
1982	22000	37900	*	59900	24	
1983	49600	104100	-	153700	62	
1984	45100	124300	60	169460	61	
1985	29000	87700	-	116700	40	
1986	36600	93500	-	130100	36	
1987	66400	79800	-	146200	35	12.7
1988	25200	99400	-	124600	32	3.4
1989	41200	90400	-	131600	34	7.7
1990	56670	105660	-	162230	38	

Remarks

- indicates that landings are not available but may be greater than zero.

* indicates values less than 10 metric tons

Landings before 1979 are from Pacific Tuna Development Foundation exploratory fishing charters. Landings from other U.S. vessels fishing in 1976 to 1978 are unknown.

Values in this table for 1980 to 1985 are different than those in Doulman 1987 due to inclusion here of U.S. vessels operating out of Guam and direct exports.

Catch estimates for 1990 were provided by the National Marine Fisheries Service (Sakagawa, PRO 93/3/30, June 1991); these statistics are preliminary.

NMFS (1991) has reported 38 vessels active and 3 vessels inactive at the end of 1990.

2.2.4 Philippines and Indonesian fisheries

For the Philippines and Indonesia, the yellowfin catch statistics include other tunas such as bigeye. In addition, swordfish and billfishes are included in the Indonesian yellowfin statistics. Table 6 shows recent catches of yellowfin tuna by major fishing gear. However, a substantial part of the catch is taken by unclassified gears. Fish aggregating devices (FADs) are commonly used by both the Philippines and Indonesia to catch tunas and large pelagic fishes.

The Philippine catches of yellowfin tuna appear to be stable during the past decade, while there is a sign of recent increase of yellowfin catches in the Indonesian fishery, especially increased catch by longline gear. While not shown separately in Table 6, the handline is one of the major fishing gear used in the Philippines and Indonesia; the handline gear catches small and large yellowfin tunas.

2.3   Size of Fish in the Catch by Major Fishery

Examples of length composition of yellowfin tuna taken by the major fishing gears mentioned in the previous section are shown in Figure 3 for the Japanese purse-seine fishery, in Figure 4 for the USA purse-seine fishery, in Figure 5 for the Japanese longline fishery (in comparison with that of the Japanese purse-seine catch within major purse-seine fishing grounds), and in Figure 6 for the Philippine fisheries. In each figure, lines indicating 30 cm and 120 cm in fork length are shown for comparison. The relevant figures for the Indonesian fisheries were not available.

The length composition of yellowfin tuna caught by the Japanese and USA purse seiners is similar but there are some important differences. Percentage of the small fish (less than about 70 cm) relative to the larger fish (over 90 cm) is much higher in the Japanese catches than USA catches (Figures 3 and 4). For the small fish group, which is a major component of the purse-seine catch, the peak size of the Japanese catch is much smaller than that in the USA catch. This difference may be due to difference in fishing areas, school types, or degree of discard of the smaller fish.

In Figure 5, length composition of yellowfin tuna caught by the Japanese longline fishery is compared with that taken by the Japanese purse-seine fishery. The major size of longline-caught yellowfin tuna in the western Pacific covers from about 90 to 140 cm.

Major fishing gears of the Philippines for catching yellowfin tuna are purse seine, ringnet (small purse seiners), and handline. The catch by size of yellowfin tuna for these fisheries was calculated from the data collected by the Indo-Pacific Tuna Development and Management Programme (IPTP) tuna sampling project which started in the early 1980s (Suzuki, 1989). One of the unique characteristics of the size of yellowfin taken by the purse-seine and ringnet fisheries of the Philippines is very small fish of less than 30 cm often found in the catches (Figure 6). The other important observation comes from the size composition of yellowfin tuna taken by the handline fishery, which indicates a significant catch of probably the largest yellowfin tuna, over 140 cm, caught in a sizable amount in the western equatorial region. It should be noted that middle sized fish between 60 and 100 cm are very few in the handline catches. The lack of this size class of fish in the catch appears to be a common feature for all major fisheries in the western

Table 6. Catches of yellowfin tuna from domestic fisheries in the Philippines and Indonesia

After SPC (1991)

Unit: ton

Indonesia

YEAR	LL	BB	PS	GILL	UNCL	TOTAL
1979	–	–	–	–	17,899	17,899
1980	–	–	–	–	20,898	20,898
1981	–	–	–	–	25,239	25,239
1982	4,120	963	1,445	–	21,552	28,080
1983	–	–	–	–	26,088	26,088
1984	2,255	2,282	2,135	–	24,025	30,697
1985	2,907	2,344	2,136	–	26,743	34,130
1986	2,557	2,278	1,794	21	30,858	37,508
1987	–	2,323	1,832	21	31,530	35,706
1988	–	2,439	1,923	22	33,107	37,491
1989	13,147	4,707	2,547	122	37,472	57,995
1990[1]	13,147	4,707	2,547	122	37,472	57,995

Philippines

YEAR	LL	PS	GILL	UNCL	TOTAL
1979	–	12,301	2,027	34,896	49,224
1980	–	12,463	2,301	33,259	48,023
1981	1,073	18,182	2,655	34,266	56,176
1982	1,897	17,676	1,386	30,963	51,922
1983	–	20,779	1,260	39,997	62,036
1984	1,284	22,989	2,161	32,490	58,924
1985	1,819	21,591	2,040	38,843	64,293
1986	2,411	17,591	2,137	37,371	59,510
1987	3,774	18,087	2,161	27,788	51,810
1988	–	–	–	57,060	57,060
1989	–	–	–	62,146	62,146
1990[1]	–	–	–	62,146	62,146

KEY:
LL	LONGLINE
BB	POLE-AND-LINE
PS	PURSE SEINE
GILL	GILLNETS
UNCL	UNCLASSIFIED

SOURCE

All statistics were taken from IPTP (1991), except where noted.

and central Pacific, although the USA purse seiners catch this size class substantially in some years (Coan, 1993).

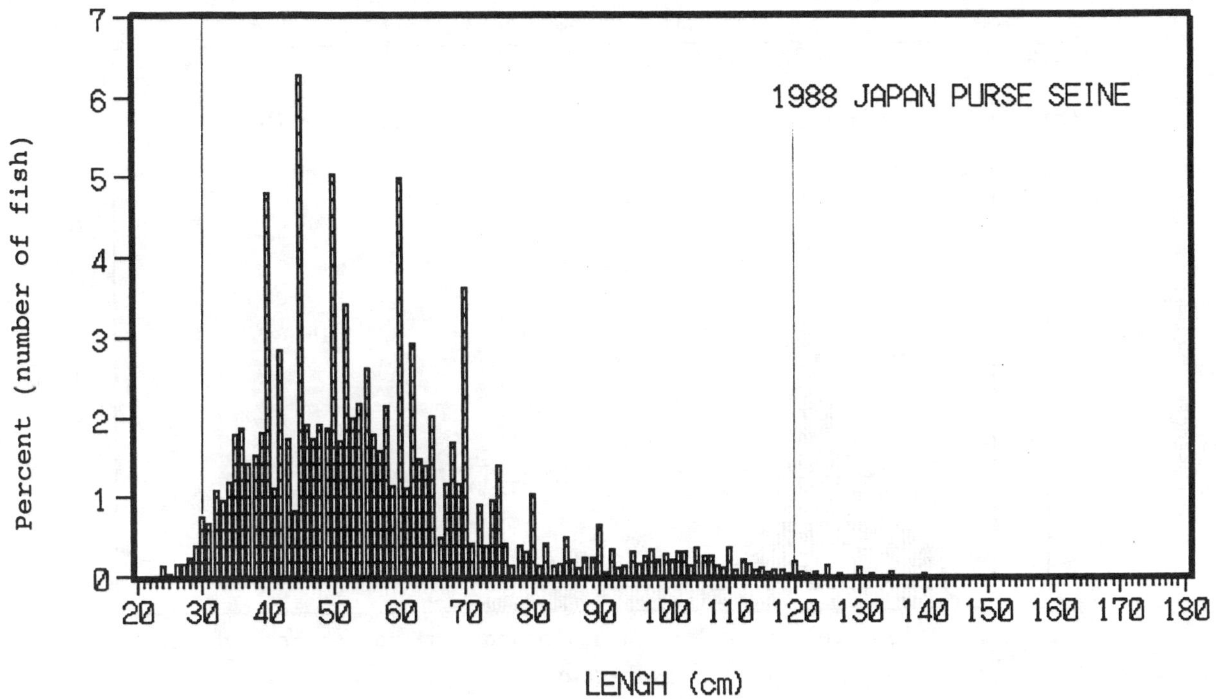

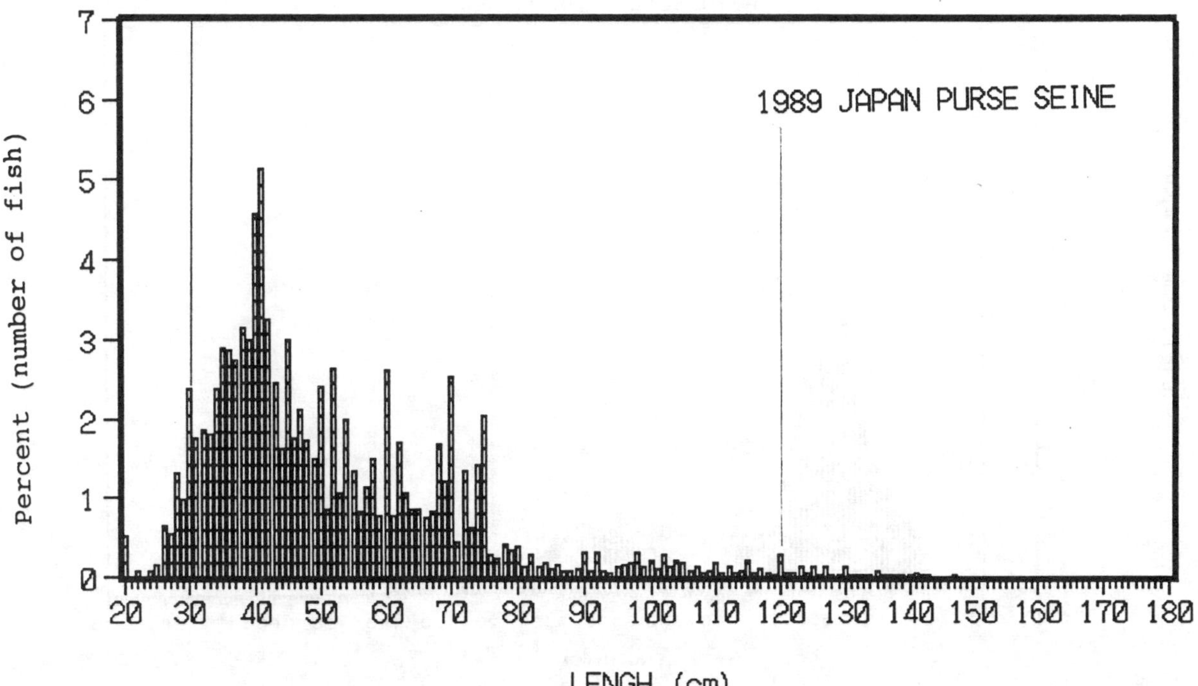

Fig. 3. Length composition of yellowfin tuna caught by the Japanese purse seiners in the western and central Pacific (west of 180, 20N-20S)

1988 U.S. CENTRAL AND WESTERN PACIFIC

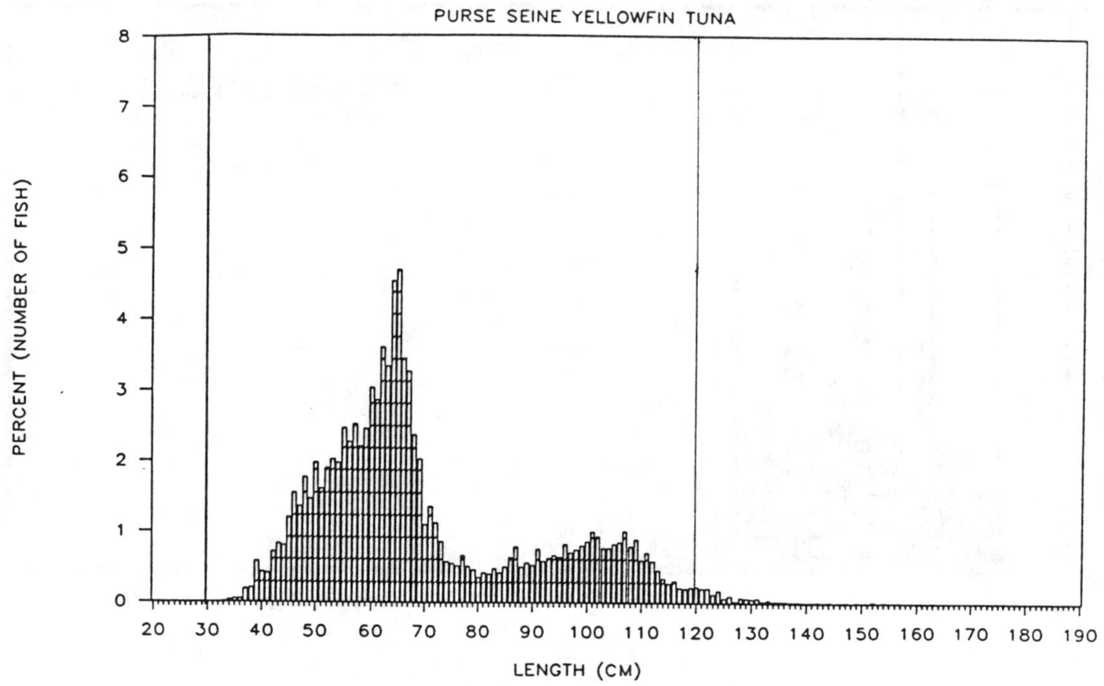

1989 U.S. CENTRAL AND WESTERN PACIFIC

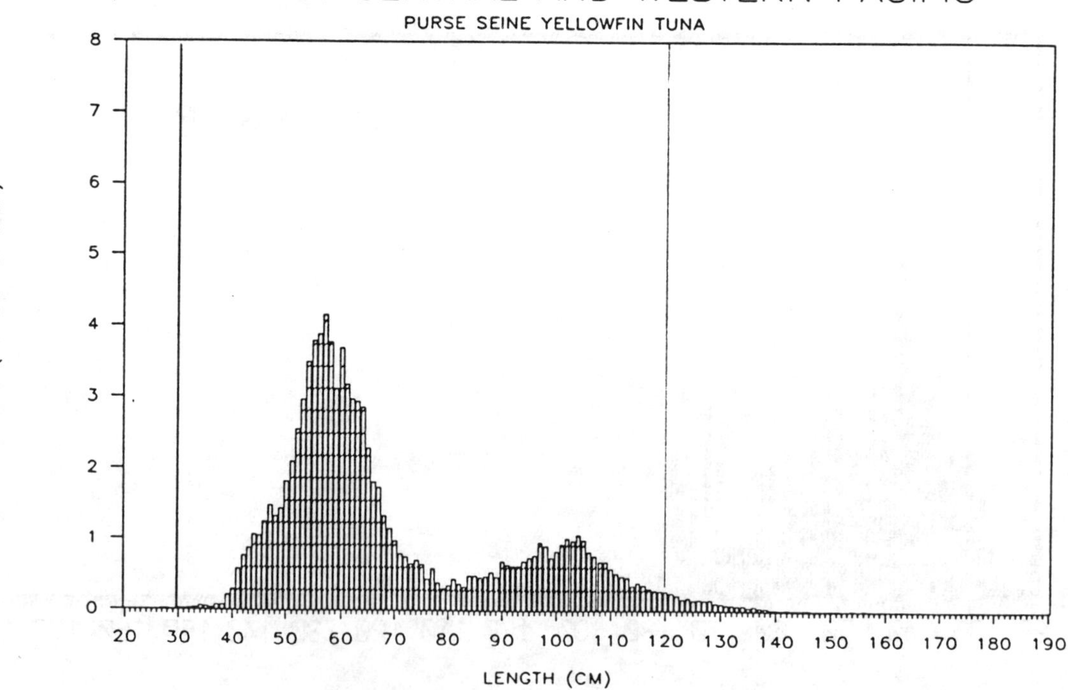

Fig. 4. Length composition of yellowfin tuna caught by the U.S. purse seiners in the western and central Pacific

After Coan (in press)

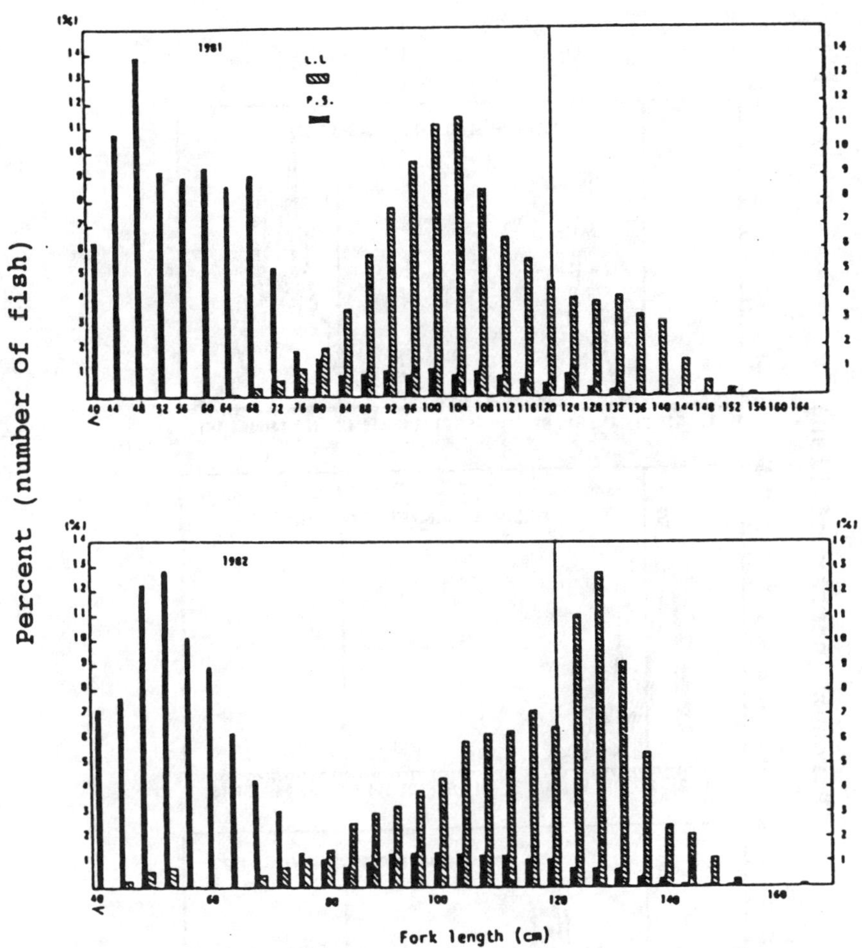

Fig. 5. Comparison of length composition of yellowfin tuna taken by the Japanese purse seine (P.S.) and by the Japanese longline boats (L.L.) in the main purse seine fishing ground (0-10N, 140-160E).

After Suzuki (1988)

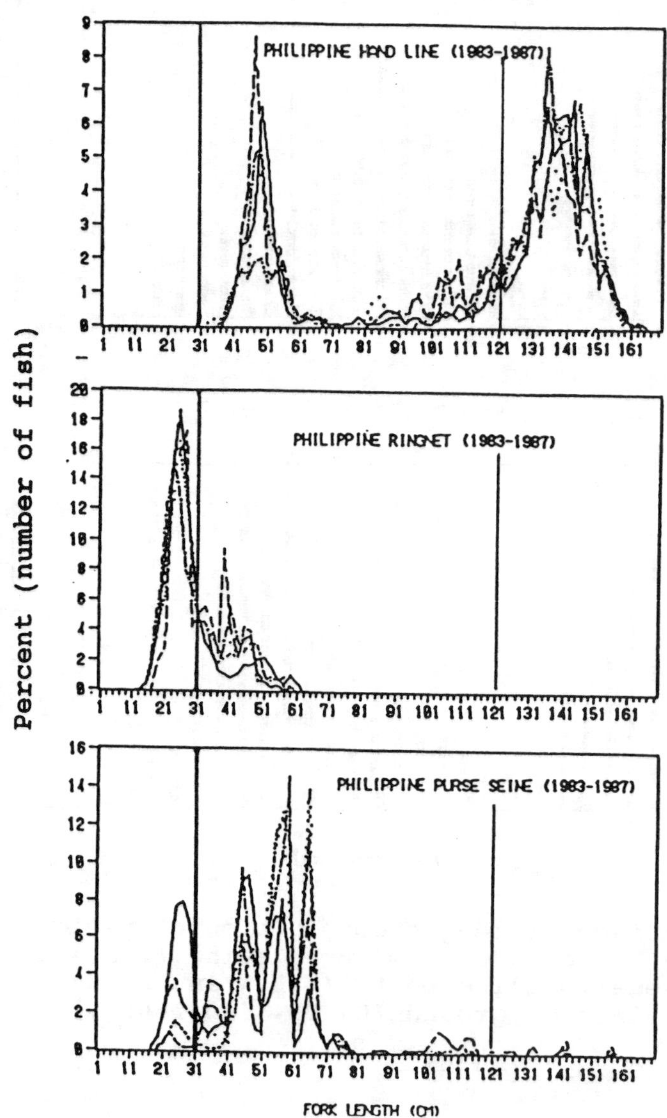

Fig. 6. Length composition of yellowfin tuna caught in the major
fisheries of the Philippines

Modified from Suzuki (1989)

3. FACTORS INTERVENING IN THE FISHERIES INTERACTION

It is obvious that there are a number of factors playing important roles in the possible interaction between purse-seine and longline fisheries other than the changes in the magnitude of the two fisheries. They include (1) changes in population size caused, independent of the fisheries, by physical environmental changes (this factor is partly correlated with the changes in catchability of the fishing gear), and (2) possible biological interaction of the yellowfin tuna population with other animal populations such as forage organisms and competing fish populations. Although these factors appear to have a great impact on the process of interaction, very little is known about them at present.

However, some work on physical environment as related to fishing performance and probably to recruitment strength, especially in connection with El Niño events, has been done (*e.g.*, IATTC, 1989). Apparent changes in catchability by purse- scine fishing as well as longlining possibly due to El Niño or quasi-El Niño events have been reported. Figure 7 shows a possible example in the western equatorial Pacific which might indicate that the purse-seine catch rate of large fish, mostly found in free schools, is affected by change in the thermocline caused by El Niño events. A dramatic decline in purse-seine CPUE from 1987 to 1988 appears to be related to a change from El Niño year to a non-El Niño year. This possible relationship should be monitored carefully in the future.

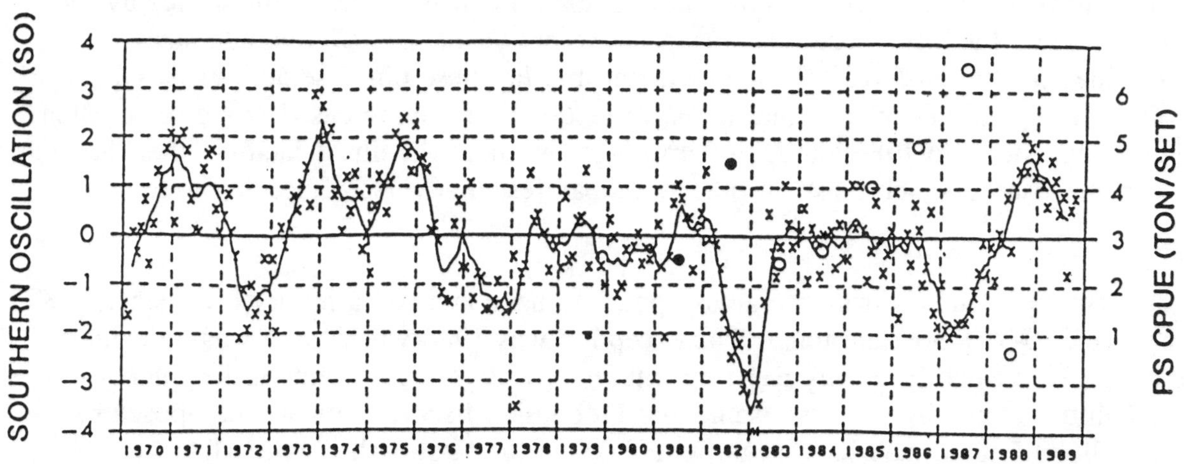

Fig. 7.   Comparison of the Southern Oscillation Index (SOI) and yellowfin CPUE for free swimming schools taken by the Japanese purse seine fishery.

After Suzuki (1991)

Open and solid circles denote the CPUE of fish above 10 Kg and 20 Kg, respectively (there was a change in data entry for weight category in 1983 from 10 Kg to 20 Kg). Cross signs show standard deviations of the observed SOI (the smoothed values are shown by the curve).

Another aspect which hinders the interaction studies is lack of fundamental knowledge on biology and fisheries of yellowfin tuna. This lack of data, prohibits quantitative investigations of this subject to a satisfactory degree. They include (1) stock structure, (2) natural and fishing mortality rates, (3) rate of mixing (migration) in both vertical and horizontal direction, (4) growth and aging, (5) possible behavioural change related to spawning, (6) catch/effort and size measurement statistics, and (7) development of reliable CPUE series. These aspects are summarized in the review paper of biology and fisheries of yellowfin tuna in the western and central Pacific (Suzuki, 1993).

One thing which appears often overlooked in the fisheries interaction studies on yellowfin and other tropical tunas is the role of payao or the FADs. They must play an important role in the process of the interactions (e.g., SPC, 1990).

4.   INFERENCE ON INTERACTION BETWEEN PURSE SEINE AND LONGLINE

4.1   Status of Stock

Although the status of yellowfin in the western and central Pacific is not known, Suzuki *et al.* (1989) suggested, on the basis of fisheries data through 1986, that the total catch of about 210 thousand tons was sustainable. Pieces of circumstantial evidence supported this inference, including (1) low but relatively stable longline CPUE despite a substantial increase in the purse-seine catches and no consistent trend in catches by the Philippines and Indonesia fisheries during the recent years, and (2) average sizes of yellowfin taken by the Japanese longliners and the Japanese purse seiners were stable. Suzuki *et al.* (1989) concluded that the adverse effect of the increased purse-seine catch was not evident up to 1986 although there might be more dominant factors other than the purse-seine fishery which mask an otherwise-manifested adverse effect on the longline fishery.

The total catches from the areas appear to have attained a higher level since 1987, of around 300-350 thousand tons. The exception was 1988 when there was a decline of the purse-seine catch, probably environmentally induced. The recent further increase in yellowfin tuna catch by the purse-seine and Indonesian fisheries, mostly composed of small yellowfin tuna, may have negative effects on the large-fish longline fishery.

4.2   Inference on the Interaction

As previously mentioned in the section on the Japanese longline trends, there was a somewhat big drop in CPUE in 1989. Figures 3, 4, and 5 indicate that the difference in dominant ages of the purse-seine catch (about 1 year old) and the longline catch (2 and 3 years old) is approximately 2 years. This might be an indication of the effect of higher exploitation of the surface fisheries since 1987. A correlation analysis between longline CPUE and purse-seine catch should be done with a longer time series accounting for a time lag.

Another way to detect the effect of increased purse-seine and Indonesian fisheries catches on longlining is to see the areal differences before and after the increase of catches by these fisheries. Polacheck (1988) compared spatially the percentage change in the yellowfin tuna catch rate by the Japanese longliners for 1984-1985 relative to 1979-1981 but could not find large declines in longline catch rates in the areas of major Japanese purse-seine catches. A similar comparison, not the percentage change, but the absolute differences of the average CPUE based on recent data between 1979-1981 and 1987-1989, were calculated by 5 x 5 degree squares. The 1979-1981 period roughly covers the period when full scale purse-seine operations had not yet started.

The results are shown in Figure 8 and Table 7. All differences in average CPUE between the two periods are listed in Table 7 and shown in Figure 8. It is clear that the largest decrease roughly corresponds to the areas of the major purse-seine catches (Figure 9) whereas in the areas north of the major purse-seine fishing grounds, there are no consistent patterns (Table 7). In the areas to the south of the major purse-seine fishing grounds, there appears to be an increase in catch rates. A large decline in a somewhat isolated area between the 170°E-180° and 0°-10°S is observed. Although more detailed analyses with additional new data are required to give a conclusion, if this result is interpreted as an indication of the adverse effect of purse seining on longline fishing, there might be some indication that the mixture of yellowfin is not homogeneous spatially throughout the western and central Pacific.

It is noteworthy that the overall decline of the longline CPUE in the western and central Pacific is remarkably small, about one-half that of the early exploitation period compared to a several-fold decline in all other stocks of yellowfin. This may suggest that the western and central Pacific yellowfin stock is significantly larger than other stocks. However, it remains an open question whether the recent higher harvest of this stock is sustainable or not in the long term.

5.   RECOMMENDATION FOR IMPROVEMENT OF THE INTERACTION STUDIES

5.1   Statistics

Establishment of a centralized data collecting body is urgently needed. The SPC Tuna and Billfish Assessment Programme (TBAP) could be a promising body for this purpose if the membership is broadened to cover all the distant water fishing nations (DWFN). There is no time to waste, at this critical time of ever increasing fishing effort on tropical tunas in the western and central Pacific, in formulating such a data collecting system, preferably with binding powers for data submission. Both coastal countries and the distant water fishing nations (DWFN) must abandon selfish speculative ideas about the common fisheries resources of this region and should cooperatively strive to establish better databases.

Data collection from the southeast Asian countries should be continued and improved since the IPTP terminates its activity at the end of 1991. Effort should be made to have catch statistics by species, especially the separation of bigeye and yellowfin tunas. Size measurements by species/fishing gears should also be improved.

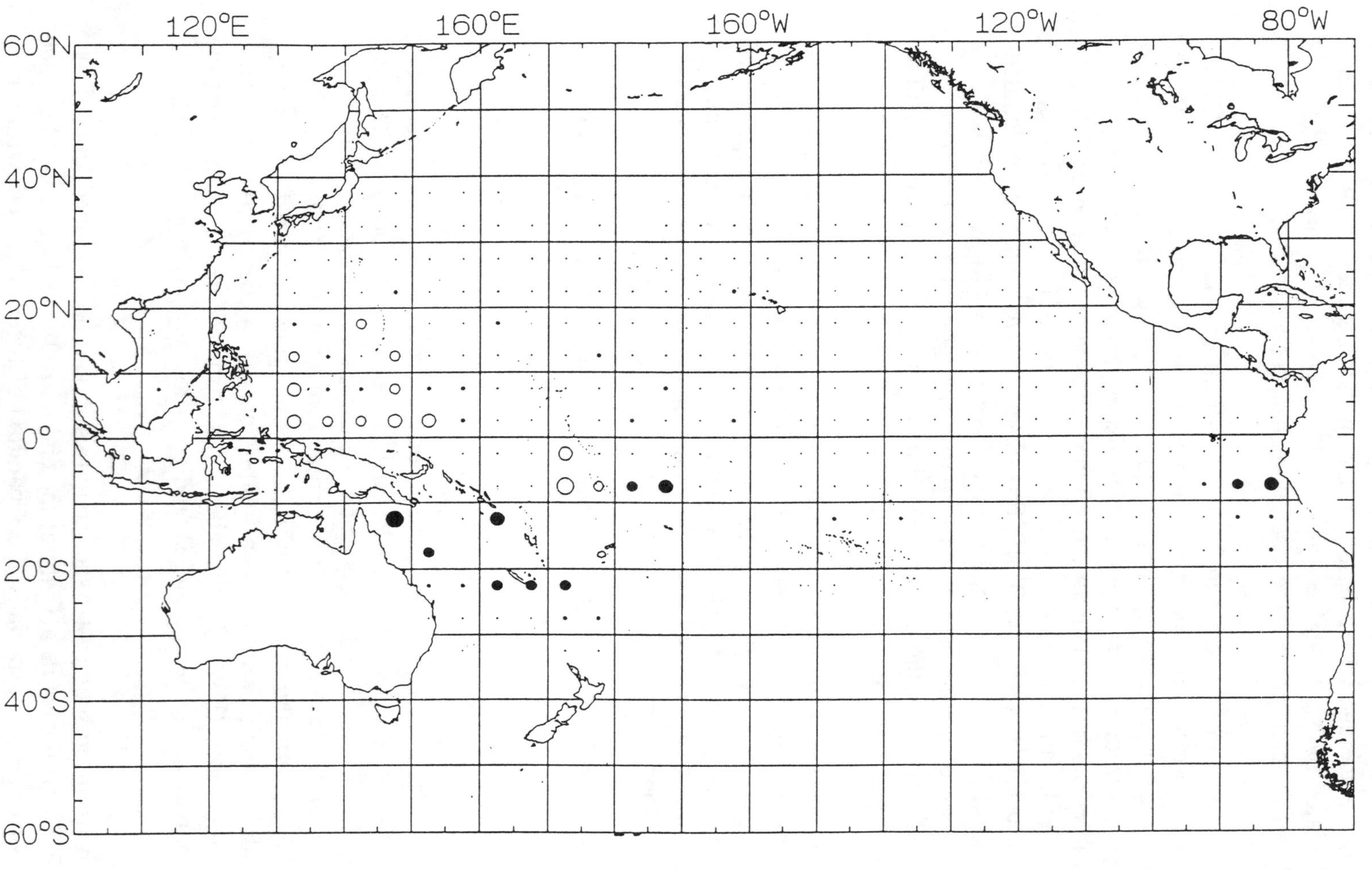

Fig. 8. Distribution of differences in the Japanese longline
CPUE (average 1979-1981 minus average 1987-1989)

Open and solid circles denote the decrease and increase
of the average CPUE during the two period. Size of the
circles show the magnitude of the difference.

See Table 7 for exact figures of the difference. Note
that the cells with small samples are not shown in the
Figure.

Table 7. Differences in the Japanese longline CPUE for yellowfin
tuna (average 1979-1981 minus average 1987-1989)

	60-°S	55-°	50-°	45-°	40-°	35-°	30-°	25-°	20-°	15-°	10-°	05-0°	0-05°	-10°	-15°	-20°	-25°	-30°	-35°	-40°	-45°	-50°	-55°	-60°N
100-105°E	.00	.00	.00	.00	.00	.00	.00	.00	.00	.00	.00	.00	.00	.00	.00	.00	.00	.00	.00	.00	.00	.00	.00	.00
105°	.00	.00	.00	.00	.00	.00	.00	.00	.00	.00	.00	.00	.00	.00	.00	.00	.00	.00	.00	.00	.00	.00	.00	.00
110°	.00	.00	.00	.00	.00	.00	.00	.00	.00	.00	.00	.00	.57	.26	.04	.29	.00	.00	.00	.00	.00	.00	.00	.00
115°	.00	.00	.00	.00	.00	.00	.00	.00	.00	.00	.00	.00	1.03	.24	.29	.31	.00	.00	.00	.00	.00	.00	.00	.00
120°	.00	.00	.00	.00	.00	.00	.00	.00	.00	.00	.00	.00	.82	.00	.12	.00	.14	.00	.00	.00	.00	.00	.00	.00
125°	.00	.00	.00	.00	.00	.00	.00	.00	.00	.00	.00	.00	.83	.76	.83	.50	-.14	-.46	.00	.00	.00	.00	.00	.00
130°	.00	.00	.00	.00	.00	.00	.00	.00	.00	.00	.00	.00	.79	.83	.73	.00	-.20	-.02	.05	.00	.00	.00	.00	.00
135°	.00	.00	.00	.00	.00	.00	.00	.00	.00	.00	.00	.85	.61	.44	.38	-.17	.09	.00	-.07	.00	.00	.00	.00	.00
140°	.00	.00	.00	.00	.00	.00	.00	.00	.00	.00	1.01	.98	.66	.37	.09	.54	.03	-.02	.01	.00	.00	.00	.00	.00
145°	.00	.00	.00	.00	.00	.00	.00	.00	-.28	-1.43	.71	1.03	.97	.65	.51	-.33	-.27	-.01	.01	-.02	-.03	.00	.00	.00
150°	.00	.00	.00	.00	-.02	.24	.13	-.41	-.66	1.37	.62	.90	.79	.36	.18	.08	-.23	-.10	.03	-.02	.00	.00	.00	.00
155°	.00	.00	.00	.00	-.06	-.18	-.23	-.41	-.66	-.22	.04	.09	.37	.32	.11	.11	.09	.02	.00	-.08	-.67	.00	.00	.00
160°	.00	.00	.00	.00	-.03	-.14	-.20	-.53	-.88	-.80	.10	.19	.06	.15	.10	-.33	-.03	.07	.00	.00	.00	.00	.00	.00
165°	.00	.00	.00	.00	-.08	-.10	.02	-.70	.00	-.07	.03	.20	.09	-.19	.03	-.17	.10	-.02	.00	-.02	.03	.00	.00	.00
170°	.00	.00	.00	.00	-.08	-.05	-.30	-.52	.00	.00	1.06	.76	.12	.00	-.07	.01	-.13	-.01	.01	-.03	.00	.00	.00	.00
175°	.00	.00	.00	.00	.02	-.08	.00	.00	.00	.00	.75	-.13	.12	.01	.34	.00	-.01	.05	.07	-.04	.00	.00	.00	.00
175°	.00	.00	.00	.00	.00	.13	-.50	.00	.00	.00	-.63	-.09	.38	.16	.02	.11	.02	.17	-.02	.01	.00	.00	.00	.00
170°	.00	.00	.00	.00	.25	.00	.00	.00	.00	.00	-.84	-.01	-.07	.40	-.06	.21	.07	.02	.02	-.07	.00	.00	.00	.00
165°	.00	.00	.00	.00	.27	-.09	.00	.00	.00	1.37	-.20	-.09	-.07	.16	-.07	.05	.09	.03	-.01	-.05	.00	.00	.00	.00
160°	.00	.00	.00	.00	-.25	.01	.00	.00	.00	-3.89	-.59	-1.18	-.26	.19	-.04	.00	.28	.01	-.03	-.08	.00	.00	.00	.00
155°	.00	.00	.00	.00	-.19	-.14	.00	.00	-.27	1.24	1.22	-.59	-.60	.05	-.02	-.03	.16	.01	-.05	-.09	.00	.00	.00	.00
150°	.00	.00	.00	.00	.00	.00	.29	-.24	-.32	.66	.18	.16	.04	-.03	.02	-.06	.08	.05	-.04	-.11	.00	.00	.00	.00
145°	.00	.00	.00	.00	.00	.00	-.10	-.15	-.52	-.27	-.01	.07	-.12	.08	-.02	-.03	.05	.01	-.02	-.06	.00	.00	.00	.00
140°	.00	.00	.00	.00	.00	.00	.00	.00	.00	.09	.21	.02	.06	-.07	.03	-.01	.02	.04	-.01	-.02	.00	.00	.00	.00
135°	.00	.00	.00	.00	.00	.00	-.26	.08	.24	.26	.16	-.01	-.09	.00	-.14	.04	.06	-.04	.00	-.15	.00	.00	.00	.00
130°	.00	.00	.00	.00	.00	.00	.04	-.20	.01	.03	.08	.05	-.08	-.13	-.03	.03	.09	-.07	.02	.00	.00	.00	.00	.00
125°	.00	.00	.00	.00	.00	.00	.00	.12	.10	-.05	.06	-.04	-.05	.03	-.08	.00	-.18	-.39	.00	.01	.00	.00	.00	.00
120°	.00	.00	.00	.00	.00	.00	.17	.07	.18	.05	.01	.03	.02	-.02	-.01	.03	.10	.11	.01	.00	.00	.00	.00	.00
115°	.00	.00	.00	.00	.00	.00	.11	.00	.07	-.05	.03	.11	-.03	-.11	-.30	-.08	-.07	-.30	.00	.00	.00	.00	.00	.00
110°	.00	.00	.00	.00	.00	.00	.04	.11	-.02	.03	.02	.08	-.03	-.09	.01	-.05	-.06	-.10	.00	.00	.00	.00	.00	.00
105°	.00	.00	.00	.00	.00	.00	-.05	-.02	-.10	.02	.05	-.02	-.05	-.09	-.10	-.15	-.04	.00	.00	.00	.00	.00	.00	.00
100°	.00	.00	.00	.00	.00	.00	.00	-.20	-.02	-.04	.00	.04	-.06	-.11	.10	.00	.00	.00	.00	.00	.00	.00	.00	.00
95°	.00	.00	.00	.00	.00	.00	.05	.00	-.06	-.01	-.04	-.13	-.11	-.05	.12	.00	.00	.00	.00	.00	.00	.00	.00	.00
90°	.00	.00	.00	.00	.00	.00	.03	.25	-.09	-.10	-.26	-.20	-.07	-.05	.02	.00	.00	.00	.00	.00	.00	.00	.00	.00
85°	.00	.00	.00	.00	.00	.00	.23	.08	.03	-.41	-.75	.00	-.01	.04	-.97	.00	.00	.00	.00	.00	.00	.00	.00	.00
80°	.00	.00	.00	.00	.00	.01	.09	.19	.26	-.38	-.99	-.03	-.02	.00	.00	.00	.00	.00	.00	.00	.00	.00	.00	.00
75°	.00	.00	.00	.00	.00	.00	.02	.41	.05	-1.29	.00	.00	-.14	-.70	.00	.00	.00	.00	.00	.00	.00	.00	.00	.00
-70°W	.00	.00	.00	.00	.00	.00	.00	.00	.00	.00	.00	.00	.00	.00	.00	.00	.00	.00	.00	.00	.00	.00	.00	.00

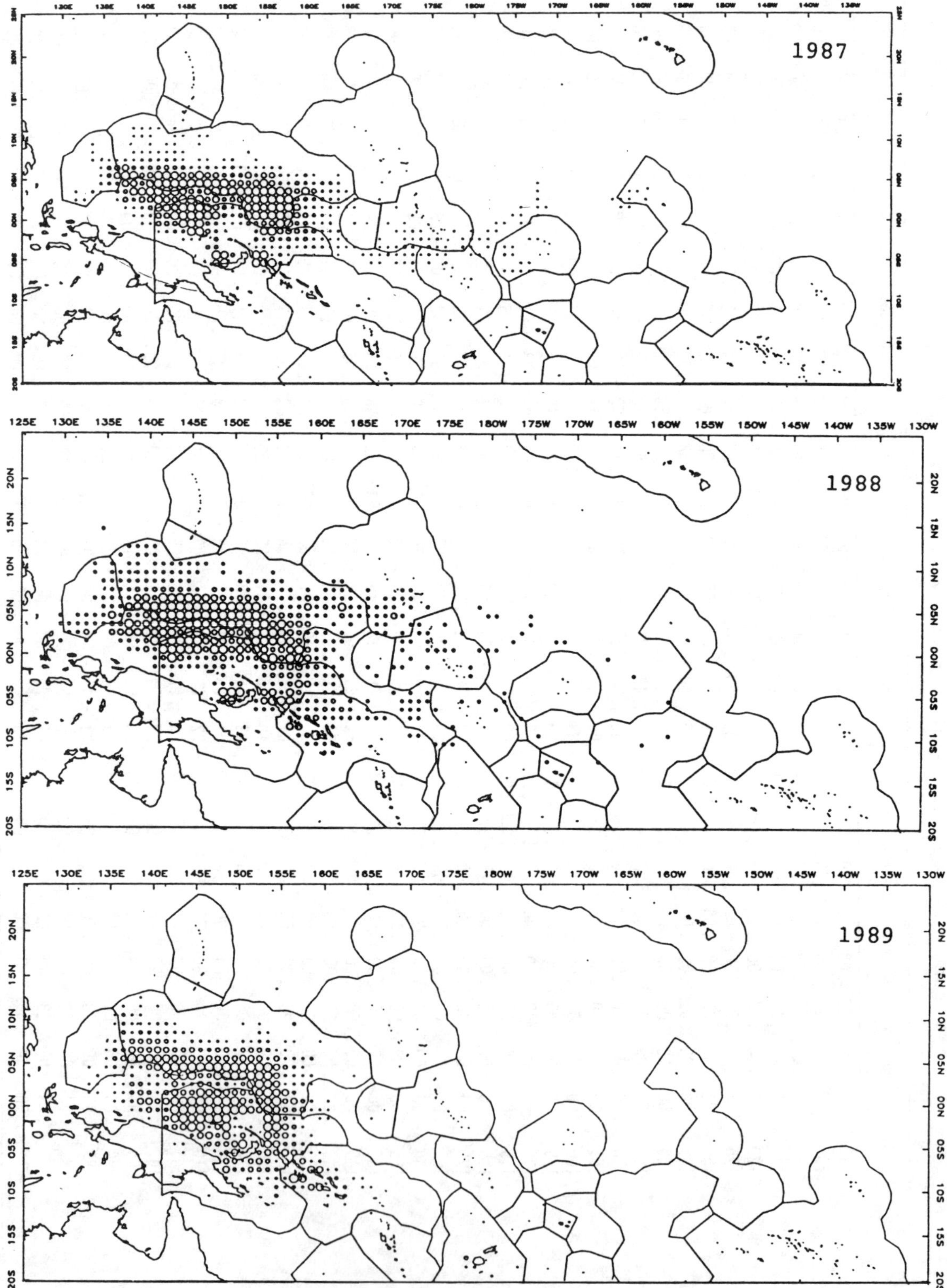

Fig. 9. Distribution of purse seine fishing effort in the SPC
areas, 1987-1989.

After SPC (1988, 1989a and 1989b)

5.2 <u>Research</u>

Movement and migration studies as well as developing movement models should be intensified, including the use of results obtained from the large-scale tagging experiments by the SPC. Developing reliable abundance indices for major fisheries and application of non-equilibrium production models (*e.g.*, Prager, 1992) should be encouraged. Age-specific analyses also should be attempted.

Stock structure and growth aspects should be investigated in more detail. Effects of the environment on fishing efficiency and recruitment should be studied.

6. REFERENCES CITED

Coan, A. Jr. 1993. USA distant-water and artisanal fisheries for yellowfin tuna in the central and western Pacific. *In* Proceedings of the FAO Expert Consultation on Interactions of Pacific Tuna Fisheries, edited by R.S. Shomura, J. Majkowski, and S. Langi, 3-11 December 1991, Noumea, New Caledonia [See this document.]

Deriso, R.B., R.G. Punsly, and W.H. Bayliff. 1991. A Markov movement model of yellowfin tuna in the Eastern Pacific Ocean and some analyses for international management. *Fish.Res.*, 11(3-4):375-95.

Doulman, D.J. 1987. Development and expansion of the tuna purse seine fishery. *In* Tuna Issues and Perspectives in the Pacific Islands Region, edited by D.J. Doulman. Honolulu, East-West Center, pp. 133-60.

FAO. 1991. Catches and landings, 1989. *FAO Yearb.Fish.Statist.*, (68):516 p.

Habib, G. 1984. Overview of purse seining in the South Pacific. Paper presented at the Workshop on National Tuna Fishing Operations, Tarawa, Kiribati, 1984, Forum Fisheries Agency, TW/15:26 p.

Hilborn, R. 1989. Yield estimation for spatially connected populations: an example of surface and longline fisheries for yellowfin tuna. *J.North Am.Fish.Mgt.*, 9(4):402-10.

Honma, M. 1974. Estimation of effective overall fishing intensity of tuna longline fishery - yellowfin tuna in the Atlantic Ocean as an example of seasonably fluctuating stocks. *Bull.Far Seas Fish.Res.Lab.*, (10):63-75.

Honma, M., and Z. Suzuki. 1978. Japanese tuna purse seine fishery in the western Pacific. *S Ser.Far Seas Fish.Res.Lab.*, (10):66 p.

IATTC. 1989. Annual Report of the Inter-American Tropical Tuna Commission, 1988. *Annu.Rep.I-ATTC*, (1988):288 p.

ICCAT. 1992. Meeting of the Working Group on Western Atlantic Tropical Tunas. *Collect.Vol.Sci.Pap.ICCAT*, 38:285 p.

IPTP. 1991. Indian Ocean and Southeast Asian tuna fisheries data summary for 1989. *Data Summ.Indo-Pac.Tuna Dev.Mgt.Programme, Colombo, Sri Lanka,* 11:96 p.

Lenarz, W., and J.R. Zweifel. 1979. A theoretical examination of some aspects of the interactions between longline and surface fisheries for yellowfin tuna, *Thunnus albacares. Fish.Bull.NOAA-NMFS,* 76(4):807-25.

NMFS. 1991. U.S. Tuna Fleet Quarterly Report- First Quarter 1991. *Quart.Rep. Southwest Reg., NOAA-NMFS,* 8 p.

Polacheck, T. 1988. Yellowfin tuna, *Thunnus albacares,* catch rates in the western Pacific. *Fish.Bull.NOAA-NMFS,* 87(1):123-44.

Prager, M. 1992. ASPIC- A surplus-production model incorporating covariates. *Collect.Vol.Sci.Pap.ICCAT,* 38:218-29.

SPC. 1988. Regional Tuna Bulletin, First Quarter 1988. *Reg.Tuna Bull.Tuna Billfish Assess.Programme, S.Pac.Comm.,* 28 p.

SPC. 1989a. Regional Tuna Bulletin, Fourth Quarter 1988. *Reg.Tuna Bull.Tuna Billfish Assess.Programme, S.Pac.Comm.,* 38 p.

SPC. 1989b. Regional Tuna Bulletin, Fourth Quarter 1989. *Reg.Tuna Bull.Tuna Billfish Assess.Programme, S.Pac.Comm.,* 34 p.

SPC. 1990. Preliminary analysis of Solomon Islands in-country tagging project data. Paper presented at the SPC Third Meeting of the Standing Committee on Tuna and Billfish, 6-8 June 1990, Noumea, New Caledonia, SCTB/WP4:10 p.

SPC. 1991. Status of tuna fisheries in the SPC area during 1990, with annual catches since 1952. Paper presented at the SPC Fourth Meeting of the Standing Committee on Tuna and Billfish, 17-19 June 1991, Port Vila, Vanuatu, SCTB/WP:75 p.

Suzuki, Z. 1988. Study of interaction between longline and purse seine fisheries on yellowfin tuna, *Thunnus albacares* (Bonnaterre). *Bull.Far Seas Fish.Res.Lab.,* (25):73-143.

Suzuki, Z. 1989. A consideration on stock structure of yellowfin tuna in the western pacific Ocean. *Summ.Rep.Jap.Soc.Fish.Oceanogr.,* A2:7-8.

Suzuki, Z., N. Miyabe, and S. Tsuji. 1989. Preliminary analysis of fisheries and some inference on stock status for yellowfin tuna in the western Pacific. Report submitted to SPC Second Meeting of the Standing Committee on Tuna and Billfish, 19-21 June 1989, Suva, Fiji, SCTB/WP9.

Suzuki, Z. 1991. Status of world yellowfin tuna fisheries and stock. *Tuna Newsl.,* Issue 101:5-8.

Suzuki, Z. 1993. A review of the biology and fisheries for yellowfin tuna (*Thunnus albacares*) in the western and central Pacific Ocean. *In* Proceedings of the FAO Expert Consultation on Interactions of Pacific Tuna Fisheries, edited by R.S. Shomura, J. Majkowski, and S. Langi, 3-11 December 1991, Noumea, New Caledonia [See this document.]

ESTIMATING THE IMPACT OF PURSE-SEINE CATCHES ON LONGLINE

Paul A.H. Medley
Marine Resources Assessment Group
8 Prince's Gardens
London, UK

ABSTRACT

The longline monthly catch per unit effort time series of the southwest Pacific yellowfin tuna stock (1980-1990) is analysed using a generalised linear model of the stock dynamics. The model accounts for changes in the catch rate which may be attributable to oceanographic effects, fishing location, and previous purse-seine catches. A simple example is given of how the parameter describing the impact of purse seine on longline might be used in the analysis of a decision to allocate the stock between two conflicting gears taking into account the uncertainty in the parameter estimation.

1.   INTRODUCTION

The status of yellowfin tuna (*Thunnus albacares*) continues to be an important issue to the southwest Pacific tuna fishery. Although yellowfin tuna is the second largest single fishery resource in the region, there is still uncertainty regarding the potential sustainable yield. Concern over this problem has been increased by the experiences of other fisheries where yellowfin have proved vulnerable to overexploitation by purse seiners (IATTC, 1979, 1980, 1981, 1982; Fonteneau and Diouf, 1983; Au, 1983).

While longline effort has remained relatively constant, there has been a marked increase since 1979 in purse seine effort in the southwest Pacific. Purse seiners capture yellowfin, most tending to be small, young fish in association with skipjack and logs. Longliners take larger yellowfin (Cole, 1980), which form the biggest component of their catch. If purse seiners and longliners are fishing the same stock, which seems likely, and longline catch per unit effort (CPUE) data reflect changes in stock size, the longline CPUE should decrease as the purse-seine catch increases, after allowing for a time delay based on their different selectivities.

Several papers have covered the potential interaction between purse seine and longline. Polacheck (1988) and Suzuki (1985) found no significant interaction between the surface fishery and longline using data available for a region of the southwest Pacific. Suzuki (1988) found a significant interaction, however, between the surface and longline fisheries in the eastern Pacific and eastern Atlantic oceans, and surmised that an interaction would in time become apparent in the southwest Pacific even though none was detectable at that time. In addition, the South Pacific Commission (SPC, 1988) reported longline CPUE falling in areas where longline operations coincided with purse seine.

The aim of this study is to derive a method whereby the potential interaction can be included in economic models of the fishery. By reducing the problem to a single measurable variable, it should become much easier to include this potential interaction in the decision-making process. The final section gives a simple example of how such an estimate might be used even where significant doubt exists that it is accurate.

2.   THE CPUE TIME SERIES

For there to be a significant economic interaction between longline and purse seine, two assumptions need to be satisfied. Firstly, the longline yellowfin catch rate must reflect the abundance of fish. While there may be reservations about the exact form of this relationship, it seems reasonable to assume this is the case on theoretical grounds (Medley, 1990). Secondly, longline and purse seine must be fishing the same yellowfin stock. To claim this is not the case requires that stocks occupying the same geographical area are separated by depth, which has no accepted theoretical basis, or that the gears are fishing stocks in different geographical areas. In the latter case longline effort is distributed more widely than purse seine, so that longline catch rates might be expected to be falling more rapidly in those areas also used by purse seiners.

The main argument regarding the interaction is over the degree rather than whether or not it exists. The interaction might be insignificant for a number of reasons, a high natural mortality or large stock size and relatively low fishing mortality being the most likely. However if yellowfin catches are already close to their maximum sustainable yield (MSY), further increases in catch should lead to interactions becoming much more significant.

Figure 1 shows the monthly average catch per hook of longline from published Japanese Yellow Book data (1962-1980) and logsheet data from the SPC data base (1978-1989) combined for a sub-region 10°N-20°S, 125°E-175°E. This subset of the data set was chosen to avoid noise generated by different fishing practices further south, most notably for albacore, as well as to reduce the data set to manageable size. Although from different sources, the logsheet and Japanese Yellow Book data show a similar time series and combine reasonably well.

The important pattern in the data is the decline in longline catch per hook since 1980, coinciding with the increasing catches of purse seine. This does not constitute proof that one causes the other. All the trends can be explained by a number of hypotheses including changes in gear, targeting, and external oceanographic effects. In particular there is a similar negative trend during the period 1962-1975, before purse seine came into operation. It has been suggested that these trends coincide with the rise and decline of the pole and line fleets (Wright, pers. commun.), and the interaction between these two fleets is certainly worth investigating. In the meantime it should be assumed that declines in longline catch per hook have been observed before, but not necessarily related to activities in the surface fisheries.

There is no Box-Jenkins time series model which provides a very good fit to the data. The data are clearly nonstationary with changes in variance and trends throughout the series. Standard methods, which require transforming and differencing the data, produce models that fit the data poorly. In fact the behaviour of the series 1962-1990 is close to a random walk, with positive and negative trends even in the absence of purse seine fishing. In order to proceed with estimating the interaction, the simplest hypothesis is to assume that there would have been no trend without purse seiners, so the best estimate for the mean catch rate is the long-term mean of the series. That is, the expected behaviour would have been a horizontal straight line.

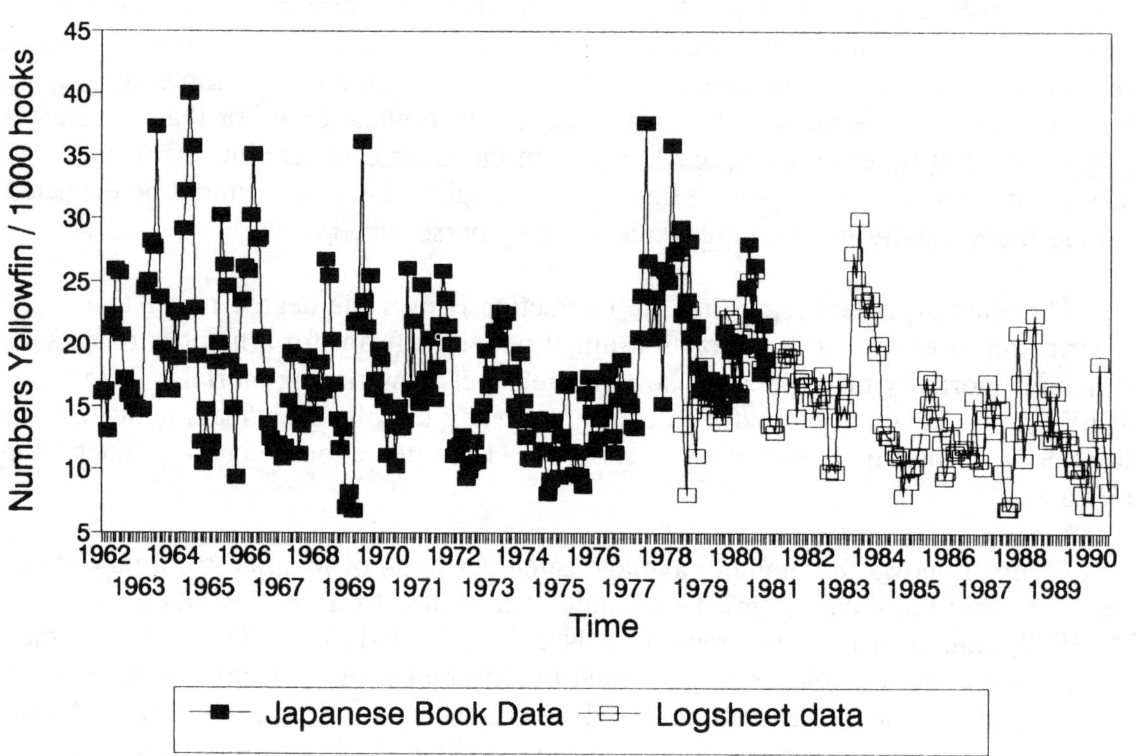

Figure 1.   Monthly catch per hook time series combining Japanese Yellowbook data and SPC logsheet data.

Even without proof of causation, it is still useful to estimate the degree of impact, assuming the observed decline in catch per hook is entirely due to purse seiners. This potential impact can be estimated by finding out how much past purse seine catches can explain the decline in longline catch rates. The estimate can be used to set the likely maximum impact of purse seine on longline, assuming the minimum impact to be zero.

3.   MODEL OF LONGLINE YELLOWFIN CATCHES INCORPORATING PURSE-SEINE CATCH

3.1   <u>Time Lag</u>

Before constructing a model it is necessary to estimate the approximate delay between fish being exploited by purse seiners and becoming available to longline. Figure 2 represents the estimated total catch of yellowfin by the different gears based on the same data set described in section 3.3, where the weight has been converted to age using a formula developed for the eastern Pacific (Wild, 1986; no such formula has been estimated for the southwest Pacific).

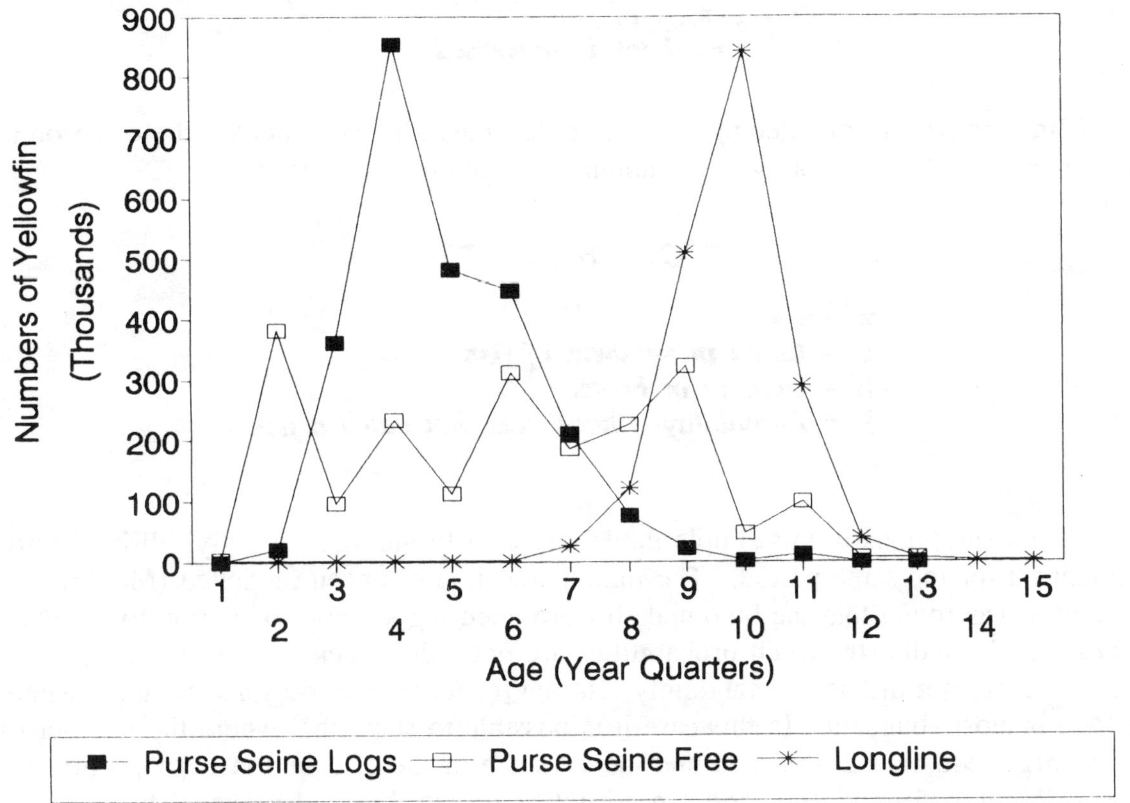

Figure 2.   Age frequency of purse-seine and longline catches determined using a growth model from the eastern Pacific (Wild, 1986).

An important point to notice with the purse seine age distribution is that whereas schools associated with logs have a relatively tight age distribution, free swimming schools (not associated with logs or flotsam) cover very young to very old fish. The age distribution for longline catches is also very tight, almost all of the reported catch appears to be between 2 and 3 years old. In particular the downward slope after the mode is very steep, implying very high mortalities if catchability of older fish remains high. It is perhaps more likely that longline mainly selects only one age class of the entire stock, and hence age-structured models even for this gear are not likely to be particularly useful. The lag between the peak purse seine age and peak longline age is about 1.5 years.

3.2   The Model

If fish encountering a hook can be described as a Poisson process, the probability a particular hook does not catch a fish can be defined as :

$$P = Exp(-\alpha\,D\,t)$$

where
***P** = Probability a hook has NO fish*   (1)
α = constant
***D** = density of fish*
t = time hook is immersed

If the probability of catching a fish is independent between hooks, the catch on a particular longline set will follow the binomial distribution with mean:

$$C = H\,(1 - P)$$

where   (2)
***C** = Catch in numbers of fish*
***H** = Number of hooks*
***P** = Probability a hook does not catch a fish*

Two assumptions in this simple model are clearly violated. Firstly, all hooks are not immersed for the same period. The immersion time has been modelled (Medley, 1990) and it was found that the binomial still provided a good approximation to the catch distribution. Secondly, the catch probabilities are not independent. This is mainly because tuna are not distributed randomly, but aggregate into schools and to areas where their food is most abundant. In this case it is possible to show that where the number of hooks is large, equation 2 is correct and the variance of the distribution is proportional to the mean (Medley, 1990). Equation 1 need not be correct but will probably be a reasonable approximation. Although the probability a hook catches a fish increases as the abundance of fish increases, the functional shape of that relationship may be different from equation 1.

The model can be further developed to include the effect of bait loss and a multi-species catch. However, much data on by-catch and bait loss are not available, so it is more sensible to propose an approximation based on the above model.

$$\mu = Exp(z\ln(H) + \beta)$$

where   (3)
μ = mean catch (Poisson parameter)
z = hook interaction term
β = linear predictor of other factors

Equation 3 should provide a reasonably good description of the data if the longline gear is not being saturated by fish. With an average catch of only 15 yellowfin per 1,000 hooks, this would seem to be a reasonable assumption. Since we are dealing with numbers of fish, which cannot fall below zero, the obvious error distribution to use is the Poisson, which has the additional feature of fixing the variance equal to the mean. This can be relaxed without difficulty to the more realistic quasi-likelihood assumption that the variance is proportional to the mean (McCullagh and Nelder, 1983).

A simple yellowfin population model can be used to describe the effect of purse seine catches on longline catch per hook. Fish available to purse seine, after a delay period, become available to longline. For simplicity it is assumed that during the delay period they are not available to any gear. Hence to calculate how many of the purse seine-caught fish would have been recruited to the longline fishery, the purse-seine catch must be discounted by the attrition rate over the appropriate delay period. This delay period, according to the catch size distribution, should be around 18 months.

Equation 1 can be expanded to refer to numbers of fish rather than density, and so include numbers of fish removed by purse seiners.

$$P = Exp(-\frac{\alpha}{V}(N - S_0 e^{-mx})t)$$

where

$$
\begin{aligned}
N &= Population\ size\ (constant) \\
V &= Volume\ of\ water\ holding\ the\ population\ N \\
S_0 &= actual\ purse\ seine\ catch \\
m &= attrition\ rate \\
x &= time\ lag\ in\ months \\
t &= hook\ immersion\ time
\end{aligned}
\tag{4}
$$

The basic population, N, might be modelled as a stationary stochastic process, with longliners sampling the population. This leads to a non-linear model and a great increase in the complexity of the fitting procedure. A simpler alternative approach, and that adopted here, is to treat the catch itself as a stochastic process. Although theoretically less appealing, such a model does take account of the high autocorrelations in the time series while keeping the model simple. A linear model can be used to account for changes in the basic population size and in catchability.

$$\mu_i = Exp(z\ln(H_i) + \beta_i - q_A S_{i-x} + \theta B_{i-1})$$

where

$$
\begin{aligned}
q_A &= purse\ seine\ impact\ coefficient \\
\theta &= auto\text{-}regressive\ parameter \\
i &= time\ in\ months
\end{aligned}
\tag{5}
$$

$$B_t = \ln\left[\frac{y_i}{Exp(z\ln(H_i) + \beta_i - q_A S_{i-x})}\right]$$

$$y_i = data\ values$$

Equation 5 presents the time-series model based on one described by Zeger and Qaqish (1988). It is equivalent to a Box-Jenkins autoregressive model with one parameter [AR(1)] model, which provided the best fit to the longline catch time series. As for equation 3, the model estimates catch rather than empty hooks. The model was fitted to the catch in numbers of fish by minimising the Poisson log-likelihood (deviance) and is equivalent to the log-linear model commonly used on contingency tables (McCullagh and Nelder, 1983). The logarithm function was used to link the mean to the linear predictor, so the model terms are multiplicative.

3.3 The Data Set

The model proposed here assumes there are two populations with fish moving from the population being exploited by purse seine to the one available to longline after an 18-month delay. It was applied to separate 10° squares assuming no movement between them.

The model uses only logsheet data because there are a number of problems with the Japanese Yellow Book data in the analysis stemming from the lack of other factors and covariates for the period prior to 1980. While inclusion of these data might be useful to improve estimates of some parameters in the model, it will not change the interaction estimate since no purse seine fishing occurred during this period. However if all yellowfin catches separated into size classes were available, the model could be extended over the entire time series and the interaction estimate probably considerably improved.

The data set consisted of the yellowfin caught and hooks set for each 10° square and month. This division into areas was done for two reasons. Firstly, area accounts for a high degree of variability in the catch rate, mainly because vessels fishing further south tend to be targeting albacore and have a lower yellowfin catch rate. Secondly, purse seiners operate predominantly in a band 10° either side of the equator. If purse seiners are affecting longliners, it might be expected that there will be a more pronounced fall in longline catch rates in this set of 10° squares.

Figure 3 shows the total purse catch for each month since 1980 when purse seiners began to move into the area. There is also an example catch from the square 0°4-10°N, 135°E-145°E, the data which form the basis of the model's discounted catch. The purse seine catch in numbers of fish is calculated by dividing the weight of yellowfin in a set by the average fish weight estimated by the fishermen. The purse seine catch parameter, S_x, depended in addition upon the estimated attrition rate (m). The attrition rate depended upon the natural mortality rate and migration rate. The migration rate included fish moving into adjacent areas and could be incorporated into the model explicitly, but there were no data on how great this migration might be or the form it would take. The attrition rate was assumed to be twice the natural mortality rate alone, representing the diffusion of yellowfin out of the range of the longliners. A low attrition rate tends to smooth the fluctuations evident in Figure 3.

Other factors and covariates explaining variation in the longline catch rate include the previous month longline mean catch rate for the whole area (20°S-10°N, 125°E-175°E), depth of 24°C thermocline, and the latitude and longitude position in 10° units.

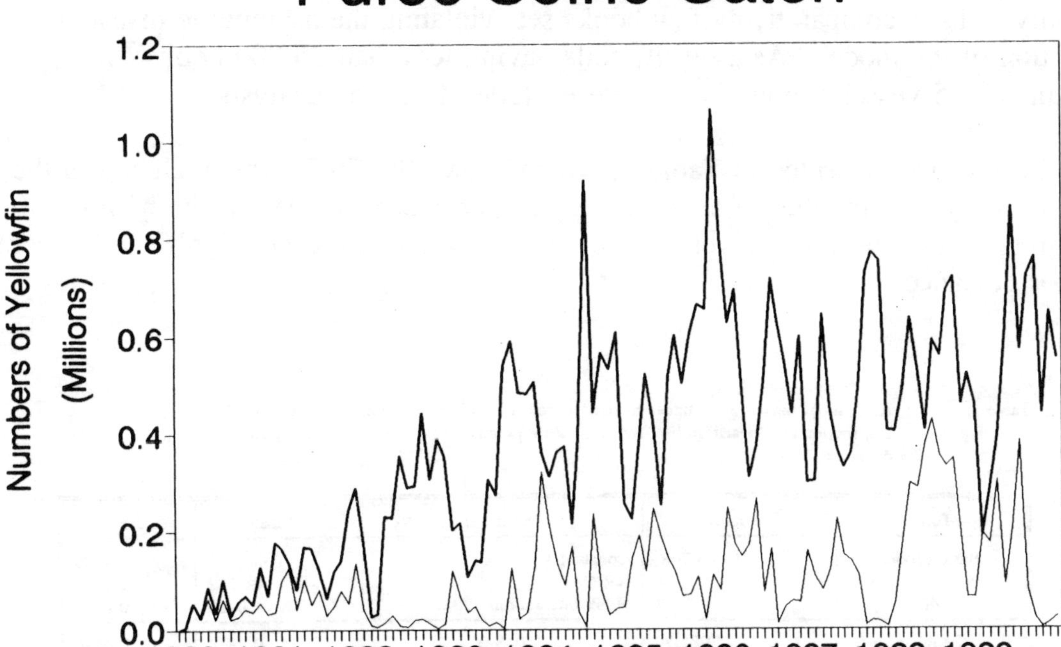

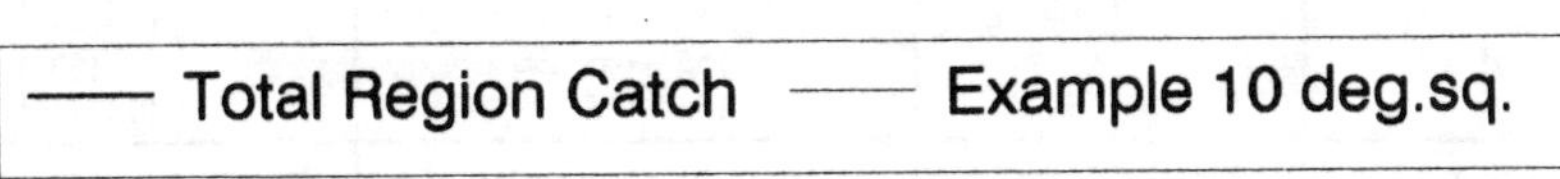

Figure 3.   Numbers of yellowfin caught by purse seiners, 1980-1990.

3.4   Tests

A number of tests can be carried out to make sure the model assumptions are not violated. The hook interaction parameter, z, should be close to 1.00 if the remaining linear predictor estimates the probability a hook catches a fish and this estimate remains constant within the stratum. If z is much smaller than 1.00, longline catches are probably significantly affecting the stock size. If z is greater than 1.00, it is likely vessels are being attracted to areas where catch rates are higher. Standard residual tests were also carried out.

3.5   Results

The model fitted the data reasonably well, with none of the assumptions being unacceptably violated. Alternative link functions to the logarithm were found to perform less well, showing effects are multiplicative. Although the data were much too dispersed to follow the Poisson distribution, the variance was found to be approximately proportional to the mean, fulfilling the quasi-likelihood requirements. Usually changes in deviance can be assumed to be distributed as chi-squared, but with the levels of dispersion present in this data set, this assumption is dangerous, hence no F statistics are given. However the large data set involved suggests that the parameter estimates will be approximately normally distributed, so their standard errors were used for additional guidance as to the inclusion of terms. In the first stages of analysis, outliers were found invariably to have comparatively few hooks set, violating the asymptotic distribution assumption of the model. As a result, cells having less than 10,000 hooks (*i.e.* approximately 5 vessel fishing days) were excluded from the analysis.

Results are presented in Table 1, which shows the final terms included in the model after trying a number of alternatives. In particular, increasing the number of autoregressive parameters or levels of the interaction terms did not significantly decrease the error deviance.

Table 1.　　　Analysis of deviance for linear model. Based on 1,151 data values, Poisson probability distribution and logarithm link function; scale parameter is estimated by the mean deviance.

Term	Deviance	Parameter	Estimate	S.E.
Basic Model	1,006,768	Constant	-5.1400	0.4826
		AR (1)	0.7282	0.0393
df	3	Hook Interaction	1.0460	0.0138
Latitude	298,630	10°S-0°	0.4814	0.4626
		0°-10°N	0.1031	0.4617
df	2			
Longitude	19,677	135°-145°E	1.1720	0.6533
		145°-155°E	0.6761	0.4698
df	4	155°-165°E	0.7153	0.4698
		165°-175°E	0.2714	0.0819
Latitude-	12,484	10°S-0°,135°-145°E	-0.8330	0.6527
Longitude		10°S-0°,145°-155°E	-0.3583	0.4638
		10°S-0°,155°-165°E	-0.5042	0.4640
df	6	10°S-0°,165°-175°E	0.0000	aliased
		0°-10°N,135°-145°E	-1.1430	0.6502
		0°-10°N,145°-155°E	-0.3562	0.4628
		0°-10°N,155°-165°E	-0.4231	0.4627
		0°-10°N,165°-175°E	0.0000	aliased
Depth of 24°C (m)	2,621	Thermocline depth	-9.805E-04	4.63E-04
df	1			
Discounted Purse-seine catch Lag 18 (S_{18}) df	22,082 1	Purse-seine catch impact (q_A)	-3,805E-06	4.07E-07
Error df	651,274 1,134			

The basic model and error terms give the deviance before and after fitting. The basic model uses the previous catch and number of hooks set in a stratum to explain the catch. Both terms are highly significant and are automatically included. The remaining terms give the change in deviance through fitting their parameters.

Although the hook parameter is close to 1.0, the standard error suggests it is still significantly different ($t_s = 3.33$***, 1,134 df). Its value indicates the catch per hook increases with the number of hooks set, probably because fishermen are concentrating effort where catch rates are highest in a particular month.

Latitude (referencing each 10° band) is clearly the most important factor after the basic model. For all discrete factors, the estimate indicates the difference from the first factor class. So for latitude the estimate measures the difference from the 20°S-10°S band, for longitude the 125°E-135°E band. The longitude and latitude-longitude interaction terms are much less important. These three factors together estimate the mean catch for each 10° square.

The depth of the 24°C thermocline, estimated for each 10° square month from data held at SPC, explained little of the total deviance. Other oceanographic variables such as depth of alternative thermoclines and sea surface temperature were looked at. These variables were all heavily correlated, so only the depth of the 24°C thermocline, which gave the largest change in deviance when fit, was used.

When free-swimming schools were used in the analysis, the model did not fit the data very well, perhaps because the impact delay is spread out much more between 24 and 0 months. Catches from free-swimming schools are excluded from the analysis. Probably the only way to include data on free-swimming school catches is to split up the catches explicitly into size classes.

The purse seine log-associated catches from 18 months previously are clearly correlated with the longline CPUE. Importantly, these catches explain significantly more of the longline catch rate variation than a simple linear decline. This is because this decline is greatest in the region 10°N-10°S, where purse seiners operate. There is a slight increase in catch rate outside this region. This result adds some weight to the argument of causation.

The estimate indicates that for every 10,000 log-associated yellowfin that purse seiners catch in a 10° square, longline catch per hook will fall by between 0.3 and 0.4% 18 months later. The longline catch rate will continue to be affected, but by a decreasing amount, in all months subsequent to the 18-month period.

The lag between fish being available to purse seine and then longline was tested by looking at the change in deviance using purse seine catches at different lags. The peaks at lags 3, 18 and 32, shown in Figure 4, may relate to different sizes of fish being exploited by purse seine. The first peak at 3 months would relate to the catch of older fish, which has a more immediate impact on longline, the larger peak at 18 months the impact from younger fish caught in association with logs. The final peak at 32/33 months, however, is not easy to explain in this way, since the sizes caught, assuming the growth curve is correct, indicate this lag is too great to be explained by an exploited cohort passing from the surface fishery to longline. The only other explanation within this framework is that a stock-recruitment relationship is increasing the impact at a later date. Purse seine catches at different lags were highly correlated, and it was not possible to fit them simultaneously with any confidence. There is little real evidence that this pattern is the result of the stock dynamics and not just spurious correlations in the time series, which makes interpretation of Figure 4 with any certainty impossible.

The results are more robust to the choice of the attrition rate than might first appear. The attrition rate has two effects. As it increases, it reduces the number of fish which finally move into the population vulnerable to longline. This part is closely related to the purse-seine catch parameter (q_A), hence in this respect the result is not sensitive to size of the attrition rate. The attrition rate also controls how any month's purse seine catch continues to affect longline catch rates. If the attrition rate is zero, any one month's catch would have an undiminishing effect and the purse-seine catch would simply accumulate in S_x. Effectively, this induces a trend in purse-seine catch rates and dampens fluctuations. A high attrition rate causes purse-seine catches to have an impact which dies out quickly.

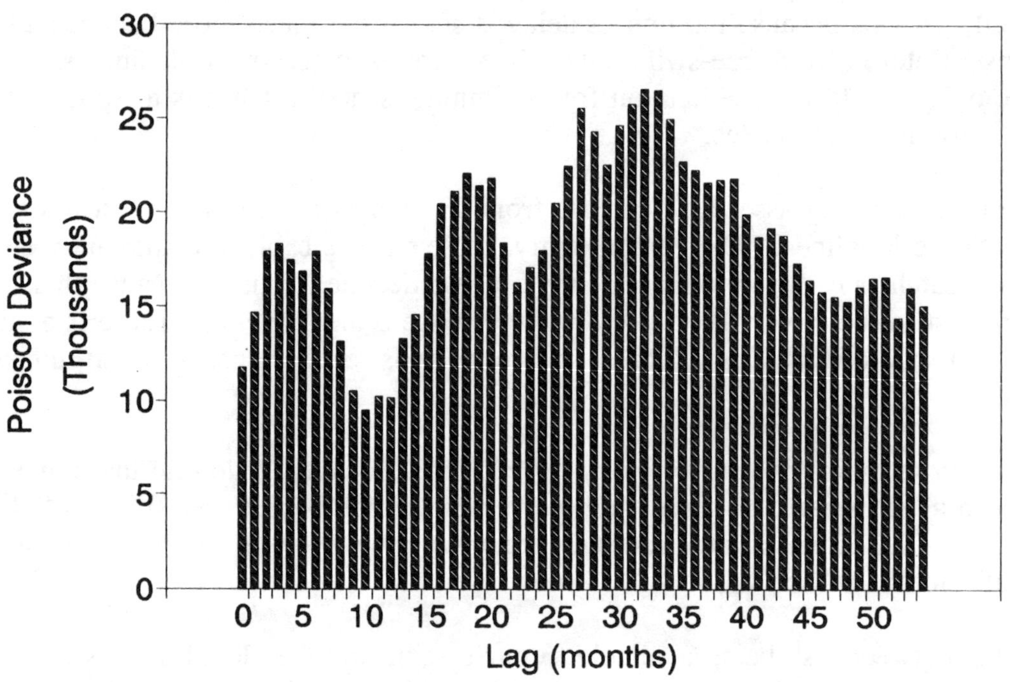

Figure 4.   Poisson deviance at different lags for the time series model.

4.   USING THE IMPACT ESTIMATE

The model has a variety of theoretical and empirical evidence supporting it as a reasonable description of longline catches. The model provides an estimate of the impact of purse seine catches based on the catch-and-effort time series for both gears. The question still remains whether the measurement being made is the correct one.

The conclusion is entirely dependent on the null hypothesis. The assumption that longline CPUE would have been constant since 1980 if purse seine operations had not started has little support from the time series analysis. The main argument that the decline is at least partly due to purse seine is based on the assumption that these gears are fishing the same stock and that longline CPUE will reflect stock abundance. Both these assumptions are reasonable. However it is still necessary to deal with an estimate that may be accounting for effects other than the one the model is trying to measure.

From the scientific point of view, the best way to proceed is to use more direct methods such as tagging to demonstrate a link between the two gears. Management needs are different. In reality management decisions are always made under uncertainty, and while this uncertainty can be reduced, it can never be eliminated entirely. Decision theory can help analyze the problem.

A decision analysis requires first of all the cost or benefit derived from each of a set of possible outcomes. The cost to longliners for allowing a purse seiner into a zone for a 2 month trip can be calculated using the estimated parameter with other statistics taken from the same data set. The average total discounted cost works out to be around US$24,000 (see Appendix). It is important to note that this estimate depends on the total number of hooks set as well as the average yellowfin catch in the area concerned. In different areas the cost to longliners may be greater or less than this.

In the simplest analysis, there are four possible outcomes, which are presented in Table 2. The 'states of nature' are that there is no significant impact of purse seine on longline catch rates or the impact is as predicted by the analysis. Equally there are two possible decisions, either to allow the purse seiner access to the zone, or deny access and forego any income from that vessel. The decision assumes one of those cases, but can be either right or wrong. The example losses given for each of the four outcomes are reductions in the economic rent of the fishery.

Table 2.   Example loss function from decisions based on the estimated purse seine impact on longline of $24,000 and purse seine trip profit of $10,000.

State of Nature	Decision	
Impact Parameter	No Access	Access
0	10,000	0
-3.805E-06	10,000	14,000

In the example, allowing a purse seiner access where there is an impact results in a net loss of $14,000. Preventing a purse seiner, from fishing results in a loss of $10,000, the purse seiners profitability, whether it has an impact or not. In this example, if the profitability of the purse seiner exceeds $24,000, then in either state of nature the optimal decision will be to give the purse seiner access (this decision will be stochastically dominant). It seems likely that the profitability of a purse seine trip is greater than this estimated average loss to longline.

How these losses are divided up between the fishermen and the owners of the resource depends on how fishermen pay for their right of access. If the access fee is unrelated to the profitability of fishing the resource, changes in profitability will have little impact on the fee, decreasing any incentive to minimise the interaction. Hence a low longliner access fee will not be eroded by decreasing catch rates, and the perceived loss in economic rent from allowing a purse seiner access by those charging the fee will be zero even where an interaction is occurring.

Although the example presents the basic decision to be made, it is grossly oversimplified. The method is used to demonstrate the approach and indicate how information might be used even where significant uncertainty exists as to its accuracy. A more realistic analysis would make greater use of the available data.

The likelihood of the different states of nature needs to be estimated. In some cases uncertainty can be estimated, for instance the impact estimate could be assumed to follow a normal distribution. For some uncertainty there is no *a priori* model, for instance in the case whether the downward trend in longline CPUE is caused by purse seine. In these cases the value of this type of analysis is in identifying the assumptions being made to derive any particular result.

The variety of states of nature may be very large indeed. In this case they should probably include variability in catches, prices (particularly supply and demand), and costs. The number of possible actions can also be increased, not only by considering the number of purse seiners to be given access, but also the areas in which they can fish. The simplest approach would be to repeat the decision analysis for each 10° square using the model estimates.

The simple decision analysis in Table 2 demonstrates the probable results if a more complicated analysis was undertaken. Given an equal chance for either state of nature, allowing a purse seiner to fish results in a smaller expected loss but greater variability in income. To maximise profit purse seine would dominate longline as the more profitable gear. If there is an interaction between the gears, longline would eventually be driven out of the fishery. However for a risk-aversive decision-maker, longline may be the optimum choice. Since longline exploits a different stock (separated in time) and uses a different market, longline may have an important role in providing greater stability in the economics of the fishery.

5.   REFERENCES CITED

Au, D. 1983. Production model analysis of the Atlantic yellowfin tuna (*Thunnus albacares*) fishery. *Collect.Vol.Sci.Pap.ICCAT,* 18:177-196.

Cole, J.S. 1980. Synopsis of biological data on the yellowfin tuna, *Thunnus albacares* (Bonnaterre, 1788), in the Pacific Ocean. *In* Synopsis of biological data on eight species of scombrids, p. 71-150, edited by W.H. Bayliff. *Spec.Rep.I-ATTC,* 2.

Fonteneau, A. and T. Diouf. 1983. Analyse de l'etat des stocks d'albacore de l'Atlantique au September 1982. *Coll.Vol.Sci.Pap.ICCAT,* 18:197-204.

IATTC. 1979. Annual report of the Inter-American Tropical Tuna Commission - 1978. *Annu.Rep.I-ATTC,* (1978):163 p.

IATTC. 1980. Annual report of the Inter-American Tropical Tuna Commission - 1979. *Annu.Rep.I-ATTC*, (1979):227 p.

IATTC. 1981. Annual report of the Inter-American Tropical Tuna Commission - 1980. *Annu.Rep.I-ATTC*, (1980):234 p.

IATTC. 1982. Annual report of the Inter-American Tropical Tuna Commission - 1981. *Annu.Rep.I-ATTC*, (1981):303 p.

McCullagh, P. and J.A. Nelder. 1983. Generalized Linear Models (Monographs on statistics and applied probability). Chapman and Hall.

Medley, P.A.H. (1990). Interaction between longline and purse seine in the south-west Pacific tuna fishery. PhD Thesis, Imperial College, London.

Polacheck, T. 1988. Yellowfin tuna, *Thunnus albacares*, catch rates in the western Pacific. *Fish.Bull.NOAA-NMFS*, 87:123-44.

SPC. 1988. Investigations on Western Pacific Yellowfin Fishery Interaction Using Catch and Effort Data. Paper presented at the Twentieth Regional Technical Meeting on Fisheries, S.Pac.Comm., New Caledonia, 1-5 August 1988, WP/17.

Suzuki, Z. 1985. Interaction study on yellowfin stock between longline and purse seine fisheries in the western Pacific. *Indo-Pac.Tuna Dev.Mgt.Programme*, TWS/85/26.

Suzuki, Z. 1988. Study of interaction between longline and purse seine fisheries on yellowfin tuna, *Thunnus albacares* (Bonnaterre). *Bull.Far Seas Fish.Res.Lab.*, 25:73-136.

Wild, A. 1986. Growth of yellowfin tuna, *Thunnus albacares*, in the Eastern Pacific Ocean based on otolith increments. *Bull.I-ATTC*, 18(6).

Wright, A. Research Coordinator, Forum Fisheries Agency.

Zeger, S.L., Qaqish, B. 1988. Markov regression models for time series: A quasi-likelihood approach. *Biometrics*, 44: 1019-31.

Appendix

To calculate statistics useful in assessing the impact of purse seiners on longliners, some parameters must be estimated from the data set or from other sources.

1)   Parameter Estimates

Although a great deal of effort was put into estimating the interaction parameter (q_A), all other estimates were derived as approximations to demonstrate the working.

Average number of longline hooks in a 10° square month = H = 388,419

Probability a longline hook catches a yellowfin = X = 0.015

Average number of yellowfin taken on a purse seine (2-month) trip = C_{ps} = 17,877

Average longline yellowfin weight = W_{yf} = 26.5 kg

Average longline yellowfin price = P_{yf} = US\$4 /kg

Natural monthly attrition rate = M = 1.6/12 = 0.1333

Monthly discount rate = d = $-Ln(0.9)/12$ = 0.00878   (10% interest rate)

Impact estimate from the linear model = q_A = $-3.805*10^{-6}$

Discounted purse seine catch = S_{18}

Time lag = 18 months

2)   Calculations

a) The first month's impact from the loss of C_{yf} yellowfin caught by purse seiners 18 months before can be calculated. The impact is the decline in the catch per hook of longline after the time lag (I_{18}).

$$I_{18} = 1 - Exp(-q_A * S_{18})$$

The reduction in the population size from the purse seine catch (S_{18}) is the original catch discounted by the attrition rate.

$$S_{18} = C_{yf} * Exp(-18*M)$$

For example, if purse seiners had caught 10,000 yellowfin, the impact on longline is as follows.

$$S_{18} = 10000 * Exp(-18 * 0.1333) = 907.234$$

$$I_{18} = 1 - \text{Exp}(-3.805*10^{-6} * 907.234) = 0.0034$$

This is equivalent to a 0.34% decline in longline catch rates.

b) The impact of a purse seine trip within a single 10° square catching 17,877 yellowfin (C_{yf}), assuming all fish are caught in the same month, can be calculated as follows.

Assuming the probability a hook catches a yellowfin (X) is 0.015, the total longline fleet catch value in a 10° square month (V) without any purse seiners is assumed to be:

$$V = X * H * W_{yf} * P_{yf} = 0.015 * 388,419 * 26.5 * 4 = \text{US\$ } 617,586$$

This can be combined with the previous result and an economic discount rate to obtain the total discounted loss to the longline fleet from a single purse seine trip.

$$L = V * \sum_{t=18}^{\infty} -I_t * \text{Exp}(-d*t) = \text{US\$24,520}$$

To simplify the calculation, it is assumed the entire purse seine catch (C_{yf}) occurs in a single month.

c) Alternatively, where a decision is made to protect longline, the estimate might be used to calculate a level of purse seine effort which will not push longline catch rates below some predetermined level. This is probably a good policy to adopt, at least in the short term, since it keeps options open.

In this example, it is decided purse seine effort should be limited to a level at which the longline catch per hook must lie above 0.012, perhaps the break-even point for longline. The probability a longline hook catches a yellowfin if no purse seiners were operating is assumed to be 0.02 yellowfin per hook. The purse seine catch required to decrease the initial longline catch rate from 0.02 to 0.012 can be derived from the purse seine impact equation.

$$0.02 * \text{Exp}(\sum_{t=0}^{\infty} q_A * C_{yf} * \text{Exp}(-M*(18+t))) = 0.012$$

Removing the summation and rearranging, the maximum purse seine catch can be obtained.

$$C_{yf} = Ln\left(\frac{0.012}{0.020}\right) \frac{(1-\text{Exp}(-M))}{q_A * \text{Exp}(-M*18)}$$

$$\approx 185000 \; \textit{yellowfin}$$

Assuming a purse seiner on average catches around 17877 yellowfin every 2 months, this works out as a level of effort within a 10° square of approximately 21 vessels per month. Hence no more than 20 vessels should be given access to a 10° square area if this management objective is to be achieved.

3)   Notes

It is worth setting the calculations up in a spreadsheet to see how sensitive they are to different parameter values. However, neither the delay (18 months) or the attrition rate (0.133) should be altered, since they were involved in the estimation process. It is also best to use the purse seine average catch per trip as calculated from the data set used for the estimation, otherwise the final result may be biased. If these parameters are changed, the impact term would have to be reestimated.

INTERACTION IN THE YELLOWFIN TUNA FISHERIES
OF THE EASTERN PART OF INDONESIAN WATERS

Nurzali Naamin and Sofri Bahar
Research Institute for Marine Fisheries
Jakarta, Indonesia

ABSTRACT

Possible interactions in the yellowfin tuna fisheries of the eastern part of
Indonesian waters was examined. There are two groups of interaction occur between
fisheries: (1) interaction with other species, and (2) interaction among fishing gears. The
data imply that there is interaction between stock of yellowfin tuna in the western Pacific
and the stock in northeastern Indonesian waters, and there is very weak evidence of
possible interactions between western Pacific and Indian Ocean stocks.

In eastern Indonesian waters yellowfin caught by longlining were frequently
associated with bigeye and albacore catches. Based on available data, interactions among
the four types of fishing gears used for catching yellowfin (longline, purse seine,
handline, and pole and line) could not significantly be detected.

1.   INTRODUCTION

Traditionally, the centre of the Indonesian tuna industry is the eastern part of
Indonesian waters (EIW). The region accounts for about 80% of the Indonesian tuna
catch, and about 80-95% of the tuna exported from Indonesia comes from this area. This
dominant position has changed since 1986, at least as far as fresh tuna for sashimi export
fishery is concerned, with the Indian Ocean, particularly south of Java and west of
Sumatra, accounting for most of the catch.

The EIW is a vast region of over 3 million km^2 making up about 52% of the 5.8
million km^2 of Indonesia's Exclusive Economic Zone (EEZ). The EIW region can be
broken down into four major fishing areas or fishing bases: the northern part of the
Lesser Sunda Islands (Bali, West Nusa Tenggara, East Nusa Tenggara, East Timor);
Sulawesi; Mollucas; and Irian Jaya. Four types of fishing gears exploit tuna resources in
the EIW: longline, purse seine, pole and line, and handline. Some other fishing gear
categorized as unclassified that also catch tuna incidentally are troll line, payang (danish
seine), and gillnet.

There are four species of tuna caught in the EIW: yellowfin (*Thunnus albacares*),
bigeye (*T. obesus*), albacore (*T. alalunga*), and southern bluefin tuna (*T. maccoyii*).
Among these species, yellowfin dominates the catch. The percentage of yellowfin caught
by the above-mentioned fishing gears ranges from 50 to 95%, 5 to 15%, and 25 to 75%,
for the longline, pole-and-line, and purse-seine fisheries, respectively. This paper
reviews the resources and fisheries associated with yellowfin tuna in the eastern part of
Indonesian waters and introduces the possibility of interaction among the fisheries.

2.   THE YELLOWFIN TUNA FISHERY IN EASTERN INDONESIAN WATERS

2.1   <u>Fishing Grounds and Fishing Bases</u>

There are two distinct yellowfin tuna fishing grounds in Indonesia: (1) the EIW from the western Pacific Ocean in the northeast and the Makasar Strait to the Lombok Strait in the western sector (Area 71 of FAO coded areas), and (2) the western part of the Indonesian waters (WIW) covering the eastern Indian Ocean (Area 57 of FAO coded areas). In 1990, about 42% of the yellowfin catch came from the EIW and the rest from the WIW.

Fishing areas of yellowfin in the EIW are: (1) the Flores and Banda Seas with fishing bases in Benoa on Bali; Kendari, Kolaka, Bone, and Ujung Pandang on Sulawesi; Maumere on Flores, and Ambon; (2) Tomini Bay and the Mollucas Sea with fishing bases in Luwuk, Gorontalo, Bitung, and Ternate; (3) the Sulawesi Sea with fishing bases in Bitung, Ternate, and Kendari; (4) north and west Irian Jaya waters with fishing bases in Biak and Sorong; and (5) Makasar Strait with fishing bases in Mamuju and Ujung Pandang (Figure 1).

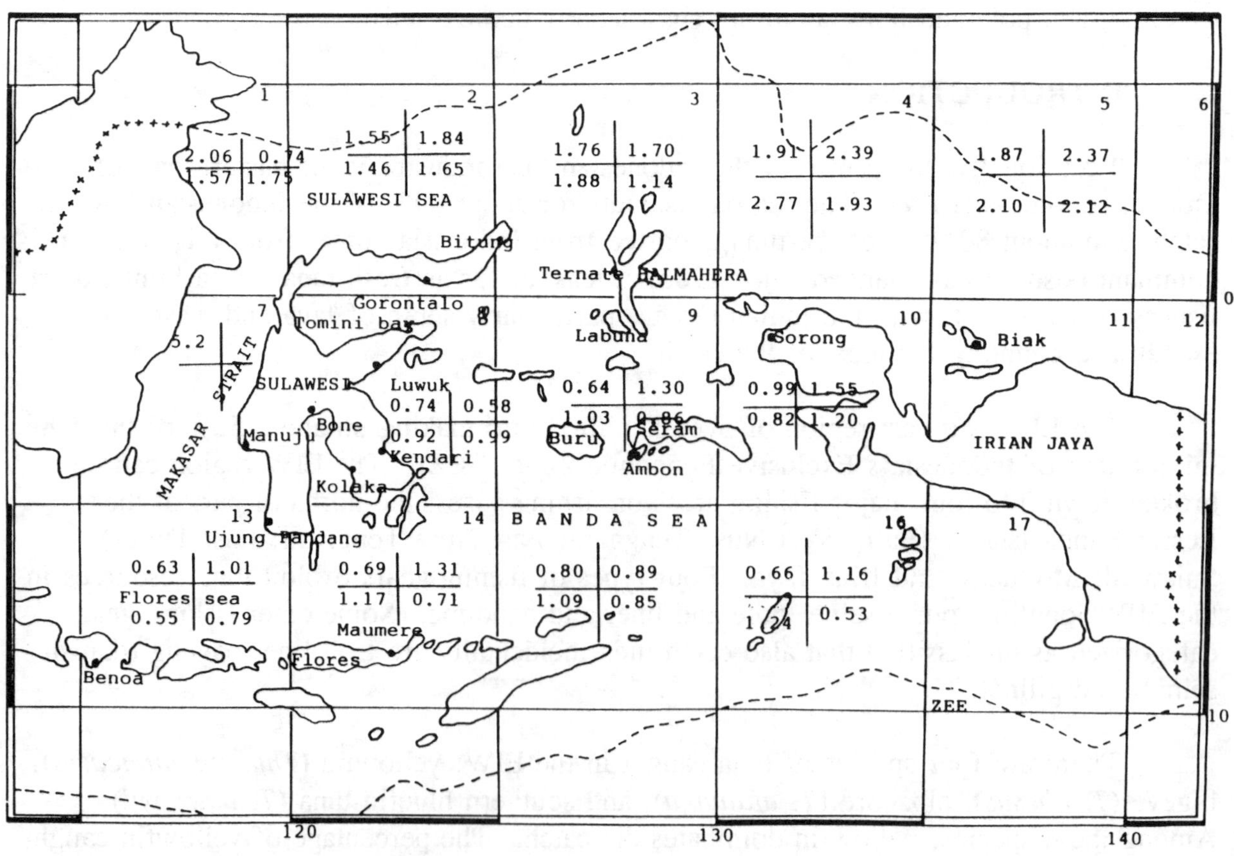

Figure 1.   Map of the eastern part of Indonesian waters showing hook rate by quarters per square degree for Japanese longliners.

2.2. <u>Fishing Methods, Fishing Gears, and Fishing Fleet</u>

There are four types of fishing gears used to catch yellowfin tuna in the EIW:
longline, purse seine, pole and line, and hand line.

2.2.1 Longline

The longline fishery for yellowfin in the EIW was started by Japanese fishermen
in 1952. Longline fishing was first carried out in the western Pacific Ocean; the fishing
grounds expanded to the Banda Sea and the Indian Ocean during the period 1952-1954.

Indonesian tuna longline fishing in EIW began in 1972. A state enterprise was set
up with a 900 mt cold-storage facility in Benoa, Bali. There were 39 longliners in 1982;
20 of these longliners were operated by the state fishing company based in Bali while the
others were newly introduced and operated from bases in Ambon and Bitung, fishing in
the Banda Sea. All of the vessels of the state fishing company are the same type; they
range in size from 110 to 115 GT and have engines of 370 - 400 HP, a 40-ton fish-hold
capacity, and a two tons/day freezing capacity. The tuna longline fishery has developed
rapidly since 1985 due to the demand for fresh tuna especially in Japan. This rapid
growth can be seen from the remarkable increase in the number of longliners from 18 in
1975 to 151 in 1990. In addition to the demand for fresh tuna, the government of
Indonesia invites foreign investors to develop tuna fisheries in Indonesia's EEZ. By the
end of 1989 there were about 598 foreign vessels operating in Indonesia's EEZ; the
vessels consisted of 341 longliners, 16 purse seiners, 79 gillnetters, 155 trawlers, and 7
trappers.

The size of boats used for large- or deep-longline fishing based in Benoa, Ujung
Pandang, Kendari, Bitung, Ambon, and Biak ranges from 90 to 400 GT. The crew size
ranges from 16 to 30 fishermen and the trip duration ranges from 14 to 140 days. The
conventional or common longline and monofilament longline fishing vessels based in
Benoa, Bitung, and Biak are from 30 to 100 GT, operate with 10-16 fishermen, and have
trip durations of around 10 to 15 days. The size of boats used for mini-longline fishing
based in Benoa and Maumere ranges from 3 to 30 GT. They carry ten fishermen and
have a trip duration of around 7 days.

2.2.2 Purse seine

Indonesia began tuna purse-seine fishing in the EIW in 1980. An Indonesian-
Japan joint-venture fishing company operated a base in Ternate from 1980 to 1982 using
a 600-GT purse seiner. A joint-venture company formed by Indonesian and French
interests was established in 1983 with Biak as its fishing base. It operated two purse
seiners of 600 GT and 750 GT and fished grounds north of Irian Jaya and in the western
Pacific Ocean. In 1990 this joint-venture company was sold to an Indonesian private
fishing company, PT Nelayan Bhakti.

Japanese purse seiners commenced operation in tropical areas of the western
Pacific Ocean north of Irian Jaya and Papua New Guinea (PNG) in 1970. The purse
seiners were mostly 500-GT vessels with a 400-ton fish-hold capacity. The catch
consisted of 70% skipjack, 6% yellowfin, and 4% bigeye. Until 1974, the fishing

grounds north of Irian Jaya and PNG were exploited only from December to March, but since 1976 fishing occurs year-round. The three types of fish schools encountered are (1) those found with drifting logs, (2) those associated with sharks or whales, and (3) schools associated with birds (Marcille *et al.*, 1984).

2.2.3 Pole and line

Although skipjack is the target species of the pole-and-line fishery, yellowfin is also caught in small amounts, making up 5-15% of the pole-and-line catch. The two pole-and-line fisheries are (1) the artisanal pole-and-line fishery based in Biak, Sorong, Ambon, Ternate, Labuha, Bitung, Gorontalo, Luwuk, Kendari, Kolaka, and Maumere; and (2) the large-scale pole-and-line fishery, run by the state enterprise and a private fishing company and based in Biak, Sorong, Ambon, Bitung, Kendari, and Maumere.

Due to the problem of availability of livebait, the pole-and-line fisheries have been using payaos as fish-aggregating devices (FADs) since 1985. The use of payaos make pole-and-line fishing more efficient, *e.g.*, it reduces the amount of livebait used and also reduces fuel consumption (Naamin and Chong, 1987).

2.2.4 Handline

Handline fishing is carried out by artisanal fishermen in Tomini Bay and Makassar Strait around "rompongs." The rompong is a traditional deep-water payao made of a bamboo raft and using rattan as line, stone as anchor, and coconut leaves as attractors. Yellowfin is a target species of this fishery. The catch consists of small-size (10-30 kg) yellowfin, which make up 25 to 75% of the total catch (Nasution *et al.*, 1986).

2.3 <u>Catch **and** Effort</u>

Catch of yellowfin tuna by type of fishing gear in the EIW (FAO statistical Area 71) is given in Table 1. Table 1 shows a trend of increasing catch of yellowfin tuna in the EIW since 1970.

Long time series of catch-and-effort data are available for the Japanese longline fishery for 1967-1981 (Table 2) and the state enterprise (PT Samodera Besar) operation for 1980-1990 (Table 3). Annual effort (number of hooks) of the Japanese longline fishery for all Indonesian waters (Areas 57 and 71) ranged between 5.6 and 24.6 million hooks; the yellowfin catch ranged from 55.4 to 76.9% or from 40,210 mt (1972) to 249,100 mt (1978). Annual effort in the Indonesian longline fishery ranged between 62,000 and 2.9 million hooks and the catch between 21 mt and 1,108 mt (Table 3). The very low effort in 1986 was due to only five longliners operating for three months.

Catch data for Japanese purse seiners in 1970-1979 are given in Table 4. The catch consisted of 26% yellowfin. Besides large-size (20-50 kg) yellowfin the purse seine also caught small amounts of small-size yellowfin (4-6 kg). Total yellowfin catches increased from only 123 mt in 1970 to 10,000 mt in 1979. Data on monthly effort (number of trips, days at sea per trip, number of effective fishing days per trip, and number of sets per trip) from May 1979 to April 1980 are given in Table 5.

Table 1. Catch of yellowfin tuna by type of fishing gear in the EIW (FAO statistical area 71).

Year	Unclassified	Pole and line	Purse seine	Longline	Handline	TOTAL
1970	5,500					5,500
1971	5,700					5,700
1972	9,000					9,000
1973	10.200					10,200
1974	10,165					10,165
1975	11,062					11.062
1976	7,530	507				8,037
1977	10,268	591				10,859
1978	8,225	1,160		1,216		10,601
1979	11,482	1,907		1,274		14,663
1980	11,626	2,269	2,177	1.478		17,550
1981	15,793	2,015	2,275	1,806		21,889
1982	17,393	1,887	1,428	3,605		24,313
1983	15,239	1,900	2,013	1,048		20,200
1984	18,140	2,282	2,108	1,670	2,250	26,450
1985	20,130	2,344	2,107	2,466	2,540	29,587
1986	25,226	2.278	1,650	2,347	2,737	34,238
1987	24,732	2.323	1,683	905	2,793	32,436
1988	26,377	2,439	1,767	576	2,899	34,058
1989	31,345	3,553	2,520	5,124	2.726	45,268
1990*	32,285	4,433	2,665	5,508	3,196	48,087

* Preliminary data

Table 2.   Sets, hooks, and catch (number of fish) by Japanese tuna longline in Indonesian waters.

Year	Sets	Hooks	BF	SBF	ALB	BE	YF	BB	SM	BuM	BaM	SF/SS	SJ	Total
1967	8773	15296550	2	4343	42194	113748	247242	5499	2166	8686	12858	7019	2184	445939
1968	11351	20640533	6	4578	30561	125252	311593	8955	7169	11874	29083	12026	3068	544165
1969	6413	11490178	5	1147	12301	63280	204201	4068	2008	6302	7780	3202	1630	305923
1970	4888	9280863	2	480	24146	65135	143540	3962	2094	4841	5101	2189	5906	257396
1971	3608	6895643	-	171	5753	28589	105729	2269	511	3552	6657	2180	391	155802
1972	2983	5610008	-	45	2693	38908	96118	1421	355	2281	1665	845	828	145159
1973	5367	9955675	6	35	7380	40168	217414	1831	927	6456	3606	3920	987	282730
1974	7990	15102811	5	222	13317	72852	217439	3337	1020	6785	5879	2612	1865	325335
1975	10789	20121569	2	387	12381	124595	231254	5711	1173	7229	7905	2653	935	394225
1976	6207	11431879	5	83	6622	65079	186531	2712	479	3193	1713	930	464	267711
1977	6041	11249151	1	40	890	91218	210739	2656	603	4209	1445	574	173	312548
1978	12230	22898564	-	20	2266	185060	452381	6647	1553	8743	4369	1580	274	662893
1979	12846	24583785	-	56	3315	122943	409196	5166	1095	6796	2988	1472	21	553048
1980	11598	22375244	-	131	7475	121133	411008	3881	1573	7147	4075	1875	80	558378
1981	9572	19244430	3	155	11063	71005	306414	3498	1083	4527	2482	1191	247	401668

Remark : BF = Bluefin tuna (*Thunnus thynnus*)
SBF = Southern Bluefin tuna (*Thunnus maccoyii*)
ALB = Albacore (*Thunnus alalunga*)
BE = Bigeye tuna (*Thunnus obesus*)
YF = Yellowfin tuna (*Thunnus albacares*)
BB = Broadbill swordfish (*Xiphias gladius*)
SM = Striped marlin (*Tetrapturus audax*)

SM = White marlin (*Tetrapturus albidus*)
BuM = Blue marlin (*Makaira mazara*)
BaM = Black marlin (*Makaira indica*)
SF = Sailfish (*Istiophorus platypterus*)
SS = Shortbill spearfish (*T. angustirostris*)
SS = Longbill spearfish (*T. pfluegeri*)
SJ = Skipjack (*Katsuwonus pelamis*).

Table 3.   Number of sets, hooks, hook rate (CPUE), catch (number of fish), and average weight of yellowfin caught by Indonesian longliners in FAO Area 71.

Year	Sets (Number)	Hooks (Number)	Hook Rate (CPUE)	Number Fish	Average Weight (kg)
1980	1,045	2,060,934	1.02	21,022	34
1981	1,334	2,121,060	1.06	22,483	34
1982	1,807	2,883,972	1.13	32,589	34
1983	691	1,131,808	1.03	11,658	30
1984	1,336	2,228,448	1.04	23,176	31
1985	1,429	2,366,424	1.17	27,687	30
1986*	38	62,320	1.04	648	32
1987**	503	784,680	0.83	6,513	36
1988**	460	705,180	0.73	5,148	31
1989**	239	359,934	0.67	2,412	36
1990**	350	532,000	1.16	6,171	34

* Only five longliners operated for three months.   ** Low hook rate because of deep longline.

Table 4. Japanese catch by purse seiners in the west equatorial Pacific Ocean, from 1970 to 1979.

Year	Total Catches (mt)	Skipjack (mt)	Yellowfin (mt)	Bigeye (mt)	Others (mt)
1970	461	338	123	-	-
1971	944	706	200	35	3
1972	782	539	188	47	8
1973	1,752	1,245	412	84	10
1974	2,261	2,159	407	36	19
1975	6,975	4,991	1,726	253	-
1976	10,539	7,509	2,756	274	-
1977	17,555	12,034	5,181	341	-
1978	32,000	25,000	7,000	-	-
1979	36,000	26,000	10,000	-	-

Table 5. Index of catch per unit effort for Japanese purse seiners (May 1979 - April 1980).

Month	T	C/T	D/T	F/T	S/T	C/S	C/F
May (1979)	2	420	32	20	15	28	22
June	5	416	45	31	26	16	14
July	6	397	40	28	23	18	14
August	9	348	38	23	20	21	15
September	5	403	38	27	26	15	15
October	6	346	48	33	26	13	11
November	4	463	53	37	31	15	13
December	4	398	56	39	30	13	10
January (1980)	8	424	38	26	26	16	16
February	8	429	43	28	24	18	16
March	8	383	33	21	18	21	19
April	10	451	34	21	17	27	21
Average value	-	407	42	28	23	18.5	15

Key: T = number of trips; C/T = catch per trip; D/T = days at sea per trip; F/T = number of effective fishing days per trip; S/T = number sets per trip; C/S = average catch per set; C/F = average catch per fishing day

2.4   Catch Rates

Catch rates in terms of hook rate (catch per 100 hooks) of yellowfin by Japanese longliners in the EIW for 1967-1981 ranged between 1.15-2.18 (average 1.62) fish per 100 hooks (Table 6) and hook rates of PT Samodra Besar (Table 3) ranged between 0.67 - 1.17 (average 0.94) yellowfin per 100 hooks. The difference between Japanese longline hook rates and the PT Samodra Besar hook rates may be due to the difference in (1) the bait used [Japanese longliners use saury (*Cololabis saira*) and squid as bait, while PT Samodra Besar longliners use small-size oil sardine (*Sardinella lemuru*)]; (2) the condition of the fishing gear and its accessories; and (3) the difference in experience of the fishing master and the skill of the crew.

The catch per unit of effort of the Japanese purse seiners in the western Pacific north of Irian Jaya and PNG was 15 mt per effective fishing day on the fishing ground. Catch per set ranged between 13 mt and 28 mt (average 18.5 mt) and consisted of 26% yellowfin. Catch per trip (trip duration ranged between 30 and 45 days; around 12 days were spent steaming to and from the fishing base and the fishing grounds) ranged from 346 mt to 463 mt and averaged 407 mt (Marcille *et al.*, 1984).

The average catch per day by the Indonesian purse seiner based in Ternate was 8.2 mt in 1980, 10.3 mt in 1981, and 14.9 mt in 1982. These catch-per-day rates are lower than those of the above-mentioned Japanese purse seiners fishing on the same fishing ground, although the Indonesian purse seiner was larger than the Japanese purse seiners.

The catch by each payao of the handline fishery in Makasar Strait was 56 mt/year or about 200 fish, with an average fish size of 23 kg.

3.   INTERACTION BETWEEN YELLOWFIN TUNA FISHERIES

3.1   Interaction Between the EIW and Indian Ocean Fisheries

A pattern of yellowfin tuna movement from the Indian Ocean to the Banda Sea was proposed based on relatively higher hook rates of longliners of PT Samodra Besar based in Benoa, Bali and fishing in the Indian Ocean and the Banda Sea during the period 1974-1984 and Japanese longliners during the period 1967-1981. Suhendrata and Bahar (1986) reported that the migration of yellowfin tuna from north to south along the west coast of Sumatra (eastern Indian Ocean) was as follows: July-September in the northern part off the west coast of Sumatra and October-December in the southern part with movement to the east. In April-June the fish appear south of central and east Java, Bali, and west Nusa Tenggara and then move to the Banda Sea in July-September. During the period April-June the fish apparently return south to the Indian Ocean (Figure 2).

A similar pattern of migration was also proposed by Warashina and Honma (M. Honma, pers. commun.; Figure 3). The yellowfin tuna move from the west coast of Sumatra in September-October to south of Bali and Nusa Tenggara in February-March. In April-June the fish move to the Banda Sea and reach its northern part around Ambon and Ceram Islands in July-September. During October-November the yellowfin return to the Indian Ocean. The above-mentioned facts imply that there is interaction between the stocks of yellowfin tuna in the Indian Ocean and the Banda Sea stocks.

Table 6.　Hook rate of yellowfin and bigeye per area by Japanese longline, 1967-1981.

Year		1	2	3	4	5	6	7	8	9	10	11	12	13	14	15	16
								Code Area									
1967	YF	2.87	2.76	1.30	2.17	1.77	1.47	2.11	1.19	1.64	1.55	1.12	2.07	2.08	2.09	2.18	2.23
	BE	0.21	0.26	0.20	0.28	0.28	0.32	0.40	0.77	0.46	0.65	0.10	0.12	0.70	0.55	0.55	0.94
1968	YF	3.95	2.98	1.34	2.18	2.12	1.75	2.30	1.52	1.62	1.74	1.32	2.31	1.34	1.60	1.49	1.70
	BE	0.21	0.41	0.22	0.46	0.33	0.36	0.07	0.60	0.41	0.44	0.29	0.18	0.52	0.32	0.54	0.46
1969	YF	2.42	2.07	1.69	1.65	1.35	2.78	-	1.83	2.16	3.12	1.37	0.90	2.41	2.04	1.90	0.62
	BE	0.22	0.27	0.18	0.16	0.39	0.39	-	0.69	0.56	0.63	0.08	0.02	0.66	0.49	0.39	0.45
1970	YF	1.41	1.60	1.86	1.11	2.11	1.95	-	1.79	1.74	1.84	0.41	1.12	2.19	2.10	1.81	2.36
	BE	0.34	0.27	0.21	0.20	0.33	0.21	-	0.21	0.29	0.45	0.17	0.04	0.82	0.45	0.39	0.27
1971	YF	-	1.89	0.96	1.80	2.85	2.90	-	1.82	1.55	1.58	2.34	2.59	1.03	1.34	1.19	1.43
	BE	-	-	0.52	0.12	0.34	0.39	-	0.33	0.36	0.40	0.06	0.09	0.61	0.32	0.35	0.39
1972	YF	-	-	1.74	1.74	1.67	1.56	-	1.24	1.74	1.34	2.32	3.18	1.75	5.19	2.24	1.33
	BE	-	-	0.14	0.30	0.40	0.73	-	0.49	0.51	0.30	0.26	0.33	0.45	0.98	0.86	0.27
1973	YF	-	-	0.93	2.76	2.22	2.20	-	0.24	1.19	3.10	3.53	3.98	0.91	0.34	0.45	0.15
	BE	-	-	0.23	0.27	0.32	0.30	-	0.07	0.28	0.55	0.33	0.45	0.19	0.34	0.45	0.15
1974	YF	0.89	1.79	1.90	3.06	2.45	1.56	-	0.43	1.01	1.25	0.33	2.44	0.93	0.77	1.44	0.88
	BE	0.17	0.30	0.18	0.29	0.28	0.38	-	0.38	0.42	0.39	0.25	0.25	0.53	0.57	0.50	0.52
1975	YF	2.04	2.35	1.34	2.01	1.74	1.60	-	0.35	0.75	1.31	-	2.98	0.64	0.46	0.72	0.86
	BE	0.41	0.46	0.42	0.26	0.28	0.33	-	0.72	0.94	0.80	-	0.13	0.80	0.82	0.98	0.98
1976	YF	1.62	1.28	1.45	2.05	2.33	1.52	-	1.35	0.91	0.96	1.69	1.74	0.65	0.38	0.40	0.67
	BE	0.33	0.47	0.40	0.32	0.41	0.54	-	0.96	1.37	1.30	0.18	0.20	1.51	1.85	1.10	1.40
1977	YF	2.66	1.79	1.58	1.63	2.01	2.39	-	0.76	1.22	0.76	1.13	3.73	-	0.68	0.85	1.02
	BE	0.34	0.39	0.59	0.61	0.58	0.54	-	1.92	1.48	1.91	0.34	0.42	-	1.33	1.24	1.47
1978	YF	1.15	1.50	1.97	3.34	2.94	2.98	-	0.76	0.57	1.85	5.48	3.48	0.61	0.71	0.65	0.54
	BE	0.65	0.82	0.71	0.37	0.34	0.34	-	1.61	1.09	1.71	0.12	0.24	2.11	1.86	1.37	1.24
1979	YF	1.95	1.62	2.28	2.19	1.74	1.71	2.52	0.79	1.00	1.82	1.83	3.01	0.98	0.73	0.92	0.82
	BE	0.38	0.33	0.35	0.32	0.28	0.36	0.60	0.90	0.99	1.21	0.25	0.18	1.31	1.19	1.00	0.72
1980	YF	2.11	1.65	1.46	2.33	2.11	2.33	-	0.74	1.12	1.16	1.23	2.53	-	0.40	0.65	1.19
	BE	0.40	0.49	0.34	0.31	0.31	0.34	-	1.20	0.86	0.89	0.17	0.28	0.13	1.69	1.22	1.22
1981	YF	1.89	1.77	1.46	1.92	1.54	1.53	-	-	-	-	3.64	2.22	-	-	-	-
	BE	0.33	0.30	0.46	0.25	0.27	0.25	-	-	-	-	0.21	0.25	-	-	-	-

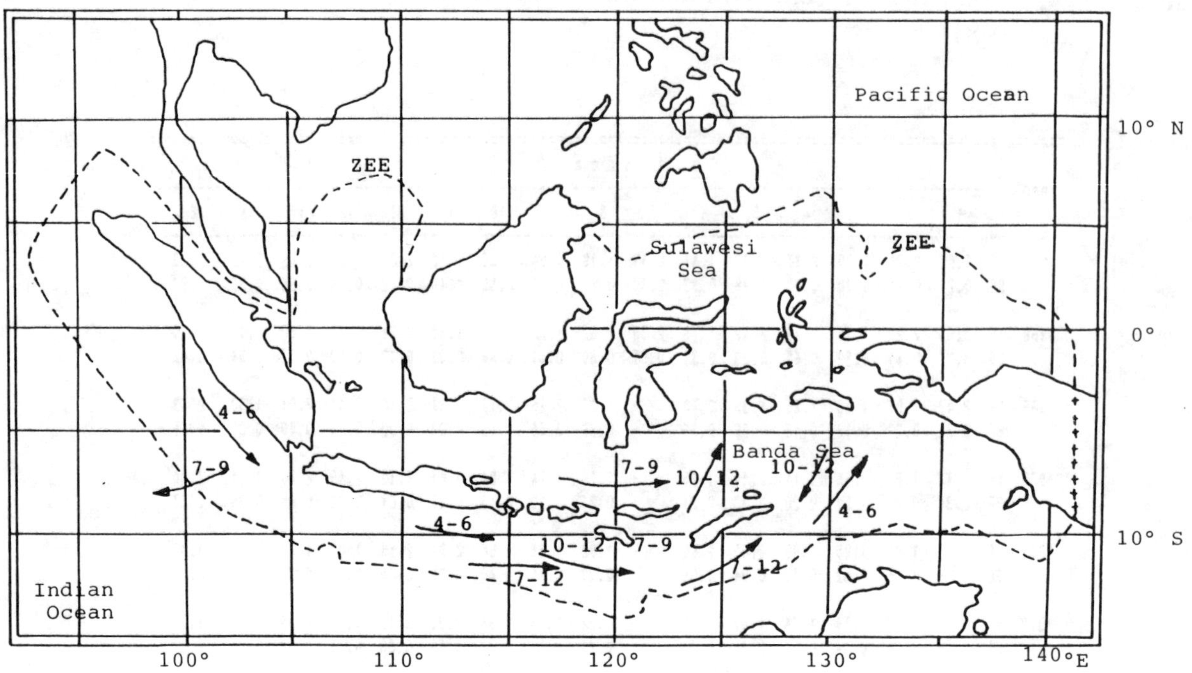

Figure 2.   The migration pattern of yellowfin tuna from the Indian Ocean to the Banda Sea and back.

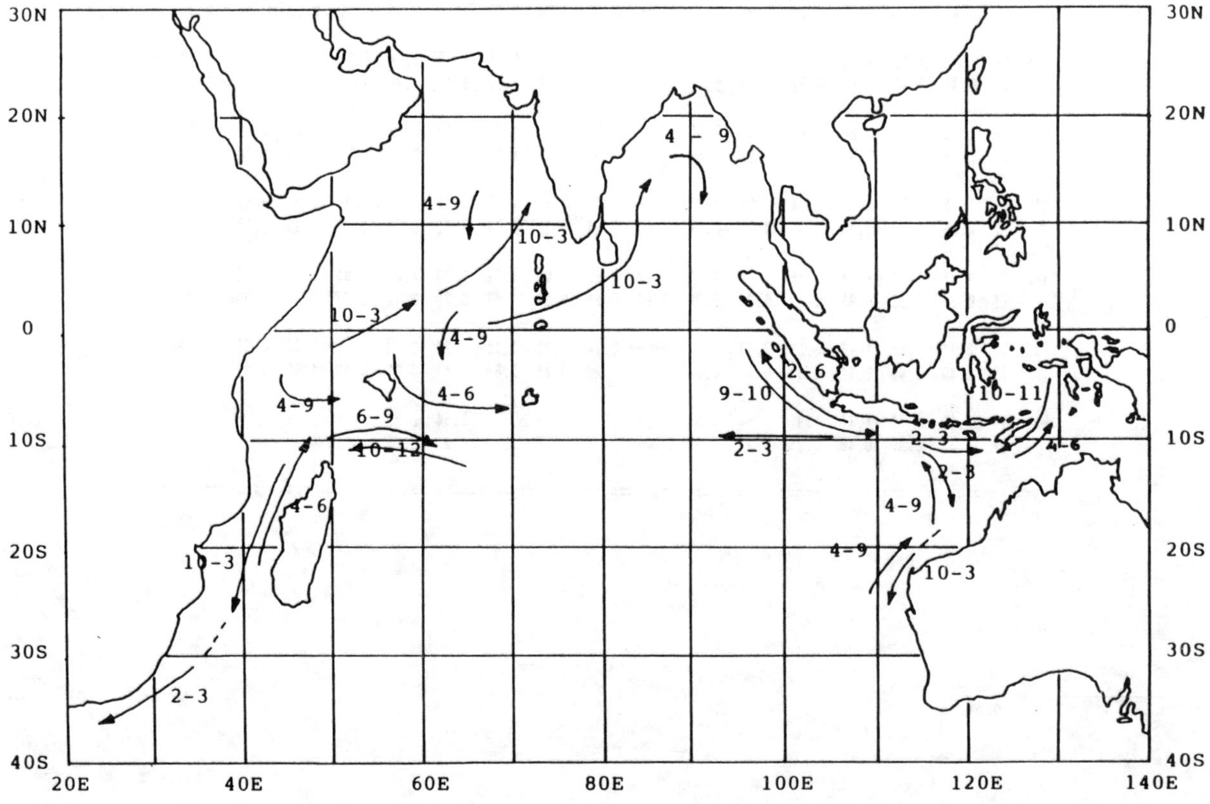

Figure 3.   Proposed migration of yellowfin tuna in the Indian Ocean and Banda Sea; numbers represent months of the year. (M. Honma, pers. commun.)

3.2   <u>Interaction Between the EIW and West Pacific Fisheries</u>

Results of tagging experiments carried out jointly by the Research Institute for Marine Fisheries and the Indo-Pacific Tuna Development and Management Programme (IPTP) in three locations (sea around Sorong, northern part of the Mollucas, and waters north of Sulawesi) in 1984-1986 show that the stock of yellowfin tuna in northeastern Indonesian waters is intermingling with the stock of yellowfin tuna in the western Pacific. Some of the yellowfin tuna tagged and released in the three tagging areas were recaptured in the western Pacific (north of PNG and near the Caroline Islands), off the southern part of Japan, and south of the Philippines.

The phenomenon described above is also supported by the results of tagging carried out by the South Pacific Commission (SPC). Tagging was done in the western Pacific north of PNG and around the Solomon Islands, and in the northeastern part of Indonesia. Some of the fish tagged and released north of PNG and around the Solomon Islands were recaptured in the northeastern sector of the EIW (north of Irian Jaya, Waigeo Island, Halmahera, and the Sulawesi Sea; Figure 4). Most of the yellowfin tuna were recaptured around payaos. Until the end of 1991 only one SPC-tagged fish was recaptured in the Indian Ocean (south of Bali Strait) and only one yellowfin was recaptured from the southern part of the Philippines. These facts suggest that yellowfin tuna migrate from east (north of PNG and around the Solomon Islands) to west (the northeastern part of Indonesian waters) and back again. These facts also imply that there is interaction between the stocks of yellowfin tuna in the northeastern part of Indonesian waters. There is weak evidence of the possible interaction between western Pacific stocks and Indian Ocean stocks.

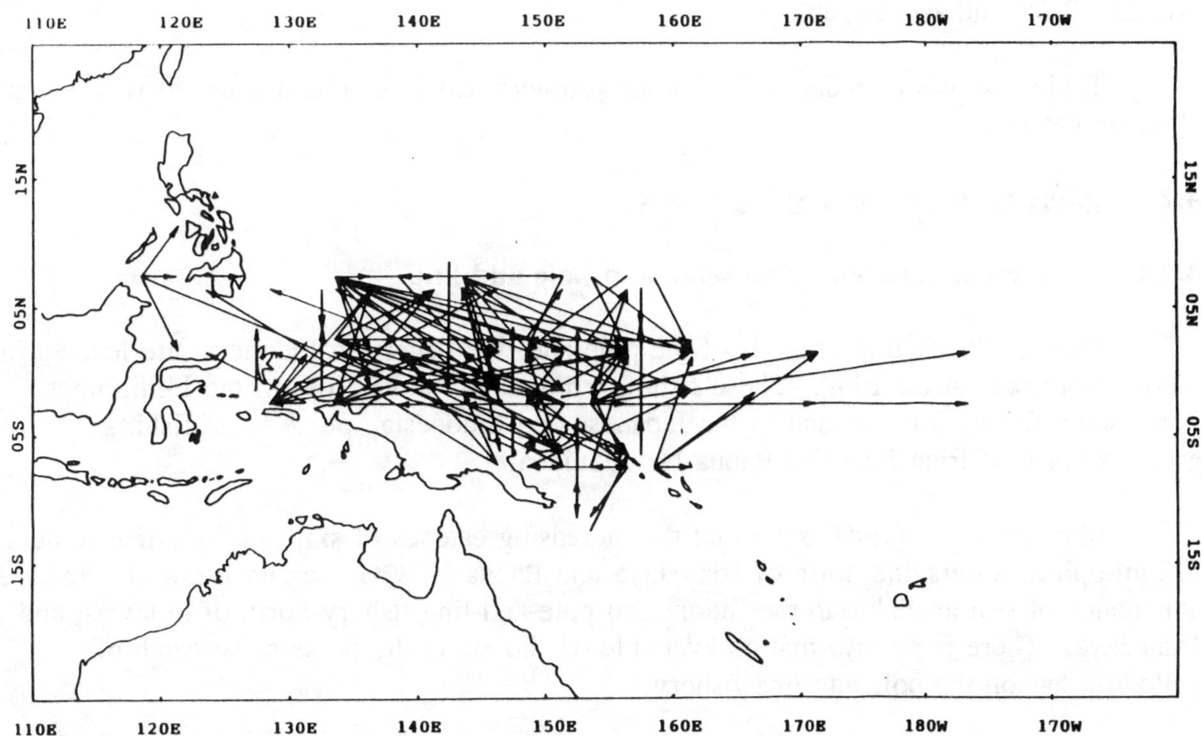

Figure 4.   Movement of yellowfin tuna based on tagging experiments carried out by the South Pacific Commission (SPC, 1991).

4. INTERACTION

The two types of interaction that occur in the region are interaction between species and interaction between fishing gears.

4.1 Interaction with Other Species

4.1.1 Bigeye

In the EIW a situation exists in which yellowfin caught by longlining are frequently associated with bigeye catches. Table 7 shows that an average of 19% of longline catches consisted of bigeye tuna. Table 7 also shows that there was a tendency for higher yellowfin catches to be associated with lower bigeye catches.

The hook rates of PT Samodera Besar longliners fishing in the eastern part of Indonesian waters in 1974-1984 did not show significant interaction between yellowfin and bigeye (Table 6).

4.1.2 Albacore

Besides yellowfin and bigeye, longliners in the eastern part of Indonesian waters also catch albacore. The catch of Japanese longliners fishing in the eastern part of Indonesian waters in 1967-1981 consisted of 55.4-76.9% (average 66.7%) yellowfin, 14.2-31.6% (average 23.4%) bigeye, and 0.3-9.5% (average 3.4%) albacore. The catch of PT Samodera Besar longliners fishing in the same area in 1975-1984 consisted of 51.4-75.9% (average 64.4%) yellowfin, 10.7-20.2% (average 15.8%) bigeye, and 2.5-15.4% (average 9.6%) albacore (Table 7).

Table 7 shows a tendency for higher yellowfin catches to be associated with lower albacore catches.

4.2 Interaction Among Fishing Gears

4.2.1 Interaction between purse seine and pole and line

Most of the fishing grounds of the pole-and-line fishery in Indonesia are located in the northern part of the EIW. These fishing grounds are quite close to the Philippines purse seine fishing grounds and to the Japanese and Indonesian purse seine fishing grounds north of Irian Jaya and Papua New Guinea.

Marcille *et al.* (1984) state that the increasing catches of skipjack by purse seiners in Philippines waters and north of Irian Jaya and Papua New Guinea have not affected the abundance of fish available to the Indonesian pole-and-line fishery north of Sulawesi and Irian Jaya. There is no information available on the effect the purse-seine catch of yellowfin has on the pole-and-line fishery.

Table 7.   Composition of the catch (%) by Japanese longline (1967-1981) and PT Perikanan Samodera Besar (1975-1984).

Code species	Year																	
	1974	1968	1969	1970	1971	1972	1973	1974	1975	1976	1977	1978	1979	1980	1981	1982	1983	1984
BF	0.00	0.00	0.00	0.00	-	-	0.00	0.00	0.00	0.00	0.00	-	-	-	-	-	-	-	*
	-	-	-	-	-	-	-	-	-	-	-	-	-	-	-	-	-	-	**
SBF	0.97	0.84	0.37	0.19	0.11	0.03	0.01	0.07	0.10	0.03	0.01	0.00	0.01	0.02	0.04	-	-	-
									0.9	0.2	0.1	0.1	0.2	0.1	0.1	0.0	0.1	0.0
ALB	9.46	5.62	4.02	9.38	3.69	1.86	2.61	4.09	3.14	2.47	0.28	0.34	0.60	1.34	2.75			
									14.1	14.4	8.7	9.5	11.9	8.2	5.7	5.5	15.4	2.5
BE	25.51	23.02	20.68	25.35	18.35	26.80	14.21	22.39	31.61	24.31	29.19	27.91	22.23	21.69	17.68			
									20.2	18.2	15.1	18.2	17.3	17.2	12.3	16.6	12.6	10.7
YF	55.44	57.26	66.75	55.77	67.86	66.22	76.90	66.84	58.66	69.34	67.43	68.24	73.99	73.61	76.28			
									51.4	55.3	67.0	60.7	59.3	64.4	73.5	71.9	64.4	75.9
SM	0.49	1.32	0.66	0.81	0.33	0.24	0.33	0.31	0.30	0.18	0.19	0.23	0.20	0.28	0.27	***		
									1.8	1.3	0.3	0.5	0.8	0.6	0.1	0.2	0.4	0.0
WM	-	-	-	-	-	-	-	-	-	-	-	-	-	-	-	-	-	-
									1.7	2.2	1.4	3.1	2.8	0.4	3.5	1.6	3.2	6.5
BaM	2.88	5.34	2.58	1.98	4.27	1.15	1.28	1.81	2.00	0.46	0.64	0.66	0.54	0.73	0.62			
									1.5	.07	0.7	1.1	1.3	0.9	0.6	0.3	0.7	0.4
BuM	1.95	2.18	2.06	1.88	2.28	1.57	2.28	2.09	1.83	1.19	1.35	1.32	1.23	1.27	1.13			
									-	-	-	-	-	-	-	-	-	-
BB	1.23	1.65	1.33	1.54	1.46	0.98	0.65	1.03	1.45	1.01	0.85	1.01	0.93	0.70	0.87			
									2.7	1.7	1.5	2.1	1.6	1.5	1.4	1.5	1.1	1.2
SF	1.57	2.21	1.05	0.85	1.40	0.58	1.38	0.80	0.67	0.35	0.18	0.24	0.27	0.34	0.30			
									2.2	2.2	2.0	1.8	1.0	1.0	1.3	0.9	1.2	1.5
SJ	0.49	0.56	0.53	2.29	0.25	0.51	0.35	0.57	0.24	0.14	0.06	0.04	0.00	0.02	0.06			
									-	-	-	-	-	-	-	-	-	-
SWR	-	-	-	-	-	-	-	-	-	-	-	-	-	-	-	-	-	-
									0.5	0.2	0.3	0.1	0.1	0.0	0.0	0.0	0.0	0.0
MR	-	-	-	-	-	-	-	-	-	-	-	-	-	-	-	-	-	-
									3.1	2.8	2.9	2.8	3.8	2.0	1.5	1.5	1.1	1.3

Remark : *) Japan
         **) PT PSB

4.2.2 Interaction Among Other Fishing Gears

Among the four fishing gears used for catching yellowfin, some small-size (3-30 kg) yellowfin are caught by purse seine and pole and line, while longline and handline catch the larger-size (30-100 kg) yellowfin. Based on available data, interactions among the above-mentioned fishing gears could not be detected.

5.   REFERENCES CITED

DGR (Directorate General of Fisheries). 1990. Fisheries statistics of Indonesia, No. 20 Stat. Prod. Departemen Pertanian, Jakarta, 77p.

Marcille, J., T. Boely, M. Unar, G.S. Merta, B. Sadhotomo, and J.C.B. Uktolseja. 1984. Tuna fishing in Indonesia. Balai Penelitian Perikanan Laut and ORSTROM. Collection Travaux et Documents No. 181, Paris, 125 p.

Naamin, N., and K.C. Chong. 1987. Technological and economic aspects of FAD-based skipjack and tuna fishing in Indonesia. Paper presented at the Fourth International Conference on Artificial Habitat for Fisheries, 2-6 November 1987, Miami, Florida.

Nasution, C., G.S. Merta, and R. Arifudin. 1986. Study on rompongs (FAD) and their aspects in Mamuju waters to develop tuna fishery in South Sulawesi. *J.Mar.Fish.Res., Jakarta*, 37:31-58.

SPC. 1991. Monthly tagging summary - August 1991. Regional Tuna Tagging Project. *Tuna Billfish Assess.Programme, S.Pac.Comm.*, Noumea, New Caledonia, 11 p.

Suhendrata, T., and S. Bahar. 1986. The tuna longline fishing grounds in Indonesian waters and the possibility of its industrial development. *J.Mar.Fish.Res., Jakarta*, 37: 79-94.

ASSESSMENT OF INTERACTION BETWEEN NORTH PACIFIC ALBACORE, *THUNNUS ALALUNGA*, FISHERIES BY USE OF A SIMULATION MODEL[1]

P. Kleiber and B. Baker
Southwest Fisheries Science Center
National Marine Fisheries Service, NOAA
La Jolla, California

ABSTRACT

Using a simulation model of a typical year in the North Pacific albacore fisheries in the 1970s, we tested for the degree to which the activity of fleets affects the performance of other fleets. The results show that rather drastic (factor of two) changes in the activity of any of the three principal albacore fleets have only a mild effect on the catch of the other fleets. With the overall exploitation rate in the model close to the exploitation rate determined from tagging results (6%), the maximum degree of interaction was a 7.5% drop in longline catch resulting from doubling the baitboat effort. The mild degree of interactions was insensitive to exploitation rate up to approximately 10% exploitation, although interaction became more severe at higher levels of exploitation.

---

[1] Published: Kleiber, P., and B. Baker. 1987. Assessment of interaction between North Pacific albacore, *Thunnus alalunga*, fisheries by use of a simulation model. *Fish.Bull.*, 85(4): 703-11.

EVIDENCE OF INTERACTIONS BETWEEN HIGH SEAS DRIFT GILLNET FISHERIES AND THE NORTH AMERICAN TROLL FISHERY FOR ALBACORE

Norman Bartoo, David Holts, and Cheryl Brown[1]
Southwest Fisheries Science Center
National Marine Fisheries Service, NOAA
La Jolla, California 92038

ABSTRACT

Observers aboard USA troll vessels fishing for albacore in the eastern North Pacific examined approximately 20,000 albacore at the time of capture for marks caused by encounters with high seas drift nets. Results indicate about 13.4% of albacore caught by USA troll vessels showed evidence of encountering drift nets prior to capture by the USA fleet. This is interpreted as direct evidence of fisheries interactions.

1.　INTRODUCTION

The Southwest Fisheries Science Center (SWFSC) in cooperation with members of the Western Fishboat Owners Association (WFOA) placed fisheries observers on USA trollers fishing in the North Pacific in 1990. This was part of a comprehensive impact assessment programme on the effects of the high-seas drift gillnet fisheries on the North Pacific albacore (*Thunnus alalunga*) stocks and other fisheries. Albacore which escape drift gillnets bear some external marks that provide direct evidence of interaction. Throughout this report, we refer to these external marks from drift gillnet encounters as damage. Damage may be minor or severe, and may include marks on the skin, bruising, cuts and broken skin (loss of scales), missing areas of skin, and scars related to any of the foregoing. This report describes the results of the 1990 observer project.

2.　BACKGROUND

Catches and catch-per-effort from the USA albacore troll and pole and line fisheries in the North Pacific have declined since the mid-1970s (Coan *et al.*, 1991; Kleiber and Perrin, in press). In addition to the USA fishery, Japan and Taiwan each have a surface fleet and a longline fleet catching albacore in the North Pacific. Three high-seas drift gillnet fleets also catch albacore in the North Pacific (Anon., 1989).

The small-mesh (90-120 mm stretched mesh) drift gillnet fleets are made up of vessels from Japan, Taiwan and the Republic of Korea (ROK). The Japanese and Taiwanese fleets fish for flying squid (*Ommastrephes bartrami*) in the North Pacific Transition Zone from about May through December. The ROK fleet operates there the entire year. The incidental catches of albacore by these drift net fisheries are not precisely known.

---

[1] Current address: Southeast Fisheries Science Center, 75 Virginia Beach Dr., Miami, FL 33149

The large-mesh (160 -180 mm stretched mesh, Bartoo and Holts, 1991) drift gillnet fleet is made up of vessels from Japan and Taiwan. Japan's fleet is currently restricted to areas east of 170°E. Albacore is a major component of that catch. Taiwan has a developing large mesh fishery for albacore. The impact of these fleets on albacore is largely unknown. Taiwan and Japan have imposed internal regulations on their drift gillnet fleets to limit fishing effort in the higher latitudes west of 170°E longitude and minimize interceptions of high-seas salmon.

Although drift gillnet catches and landings for North Pacific albacore are incomplete and unreported for some fleets, the aggregate reported annual landings for drift gillnet fisheries exceeded 20,000 mt in 1988 (Tsuji *et al.*, 1992). In the mid-1980s U. S. albacore fishermen began reporting increasing numbers of albacore injured or scarred by encounters with high-seas drift gillnets. Concerns focussed on the impact of drift gillnet fisheries on the North Pacific albacore stock and the direct interaction with the U.S. albacore troll fleet.

2.1   Troll Observer Project

In 1990, SWFSC biotechnicans observed 6 complete fishing trips aboard 5 different troll vessels. Vessels were selected to provide data over the entire season and cover the entire fishing area. The cruise tracks of the observed vessels are shown with the entire catch distribution of the troll fleet in Figure 1. The fishery followed eastward-migrating albacore beginning north of Hawaii in June, moved eastward to 135° W in July and remained east of 135° W in August and September (Coan *et al.*, 1991). Operations and observations covered a wide area of the north Pacific Ocean. The project goals were to document and estimate the interaction between high-seas drift gillnet and North American troll fisheries, and to provide biological information from which delayed mortality of albacore due to net encounters might be inferred. Specific objectives are as follows:

A.   Describe the type and severity of drift gillnet damage on troll-caught albacore

B.   Estimate the relative frequencies of undamaged albacore and damaged albacore

C.   Estimate the size-frequency of undamaged albacore and damaged albacore

D.   Determine the relative physiological condition of troll-caught albacore that show gill net damage

Here, we report results relating to the first 3 objectives.

2.1.1   Methods

Drift gillnet marks are most visible when the fish is boated. Most marks disappear or are rendered indistinguishable once the fish is frozen or dies on deck. Observers examined each fish landed and noted net scars or marks using a set of reference photographs detailing the type and severity of injuries (Figure 2). These were coded as:

Code	Damage Description
0	No gill net damage to fish
1	Minor damage along side(s) of fish, pattern of stripes due to minor scale loss where fish forces its way through or along the net.
2	Minor damage to head, chiefly forwards of pectoral fins, brush-like pattern of scale loss.
3	Severe damage with bruising or scraping away of parts of the skin, primarily in area of greatest girth and mostly on dorsal surface.
4	Old gill net damage of any degree that is partially or completely healed. This is assumed to have occurred previously.

Observers photographed, for later analysis, fish which showed damage but could not be classified on board.

In addition to damage code, observers collected information on fork lengths, maximum girth, and weights for as many fish as possible. Fishing operations usually continued all day from first light to just after dark.

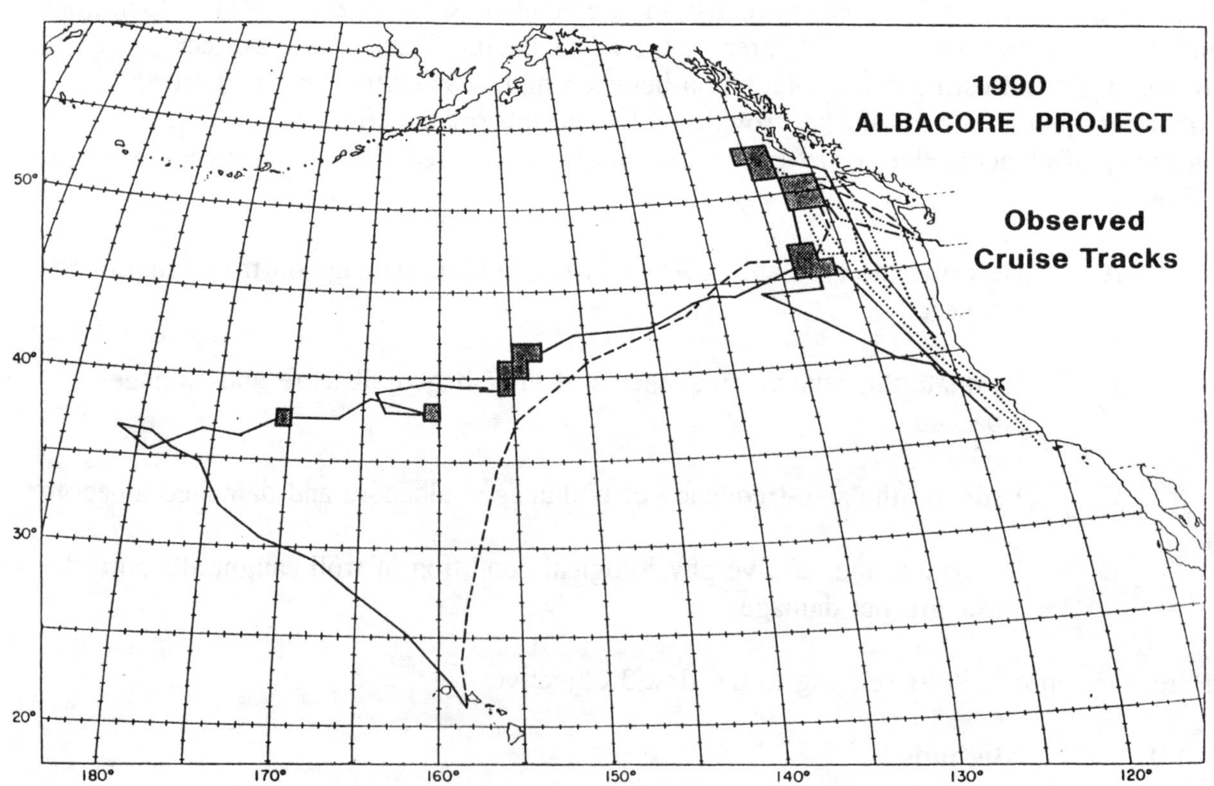

Figure 1.   General cruise track and fishing areas (shaded areas) for observed albacore troll vessels. Three vessels departed Honolulu and three departed from the mainland. All cruises terminated on the west coast.

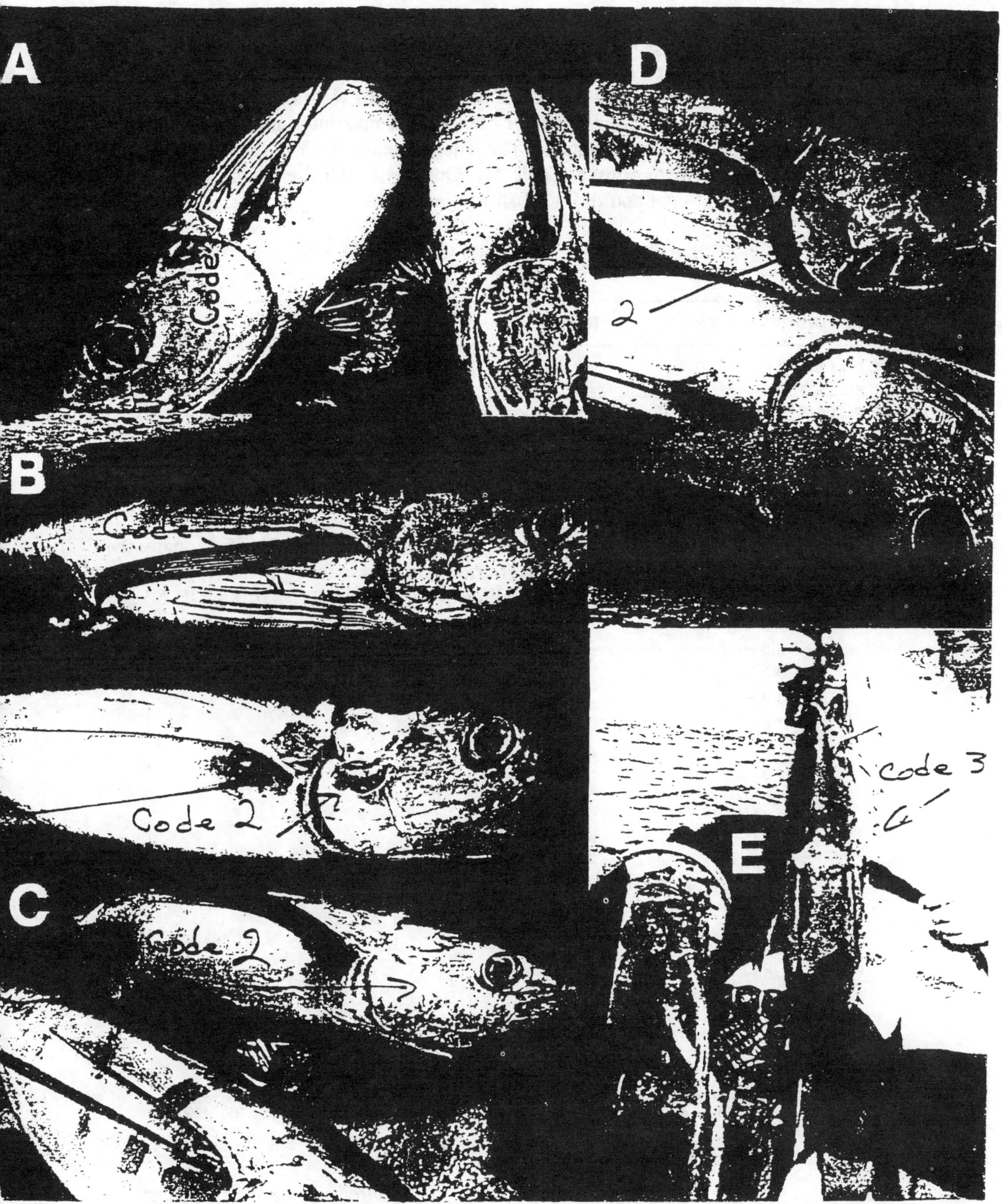

Figure 2.   Pacific albacore damaged by high-seas drift nets. Damage code 1 indicates minor scratches and scale loss (A & B). Damage code 2 shows moderate damage on the head (B. C & D).

2.1.2   Results

Six cruises, totalling 377 observed sea-days were completed between the end of May and the end of October 1990. A total of 25,177 albacore were caught and landed during the cruises. Seventy-eight percent of these (19,526) were examined for drift gillnet-related injuries and measured for fork length and maximum girth. A total of 8,720 fish were weighed to the nearest pound. The catch rate averaged 66.8 fish-per-day but varied considerably between cruises (Table 1). One vessel (Cruise 5) was on a charter with WFOA as a scout boat and did not fish routinely.

Table 1.   Albacore CPUE (fish/day) by cruise.

Cruise	Mean	Range	SD	95% C.I.
Cruise 1	108	0 - 634	146.3	76.01 - 139.9
Cruise 2	68	0 - 363	91.3	48.47 - 88.1
Cruise 3	88	0 - 291	58.5	74.02 - 101.3
Cruise 4	48	1 - 122	30.1	39.20 - 56.5
Cruise 5	10	0 - 100	22.8	2.63 - 18.3
Cruise 6	58	0 - 375	82.7	20.02 - 95.3

Three distinct size modes were caught throughout the North Pacific apparently representing age 2, 3, and 4 year old fish (Bartoo and Foreman, in press; Fig. 3). Length modes were at 54, 65 and 80 cm corresponding to weights of 3.5, 5.7, and 10.6 kg.

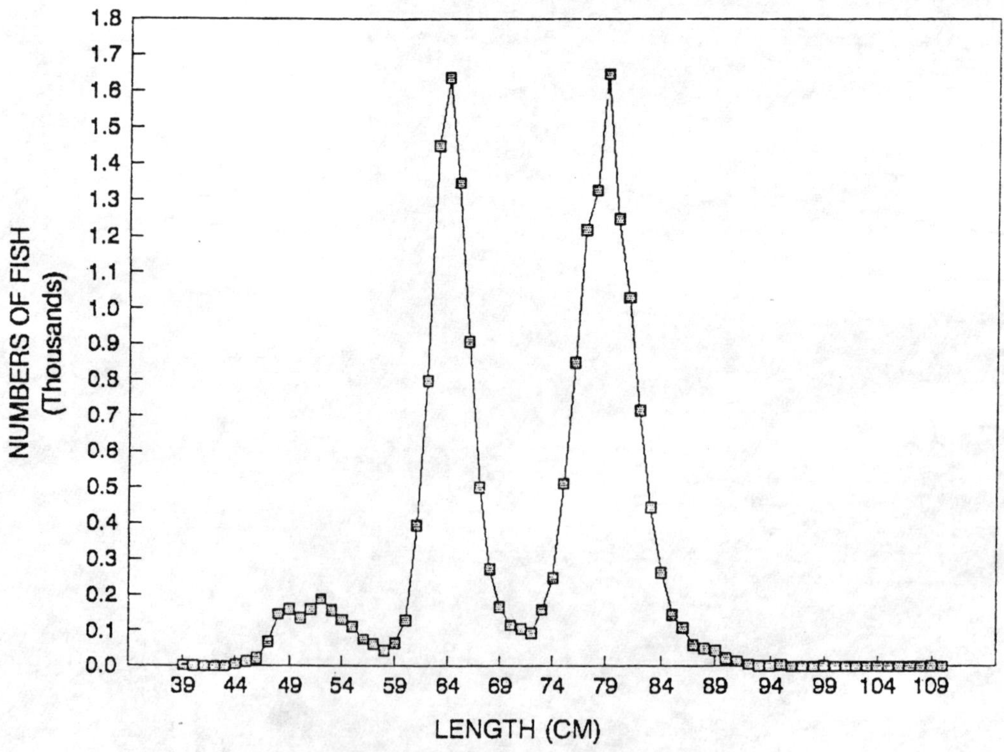

Figure 3.   Length frequency of 19,526 albacore measured during the troll fleet observer program.

Overall, 86.6% of the observed catch showed no evidence of net-related damage (Table 2). Recent gillnet damage (damage codes 1,2 & 3) totalled 9.2% while an additional 4.2% had healed scars from net encounters during the previous fishing season (code 4). Within the recently-damaged albacore, 3.5% had minor damage, 5.2% had moderate injuries on the head, gill covers and fins while 0.5% had severe bruising with significant loss of scales with skin damage to their head, sides and fins.

Table 2. Percent of albacore by fishing area damaged during encounters with drift nets and caught by troll vessels.

DAMAGE TYPE	WEST OF 160°W	160°W TO 140°W	NORTH OF 50°N	EAST OF 140°W AND SOUTH OF 50°N	μ ALL AREAS
NO DAMAGE	81.2	82.8	94.1	88.4	86.6
NEW DAMAGE					
CODE 1	6.0	4.9	0.0	3.0	3.5
CODE 2	11.4	6.3	0.4	2.7	5.2
CODE 3	0.5	0.5	0.0	1.0	0.5
TOTAL NEW DAMAGE	17.9	11.7	0.4	6.7	9.2
OLD DAMAGE	0.9	5.5	5.5	4.9	4.2
SAMPLE SIZE, n	218	7,440	4,800	7,068	19,526

Observers measured fork lengths of 11,868 albacore east of 140°W in July, August and September and 7,675 albacore west of 140°W in June and July. The catch taken west of 140°W had larger fish than the catch east of 140°W (Figure 4). These large fish averaged 9 - 11 kg (20 to 25 lbs). Although the sample size was small, injuries from recent encounters with drift gillnets were observed on nearly 18% of the catch. Old and healed scars from a previous time appeared on less than one percent of the catch west of 160°W but increased to about 5% east of 160°W (Table 2).

By the end of July, all observed vessels were fishing east of 140°W and south of 50°N, catching fish weighing 5 - 9 kg (12 to 20 lbs). These fish were caught throughout the month of August and early September. The incidence of fresh injuries remained at about 6.7% in the coastal areas and the incidence of old injuries remained fairly constant at about 5 percent.

Albacore taken north of 50°N had a higher proportion of large (>80 cm fork-length) fish than those caught east of 140°W and south of 50°N (Figure 4). No fish less than 63 cm FL were sampled north of 50°N. Fishing remained moderately good north of 50°N until seasonal storms out of the Gulf of Alaska forced many fishing vessels to return home by the end of September. Few recently-damaged fish were observed north of 50°N although the proportion of old and healed damage changed little from that seen just off Washington and Oregon (5.5% to 4.9% respectively).

The observed length frequencies varied considerably in different fishing locations. Length frequencies of injured fish did not vary greatly from uninjured fish within each

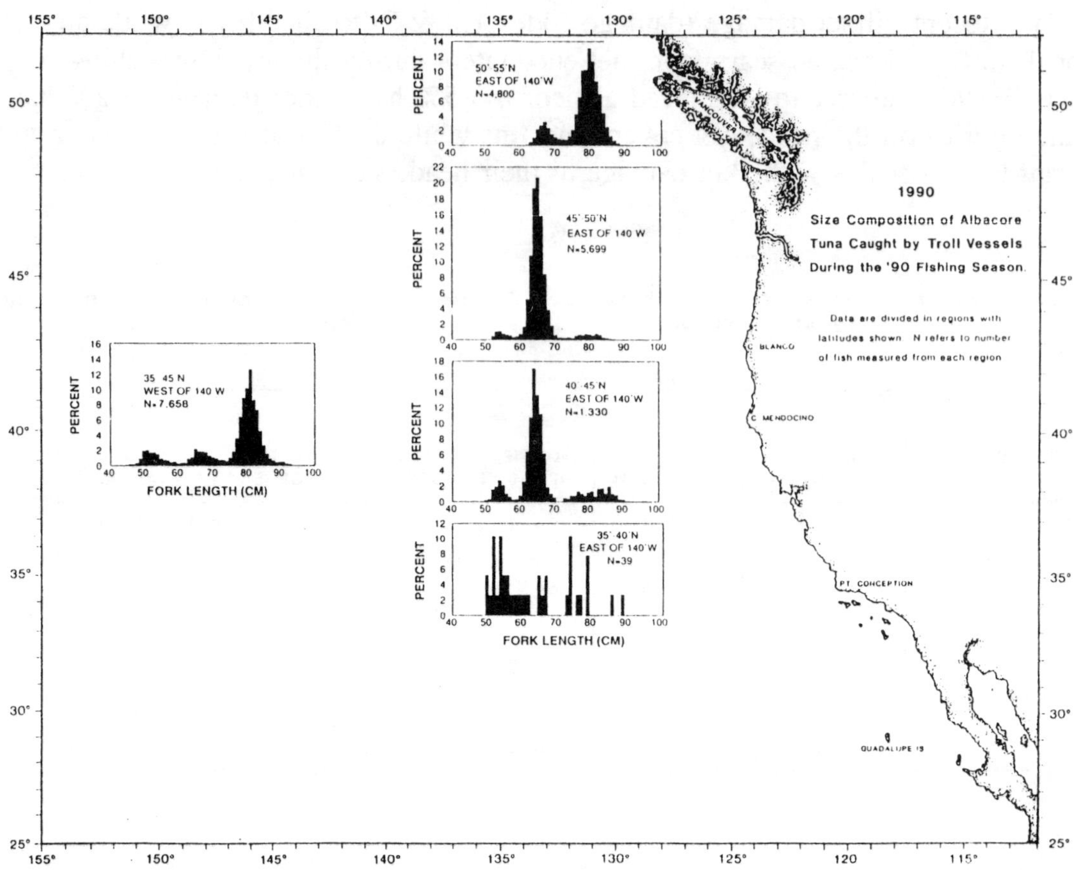

Figure 4.   Albacore length frequencies by fishing areas of the North Pacific.

area examined. The length frequencies, by area, for undamaged (code 0), recently damaged (codes 1-3) and previously damaged (code 4) fish are shown in figures 5, 6 and 7 respectively. No previous damage estimates are available from the 1989 season that could help describe survival of fish sampled (code 4) in the 1990 fishing season.

The length-weight relationship for all 8,720 weighed albacore is shown in Figure 8.

3.   DISCUSSION

Our results show that 13.4% of the albacore caught encountered drift gillnets. This proportion is less than some reports from fishermen which ranged from 40% to 90% of the catch. Our data do indicate considerable variation in damage on a daily basis, ranging from none to 100% of the catch. Our results are consistent with those reported for albacore in the South Pacific which ranged from 4.5% to 14.5% of the catch in 1989 and 1990 (Hampton *et al.*, 1991).

Our results show that 9.2% of the 1990 catch evidenced new damage and that higher proportions were found to the west. The apparent "gradient" in proportion of net marked fish from west to east may be explained by the migration pattern of albacore and

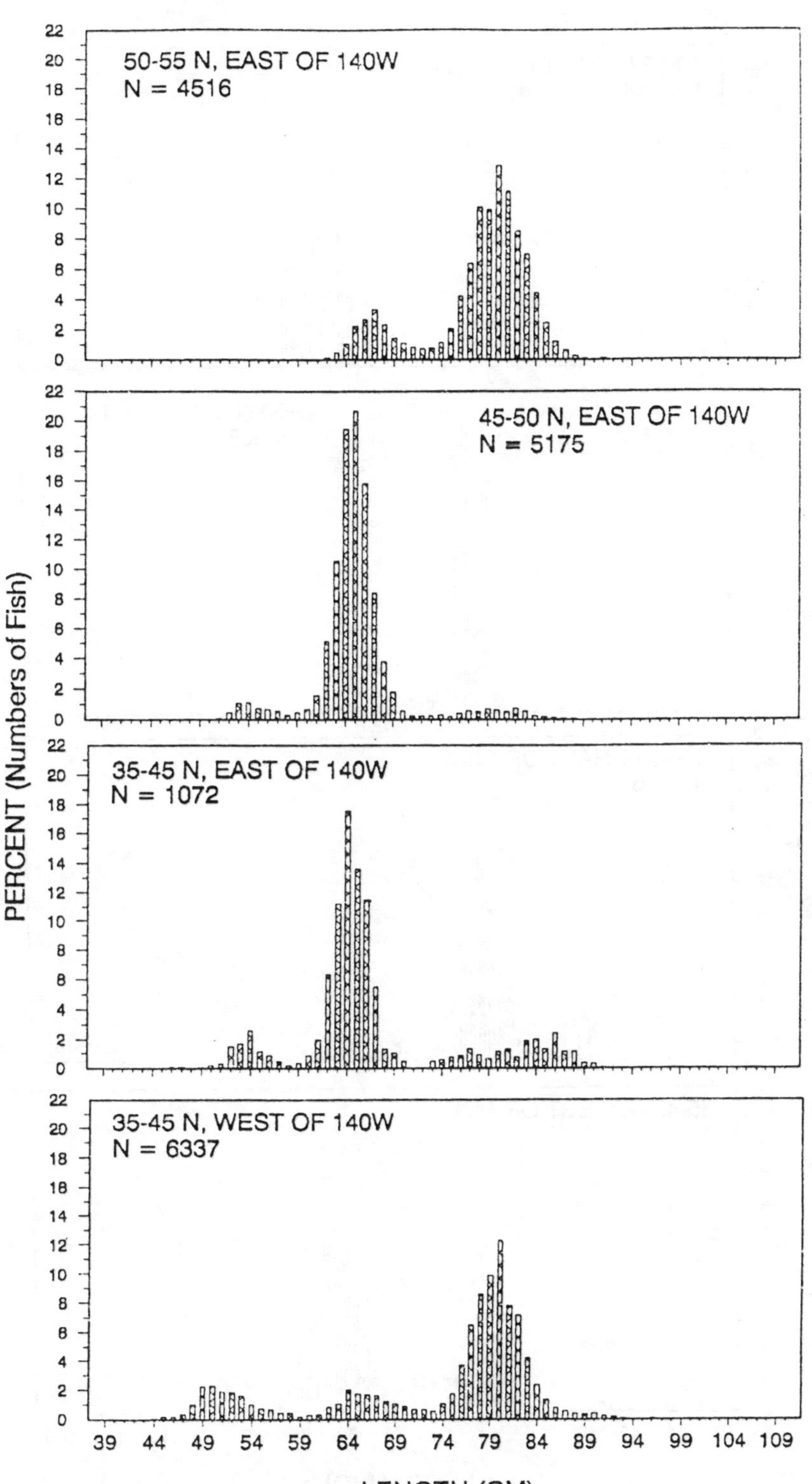

Figure 5.   Length frequencies of un-damaged albacore (Code 0) from the four fishing areas of the north Pacific.

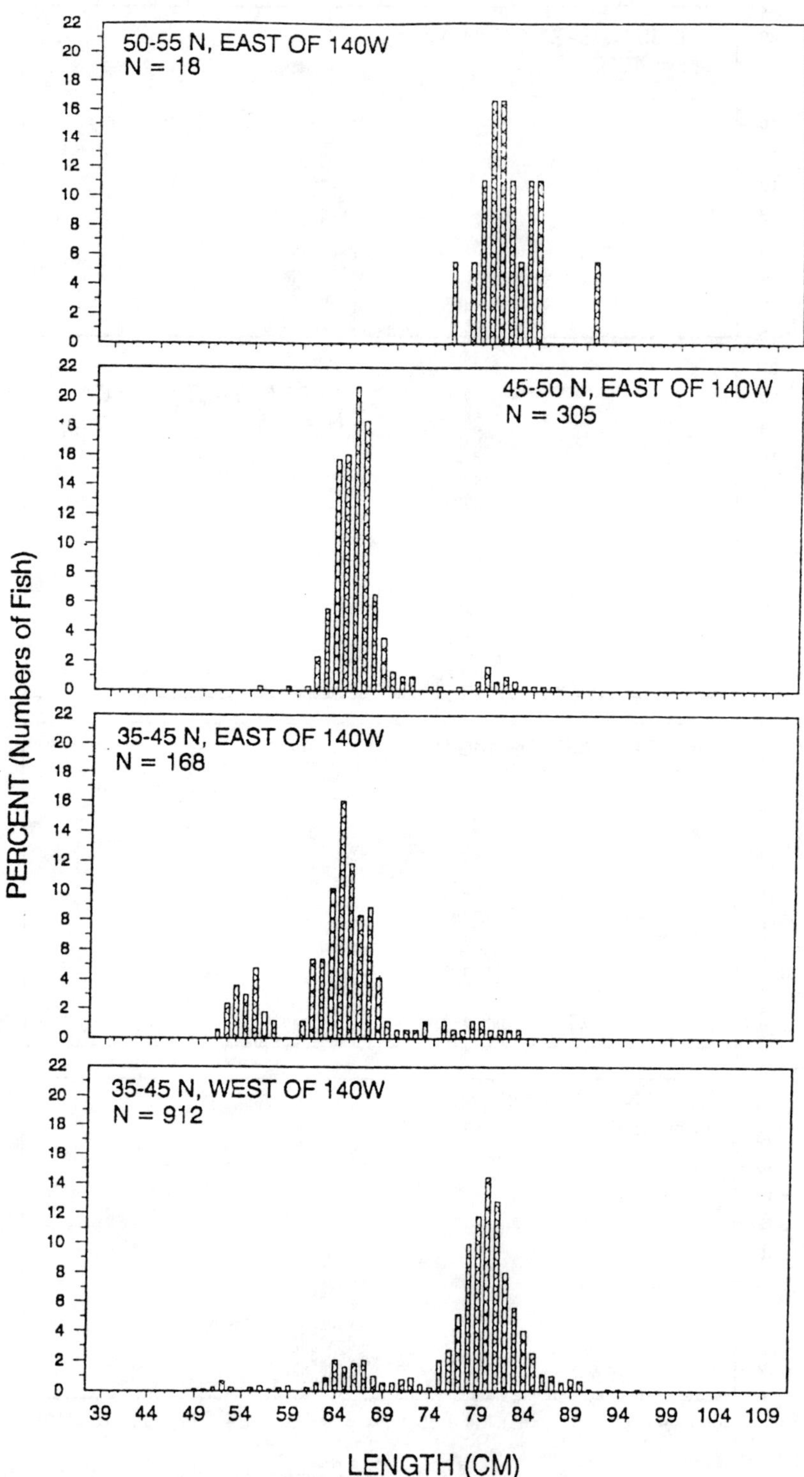

Figure 6. Length frequencies of recently damaged albacore (Codes 1, 2, and 3) from the four fishing areas of the north Pacific.

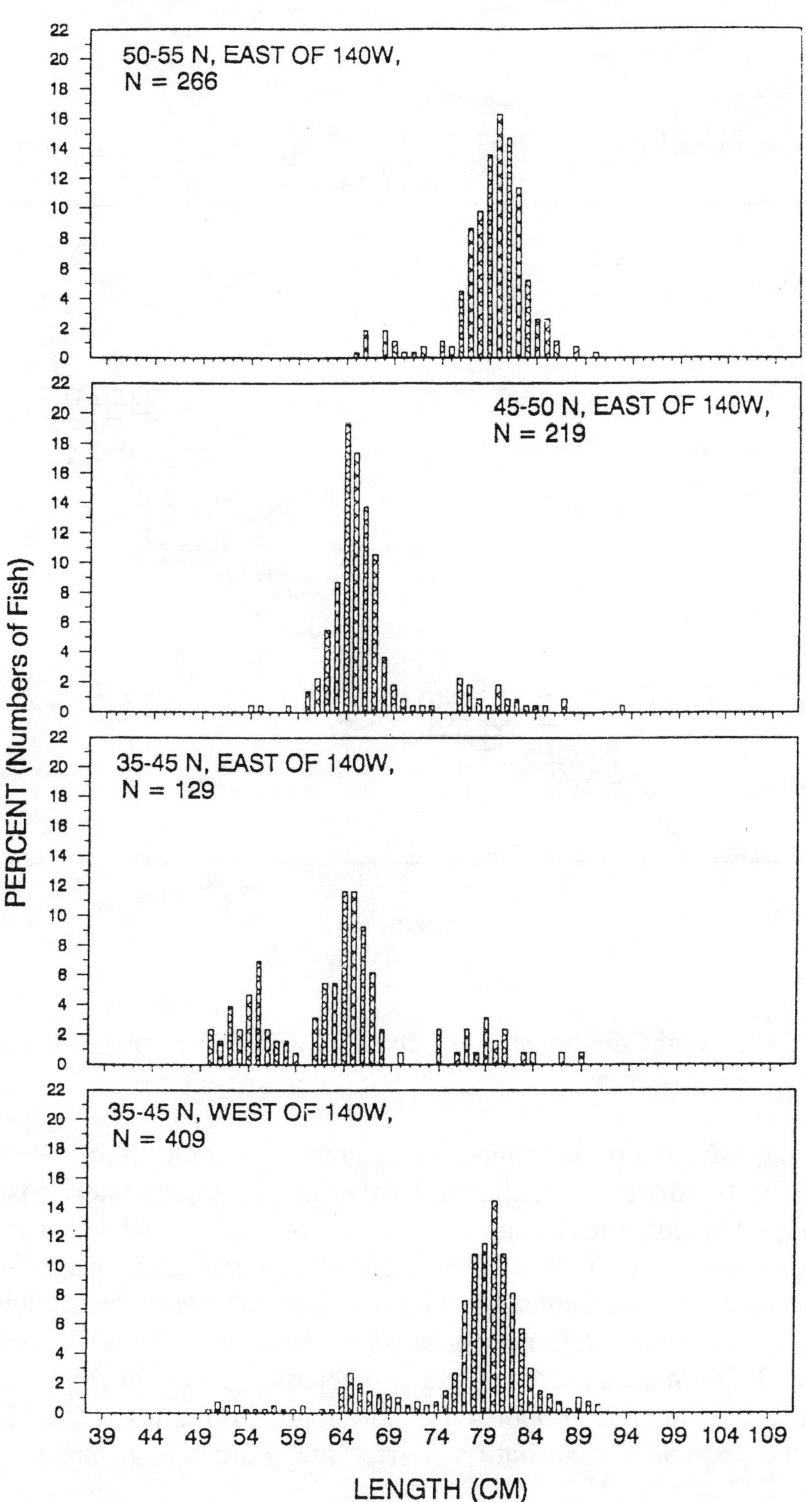

Figure 7.   Length frequencies of albacore with old injuries (Code 4) from the four fishing areas of the north Pacific.

the timing and location of the drift gillnet and troll fisheries. Additionally, increased short-term mortality due to net damage may contribute.

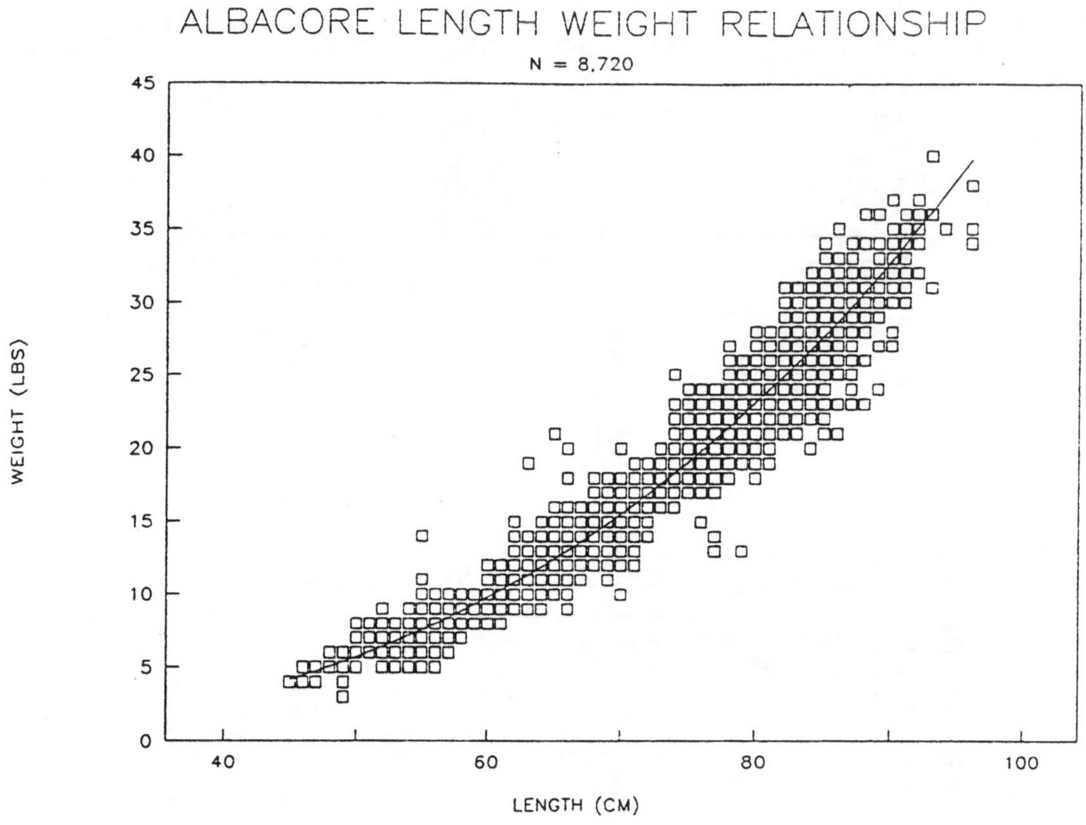

Figure 8.   Length-weight relationship of albacore weighed during the troll fleet observer program

Based on tag information the migration pattern of juvenile North Pacific albacore of the size caught by the drift gillnet and troll fisheries is predominately east-west (Otsu and Uchida, 1963; Clemens, 1961). The fish move from waters off Japan in the spring across the Pacific in the Subarctic Transition Zone (STZ) and enter the coastal waters of North America in July. In late September albacore leave the east Pacific and return to waters off Japan. This annual pattern may be repeated until age 5 or 6. Although some portion of the population makes the complete trip across the Pacific in just a few months it is apparent that a considerable proportion of the population is not available at the extreme ends of the population distribution (Bartoo and Foreman, in press).

The troll fishery as shown in Figure 1 begins near the mid-Pacific in May or June and moves with the advancing albacore to the east. The drift gillnet fisheries in 1990 were distributed as shown in Figure 9 with most of the effort centered between 170°E and 170°W and becoming less east of 170°W. These fisheries operate year round with considerably higher effort east of 170°E after May.

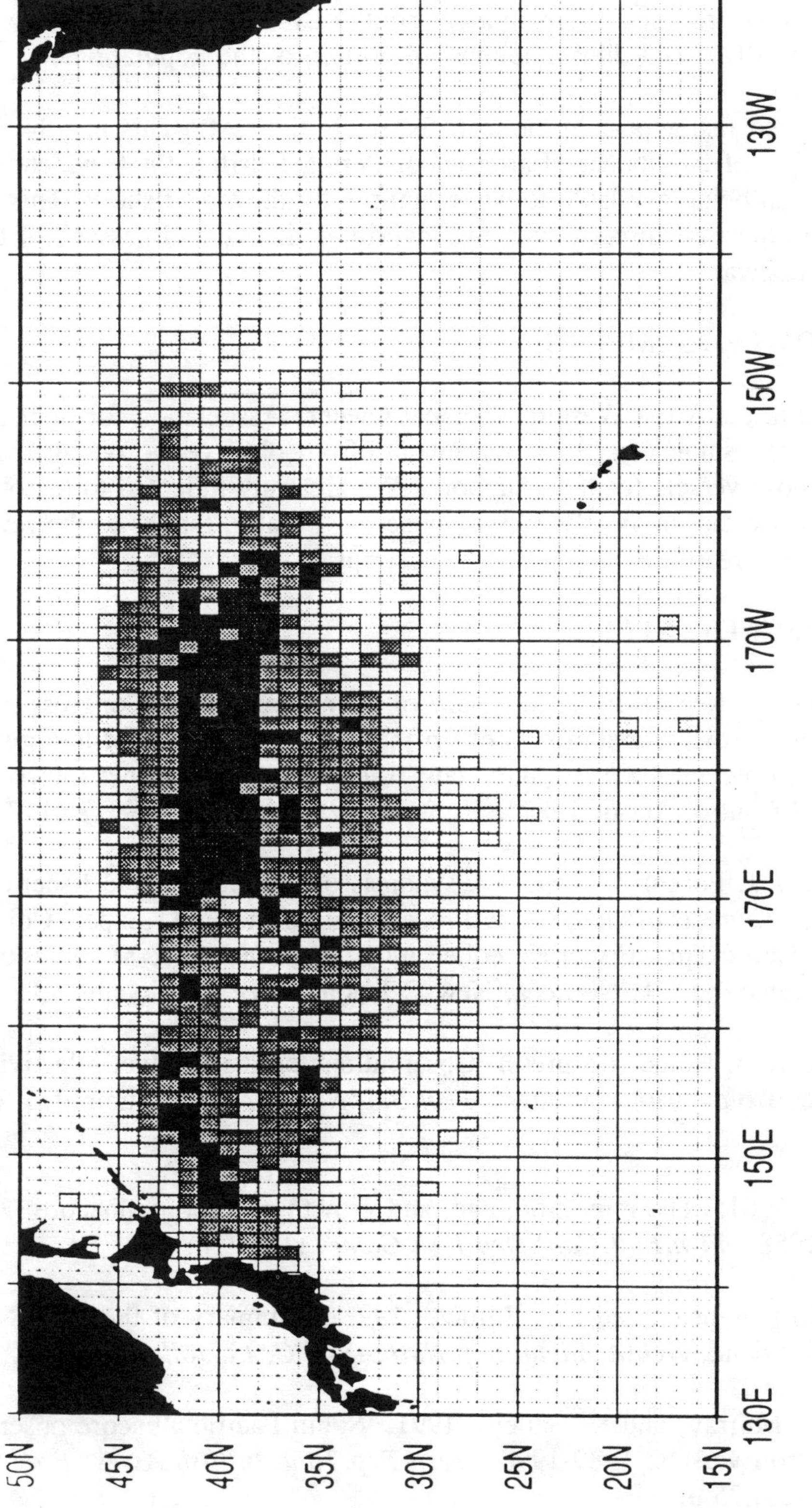

Figure 9. Distribution of observed drift net effort in the north Pacific Ocean in 1990 for both large and small mesh fisheries. Intensity of effort is shown by shading. (Pers. Comm., J. Wetherall, Nat. Mar. Fish. Serv., Honolulu, HI.)

Higher proportions of marked fish are seen west of 140°W where the drift gillnet and troll fisheries overlap in area and time. Lower proportions of marks are seen east of 140°W, outside the distribution of the drift gillnet fisheries. Additionally, some albacore may migrate into the troll fishery from areas south of the drift gillnet-fishery distribution (Laurs and Lynn, 1977) and dilute the marked portion of the population.

Less marked fish appear in the troll fishery in the east. This may be partly due to increased mortality of fish damaged after encountering a drift gillnet, which would result in less marked fish being available to be recaptured by the troll fishery. This would logically seem to increase directly with the severity of injury and passage of time (as the albacore move eastward).

4.   ACKNOWLEDGEMENTS

We wish to thank the Western Fishboat Owners Association for their help in locating owners of fishing vessels for placement of our observers. We also wish to thank the owners of those vessels for their support. We are truly indebted to the observers who examined and measured the nearly 20,000 albacore. Finally, thanks to the reviewers whose suggestions greatly improved this manuscript.

5.   REFERENCES CITED

Anonymous. 1989. Report of the Secretary of Commerce to the Congress of the United States on the nature, extent and effects of driftnet fishing in waters of the north Pacific Ocean pursuant to section 4005 of public law 100 - 220, the "Driftnet Impact Monitoring, Assessment, and Control Act of 1987". *U.S. Department of Commerce,* 147 p.

Bartoo, N., and T.J. Foreman. 1993. A synopsis of the biology and fisheries for North Pacific albacore (*Thunnus alalunga*). *In* Proceedings of the FAO Expert Consultation on Interactions of Pacific Tuna Fisheries, edited by R.S. Shomura, J. Majkowski, and S. Langi, 3-11 December 1991, Noumea, New Caledonia [See this document].

Bartoo, N., and D.B. Holts. (in press) Observations on drift gillnet selectivity for albacore inferred from various surveys. *Fish.Bull.NOAA-NMFS* [Submitted for publication].

Clemens, H.B. 1961. The migration, age, and growth of Pacific albacore (*Thunnus germo*), 1951-1958. *Fish.Bull.Calif.Dep.Fish Game,* (115):128 p.

Coan, A.L., G.M. Rensink and F.R. Miller. 1991. Summary of the 1990 North Pacific albacore fisheries data. *NOAA Admin.Rep.NMFS-SWFC, La Jolla,* LJ-91-21:33 p.

Hampton, J., T. Murray, and K. Bailey. 1991. South Pacific albacore observer programme on troll vessels, 1989-1990. *Tech.Rep.Tuna Billfish Assess.Programme, S.Pac.Comm.,* (25):25 p.

Kleiber, P., and C. Perrin. (in press) Catch per effort and stock status in the United States north Pacific albacore fishery: Re-appraisal of both. *Fish.Bull.NOAA/NMFS,* 89(3).

Laurs, R.M., and R.J. Lynn. 1977. Seasonal migration of north Pacific albacore, *Thunnus alalunga,* into North American coastal waters: distribution, relative abundance, and association with Transition Zone waters. *Fish.Bull.NOAA-NMFS,* 75(4):795-822.

Otsu, T., and R.N. Uchida. 1963. Model of the migration of albacore in the north Pacific Ocean. *Fish.Bull.U.S.Fish Wildl.Serv.,* 63(1):33-44.

Tsuji, S., H. Nakano, and N. Bartoo. 1992. Report of the Twelfth North Pacific Albacore Workshop, Shimizu, Japan July 23-25, 1991. *NOAA Admin.Rep.NMFS-SWFC, La Jolla,* LJ-92-04:15 p.

ALBACORE FISHERIES INTERACTIONS IN THE SOUTH PACIFIC OCEAN

Talbot Murray
Research Centre, MAFF Fisheries
P.O. Box 287
Wellington, New Zealand

ABSTRACT

The potential for interactions among South Pacific albacore fisheries is reviewed based on CPUE trends, recovery of tagged fish, albacore size composition in different fleets, and patterns of recovery of fish damaged by drift gillnets. Comparisons of CPUE from drift gillnet and troll fleets operating in the Subtropical Convergence Zone (1985/86-1989/90) and similar fleets operating in the Tasman Sea and off the west coast of New Zealand (1983/84-1989/90) suggest a reciprocal effect between surface fisheries. This is supported by a linear relationship between CPUE in the two components of the surface fishery (drift gillnet and troll) that accounts for over 50% of the variation in CPUE in both the Subtropical Convergence Zone and Tasman Sea. An interaction has also been postulated between surface and longline fisheries, albeit with some time lag, but there is little to support this hypothesis.

1.   INTRODUCTION

Longline fisheries for adult and surface troll fisheries for juvenile albacore have operated in the South Pacific for several decades. In the case of longlining, distant water fleets from Japan, Taiwan, and Korea have caught albacore since the early 1950s, while trolling in nearshore New Zealand waters began in the early 1960s. High seas surface fisheries for juveniles are a recent development, beginning in 1982 for the large-mesh pelagic drift gillnet fishery and 1985 for the troll fishery. The drift gill net fishery stopped in June 1991 and has not subsequently operated in the South Pacific Ocean.

The rapid development of high seas surface fisheries and the limited information available for stock assessment of South Pacific albacore raised concerns about the sustainability of harvests and the potential for interactions between fisheries. Concern over the latter, especially between surface and longline fisheries, was first raised in 1986 (Anon., 1986). The rapid expansion in drift gillnet fishing between the 1986/87 and 1988/89 fishing seasons (from about 900 mt to 25-49,000 mt) heightened these concerns (Anon., 1990, paragraphs 61-63) and extended the potential for interactions. In response, scientists from distant water fishing nations and South Pacific coastal states have worked through the South Pacific Commission's Tuna and Billfish Assessment Programme to improve catch, effort, and size composition data from all South Pacific albacore fisheries. Supplementary studies of age, growth, stock structure, and reproductive biology as well as estimates of bycatch, discarding practices and escapement have also been initiated.

While little data exists to quantify or even to demonstrate that interactions between South Pacific albacore fisheries have been strong, a range of observations suggests that some interactions occur. These include overlap in size of albacore caught in different fisheries, overlap in the time and areas fished, recovery of drift gillnet-damaged albacore in troll and in longline fisheries, tag recoveries, reports of troll vessel entanglement in drift gillnets, disruption of troll fishing operations by drift gillnet vessels, and reports of lower troll fishery catch rates in the proximity of drift gillnet vessels.

While interactions are likely to be strongest between fisheries operating in the same area, at the same time, and targeting the same stock component (*e.g.*, troll and drift gillnet fisheries), interactions with time lags of one or more season are also possible. Interactions with time lags have been discussed by Wetherall and Yong (1990). Based on tag recoveries in the longline fishery they postulate that interaction effects between surface and longline fisheries may be seen in a few months in longline fishing within or in areas immediately north of the Subtropical Convergence Zone (STCZ) or more broadly in subsequent seasons. Murray (1990) made a similar observation based on the frequency of recent drift gillnet-damaged albacore caught by longline in New Zealand waters several months after the finish of the summer surface fisheries.

2.   FISHERIES

Although the South Pacific albacore stock is broadly distributed (from the equator to 50°S and from eastern Australia to South America), fisheries do not operate throughout this range. Industrial and small scale commercial fisheries operate both in the EEZs of coastal states and in high seas areas. Artisanal catches of albacore occur throughout the South Pacific in nearshore oceanic waters of island states and small recreational catches of albacore are made in Australian and New Zealand waters.

Surface fisheries are restricted to austral summer months, primarily December to April. Fishing areas are further limited to mid-temperate latitudes where summer sea surface temperatures tend to be 16° to 21°C. In the South Pacific Ocean (including the Tasman Sea) surface fisheries operate primarily between 39° and 41°S. Within this narrow latitudinal band, commercial catches are possible from the Australian coast eastward to at least 140°E. A surface fishery of unknown size is also reported to operate off South America along the Chilean coast. The main areas of commercial surface fisheries have been the central Tasman Sea (drift gillnet), west coast of New Zealand (troll), and the STCZ (troll and drift gillnet).

In contrast, the longline fishery operates in all months, moving from north to south seasonally. These patterns are described by Wang (1988) for the Taiwanese fleet and Wetherall and Yong (1989) for the Korean fleet; both fleets target albacore. Japanese longliners catch albacore as a bycatch in fisheries directed toward bigeye, yellowfin and southern bluefin tuna. Consequently, the several longline fisheries operating in the South Pacific exhibit different operational patterns. The general pattern of fleet movement for longliners targeting albacore is southwards from January to April into subtropical waters north of the STCZ (some limited fishing occurs in the STCZ in April-May) and then northwards from July to October. Most longline fishing appears to be from 5° to 45°S west of 120°W with relatively little effort in the Tasman Sea or the waters adjacent to New Zealand. Developing longline fisheries which either target albacore (*e.g.*, Fiji, French Polynesia, and Tonga) or catch albacore as bycatch (Australia, New Caledonia, and New Zealand) do not exhibit the high mobility of distant water longline fleets and generally operate within their EEZ or adjacent waters.

Albacore recruit to surface fisheries at about 45 cm fork length (LCF) in the Tasman Sea, particularly around New Zealand and at a slightly larger size (50-55 cm LCF) east of New Zealand in the STCZ. Most of the surface fishery catch is comprised of fish smaller than 80 cm LCF in all areas and years sampled (Labelle and Murray, 1992). The size composition of the drift gillnet catch, while different, comprises the same modes and exhibits an appreciable overlap with the size of fish caught in the troll fishery (Anon., 1991). Albacore recruit to the longline fishery at about 60 cm LCF and overlap in size with those caught by surface fisheries in the range 60-80 cm LCF. The majority of the longline catch is, however, comprised of fish larger than 80 cm LCF.

3. INFORMATION AND DATA AVAILABLE FOR INTERACTION STUDIES

Data on total catches, effort and size composition has been compiled for all South Pacific albacore fisheries (Anon., 1993), although data quality, completeness, spatial and temporal resolution varies. Data on catches and effort exist for the Japanese drift gillnet fleet (Watanabe, 1990) while data on the substantial Taiwanese fleet is limited to estimates of total catch and vessel numbers. Estimates of discarding and loss of fish during landing are available for troll (Labelle and Murray, 1992) and drift gillnet fisheries (Sharples *et al.*, 1990) and for some components of the longline fishery. Aggregated size composition data are available for Taiwanese and Korean longline vessels landing in American Samoa; data from troll fisheries are available by area and time strata. Drift gillnet size composition is limited to one vessel in one season and one season's transshipment monitoring.

Growth rate has been estimated from caudal vertebrae by Murray and Bailey (1989) and from length frequency data by Hampton *et al.* (1990). Labelle (1991) has helped reduce some of the apparent discrepancy in parameter estimates between these studies reanalysing Murray and Bailey's (1989) data in light of tag recoveries since these studies. Growth rate varies with fish size and appears to be about 0.5 cm per month for juveniles. Patterns of albacore movement have also been described by Jones (1991) using tag recovery information and the spatial pattern in parasite fauna.

4. INTERACTION ISSUES

In the South Pacific discussion and data collection have focused on the potential for interactions between troll and drift gillnet fisheries for juvenile albacore and between the surface fisheries for juveniles and longline fisheries for adults. Interactions between geographical areas have also been considered (*e.g.*, the interaction between the high seas drift gillnet fishery in the western Tasman Sea and the New Zealand troll fishery; surface fisheries in the STCZ and the more northerly distributed longline fishery).

4.1 Interactions Between Surface Fisheries

The potential for interaction between troll and drift gillnet fisheries arises because they operate at the same time and in the same areas and catches have similar size compositions. That the two fisheries compete for the same fish is also evident from the frequent occurrence of albacore caught by trolling with recent drift gillnet damage (Hampton *et al.*, 1989). Similarly, reports by albacore trollers of reduced catch rates when operating in the vicinity of drift gillnet vessels suggest that interactions occur.

Comparison of the data provided by Watanabe (1990) for drift gillnet CPUE with troll CPUE from Coan and Rensink (1991) in the years when both fisheries operated in the STCZ supports this observation.

Figure 1 shows the trends in the two fisheries over the time period when these fisheries operated in the STCZ. Despite the short CPUE time series (four years), 53% of the variation in the troll fishery CPUE can be explained by a linear relationship between catch rate in the troll and drift gillnet fisheries. Low troll fishery CPUE correspond with years of high drift gillnet CPUE in the STCZ, increasing in years when drift gillnet CPUE is lower. The relationship between the performance of the New Zealand nearshore fishery in the eastern Tasman Sea and catch rate in the Tasman Sea drift gillnet fishery is shown in Figure 2. As in the STCZ surface fisheries, 52% of the variation in tonnes landed per trolling trip was explained by a linear relationship with drift gillnet CPUE in the central and western Tasman Sea over the period 1983/84 through 1989/90. Continued declines in the two troll fisheries CPUE in 1990/91 and 1991/92 may indicate that other factors also affect troll fishery performance.

4.2   Interactions Between Surface and Longline Fisheries

Several observations suggest the potential for interactions to occur between surface and longline fisheries although with some time lag. These include: overlap in albacore size composition in these fisheries (primarily fish 60-80 cm LCF), the end of the STCZ surface fishery (March-April) in temperate waters coincides with the start of the longline fishery (April-May) in subtropical waters north of the STCZ and in the STCZ, tag recovery patterns, and smaller-sized albacore caught by longline with recent drift gillnet damage. Wetherall and Yong (1989, 1990) attempted to use Taiwanese and Korean longline CPUE stratified by 5° latitudinal band to document an interaction between surface and longline fisheries following the historical peak of surface catches in 1988/89. They report that for Taiwanese and Korean longline fleets, CPUE immediately following the end of the 1988/89 surface fishery season (April-May) was the lowest of the previous 15 years. However, the variability in CPUE data over the history of this fishery, absence of more data when drift gillnet fleets operated, and changes in fishing effort by latitude were also noted as potential explanations for CPUE declines.

Figure 3 depicts the range of variation represented in albacore-targeted longline fisheries. While CPUE declines are evident in the latter years in subtropical and temperate latitudes, they are within the range of variation seen prior to the expansion of high seas surface fisheries. Hampton (1990) provides an indication of the potential severity of the interaction between surface and longline fisheries. He uses a size structured simulation model to predict the consequences of continuing surface fishery catches at the historical high 1988/89 level. Using optimistic recruitment levels he predicted that continued exploitation at the 1988/89 levels would at best, result in parental stock declines over a 5 year period equivalent to 32% of pre-exploitation levels.

A reciprocal impact of the longline fishery on surface fisheries might also be expected if surface fishery catches resulted in reduced recruitment to the spawning stock and subsequent reduced recruitment to the surface fishery. The likelihood of this occurring seems small since drift gillnet fleets have not operated in South Pacific since

Figure 1.   Comparison of CPUE trends in South Pacific albacore surface fisheries operating in the STCZ, 1985/86-1989/90.

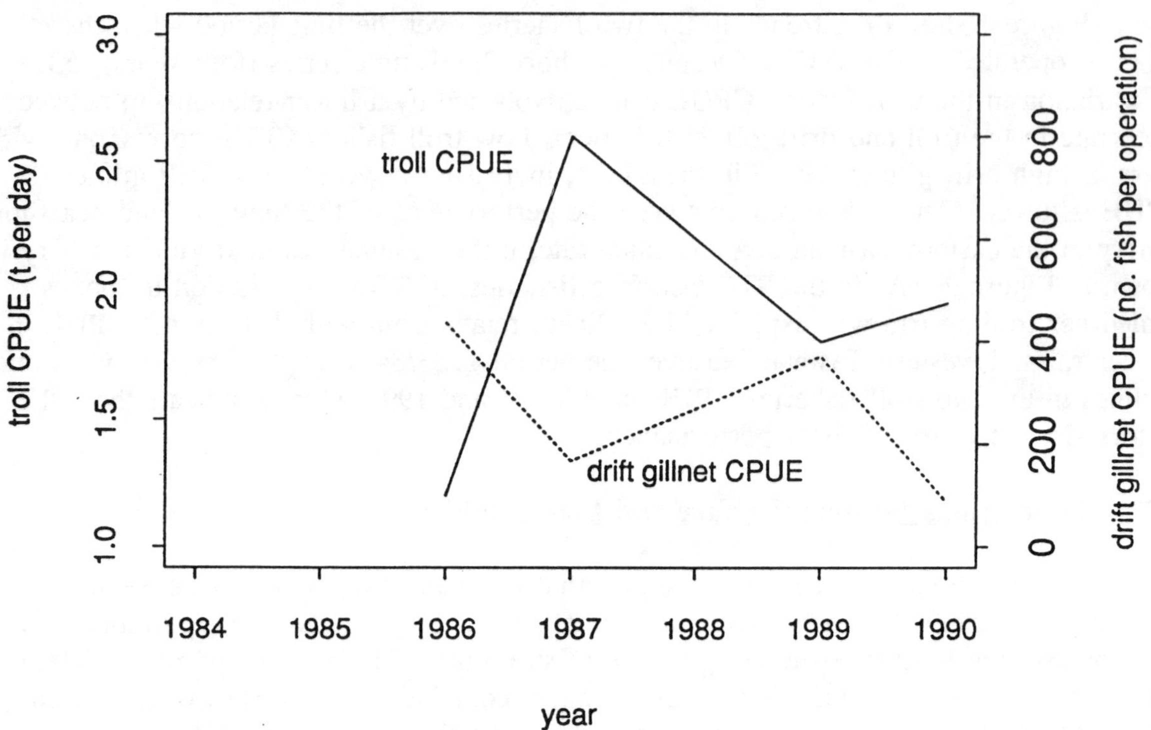

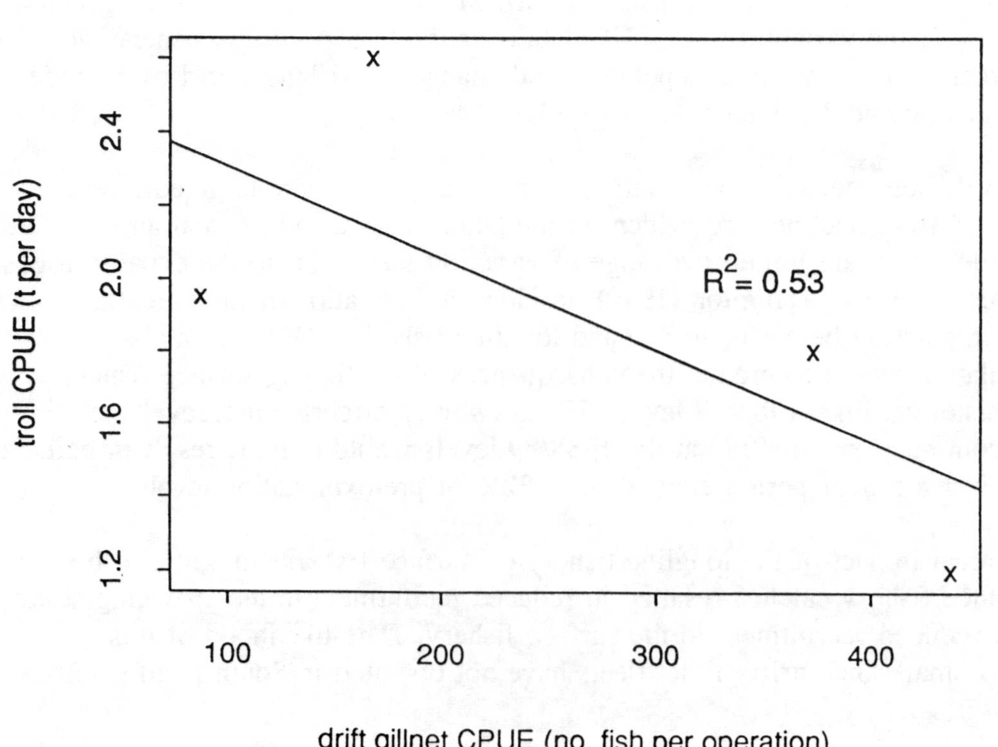

Figure 2.   Comparison of CPUE trends in South Pacific albacore surface fisheries operating in the Tasman Sea, 1983/84-1989/90.

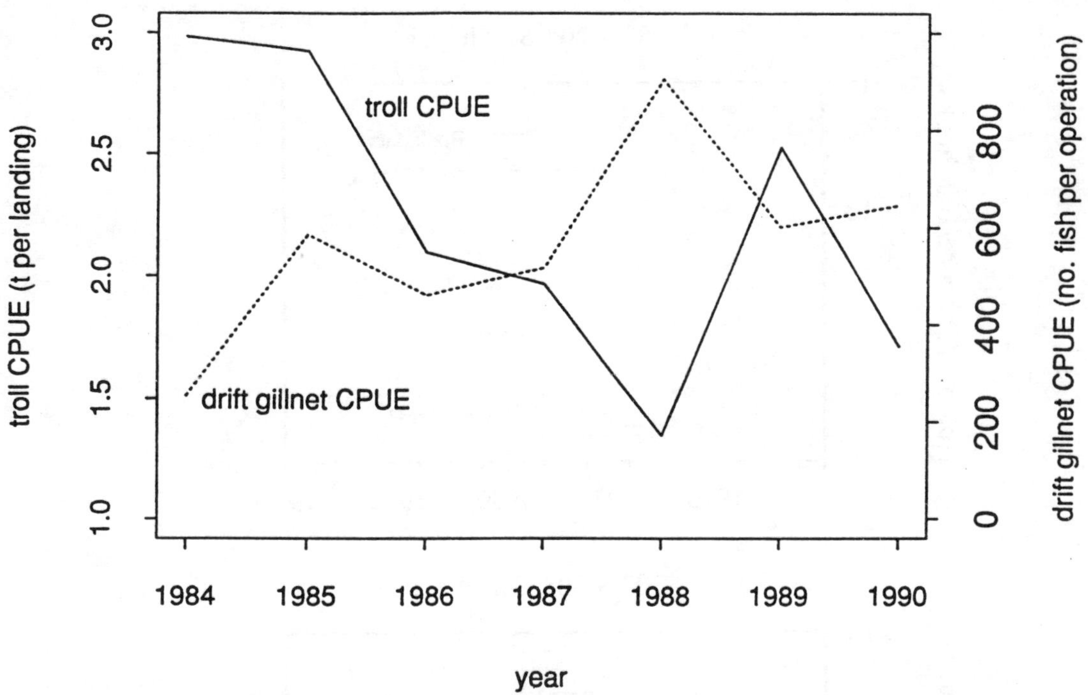

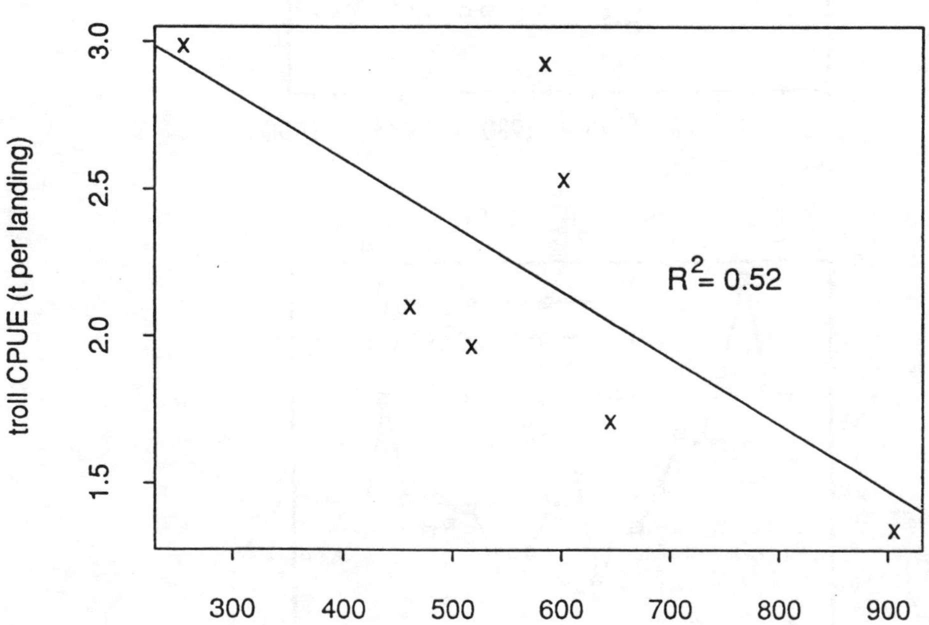

Figure 3.   CPUE trends in the Taiwanese longline fishery for South Pacific albacore stratified by latitude. Reprinted from Anon (1991).

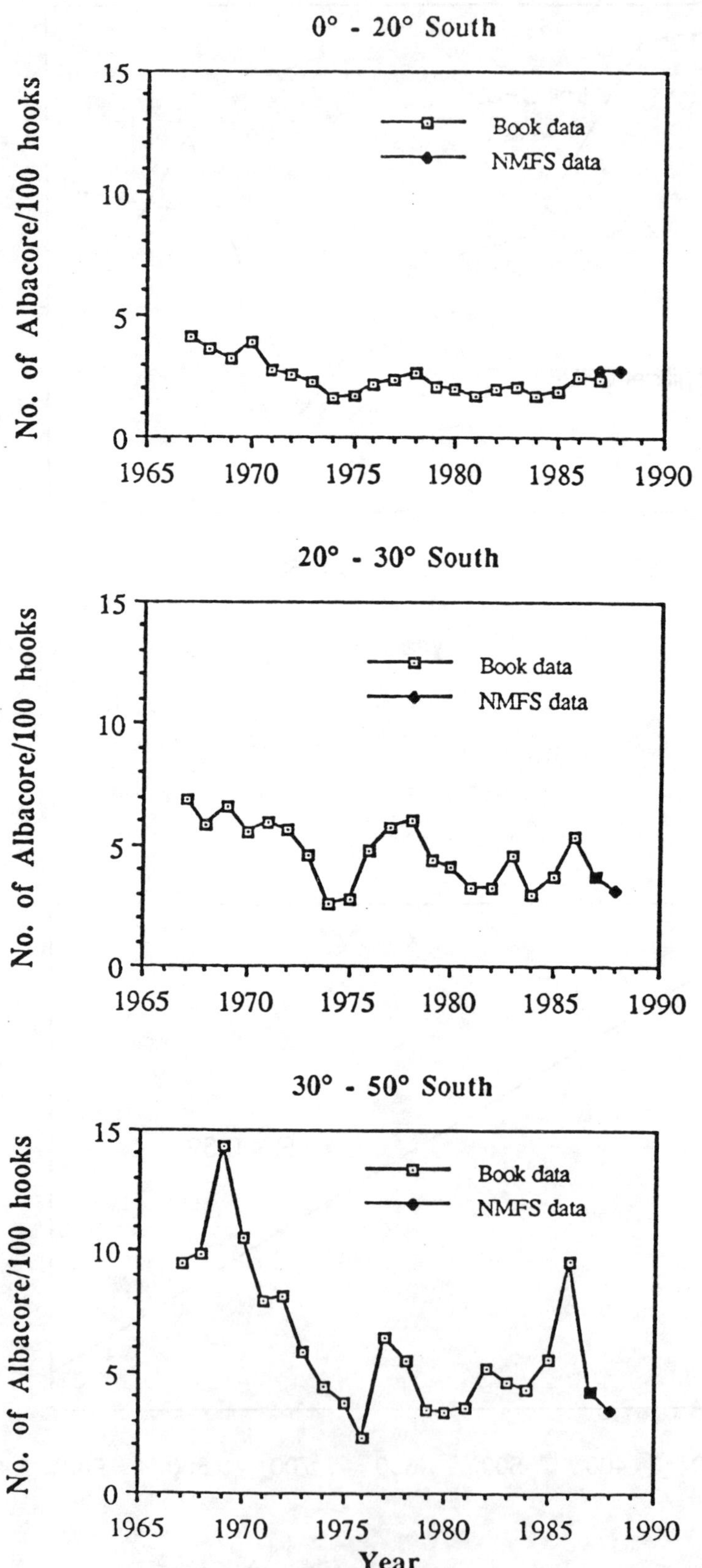

the 1991/92 season and both distant water longline fleets and troll fleets do not appear to be increasing (Richards, 1993).

4.3   <u>Interactions Between Fisheries Operating in Different Areas</u>

Interactions between albacore fisheries may operate over an appreciable distance and with time lags of one or more years. Exceptions include the drift gillnet and troll fisheries which operated in the STCZ and the small amount of longline fishing in the STCZ in March-April. The apparent interaction between the Tasman Sea drift gillnet fleet operating near Australia and the New Zealand troll fishery within a season has been discussed. The geographical scale over which this interaction was likely to operate is discussed by Murray (1990). He used the incidence of recent drift gillnet damage in troll-caught albacore to indicate that the probable geographical scale of interaction between these fisheries was on the order of 900 km. Evidence of interactions over wider areas have not been observed in South Pacific albacore.

In most cases we might expect the magnitude of an interaction to lessen as distance (or time) between two fisheries increases simply through mixing of affected and unaffected portions of the stock. The ability to distinguish an interaction given the variability in availability of a stock or vulnerability to a gear type is also expected to diminish. The possible explanations for CPUE declines, which could not be distinguished from an interaction between fisheries, noted by Wetherall and Yong (1990) highlight the difficulties in confirming an interaction between fisheries operating on different stock components in different areas.

5.   CURRENT ASSESSMENT OF INTERACTIONS

That an interaction existed between drift gillnet and troll fisheries operating in the STCZ and Tasman Sea (see Figures 1 and 2) is supported by the inverse relationship between CPUE in each fishery. Comparisons of CPUE from these fisheries in the Tasman Sea further suggests that interactions between surface fisheries can occur when fisheries are separated by several hundreds of kilometres. The appearance of recent drift gillnet damage in the New Zealand nearshore troll catch when drift gillnet fishing was restricted to the western Tasman Sea adds support to this suggestion. However, the rapid increase in drift gillnet fishing in the mid-1980s was followed by equally rapid fleet reductions and drift gillnet fleets have not operated in the South Pacific Ocean since July 1991.

Although interactions between drift gillnet and troll fisheries have stopped, there is no reason to suspect that the potential for interaction between troll fleets or other surface fisheries (*e.g.*, artisanal, recreational, *etc.*) operating in different areas separated by distances of hundreds of kilometres would not also interact to some degree. The extent to which this might occur would likely depend on distance between them and the scale of the troll fishery operating upstream of the summer migratory pattern. With the exception of the indications of a possible interaction provided by Wetherall and Yong (1990), there is little evidence at present that surface and longline fisheries interact to an appreciable extent.

Interactions between South Pacific albacore fisheries over extensive distances have not been demonstrated and the potential for interactions between industrial albacore fisheries, small scale commercial, recreational or artisanal fisheries have not been examined because of inadequate data on catch and effort in these fisheries.

6.   FUTURE STUDIES

With the cessation of drift gillnet fishing in the South Pacific from June 1991 in accordance with United Nations resolution 44/225, there is a continuing need in studies of fishery interaction to assess the potential interaction effects of the troll fishery on the longline fishery. To date these studies have been constrained by inadequate data on the size composition of the longline catch, longline catch and effort statistics for some fleets, and by low tag returns from all fisheries. There is a continuing need to improve data coverage in longline fisheries, estimate non-reporting of tags in each fishery, and increase the number of fish tagged and released (Labelle and Sharples, 1991).

Study of the relative changes in year classes affected by the 1988/89 surface catches in the longline fishery may also result in an estimate of an upper limit to the potential for surface fishery expansion. The monitoring of the longline fishery size composition in relation to the impacts of the 1988/89 surface catch should therefore continue.

7.   REFERENCES CITED

Anonymous. 1986. Report of the Meeting. First South Pacific Albacore Research Workshop, 4-6 June 1986, Auckland, New Zealand. *S.Pac.Comm.*, 33 p.

Anonymous. 1990. Report of the expert consultation on large-scale pelagic driftnet fishing, 2-6 April 1990, Rome, Italy. *FAO Fish.Rep.*, 434:78 p.

Anonymous. 1991. Report of the Meeting. Third South Pacific Albacore Research Workshop, 9-12 October 1990, Noumea, New Caledonia. *S.Pac.Comm.*, 41 p.

Anonymous. 1993. Current status of the South Pacific albacore research (SPAR) database. Paper presented at the Fifth Meeting of the South Pacific Albacore Research (SPAR) Workshop, 29 March - 4 April 1993, Papeete, Tahiti, French Polynesia. *S.Pac.Comm.*, WP1.

Coan, A.L., and G.M. Rensink. 1991. U.S. troll fishery for albacore in the South Pacific. Paper presented at the Fourth Sout Pacific Albacore Research (SPAR) Workshop, 4-8 November 1991, Taipei, Taiwan. *S.Pac.Comm.*, WP11.

Hampton, J. 1990. Simulations of the South Pacific albacore population: effects of rapid developments in the surface fishery. Paper presented at the Third South Pacific Albacore Research (SPAR) Workshop, 9-12 October 1990. *S.Pac.Comm.*, IP1.

Hampton, J., D.A. Fournier, and J.R. Sibert. 1990. MULTIFAN analysis of South Pacific albacore length-frequency data collected by observers, 1989-1990. Paper presented at the Third South Pacific Albacore Research (SPAR) Workshop, 9-12 October 1990, Noumea, New Caledonia. *S.Pac.Comm.*, WP1.

Hampton, J., T. Murray, and P. Sharples. 1989. South Pacific albacore observer programme 1988-1989. *Tech.Rep.Tuna Billfish Assess.Programme, S.Pac.Comm.*, 22: 22 p.

Jones, J.B. 1991. Movements of albacore tuna (*Thunnus alalunga*) in the South Pacific: evidence from parasites. *Mar.Biol.*, 111:1-9.

Labelle, M. 1991. Estimates of age and growth for South Pacific albacore. Paper presented at the Fourth South Pacific Albacore Reseach (SPAR) Workshop, 4-8 November 1991, Taipei, Taiwan. *S.Pac.Comm.*, WP5.

Labelle, M., and T. Murray. 1992. South Pacific albacore observer programme summary report, 1988-1991. *Tech.Rep.Tuna Billfish Assess.Programme, S.Pac. Comm.*, 26:23 p.

Labelle, M., and P. Sharples. 1991. South Pacific albacore tagging programme summary report 1990-1991. Paper presented at the Fourth South Pacific Albacore Research (SPAR) Workshop, 4-8 November 1991, Taipei, Taiwan. *S.Pac.Comm.*, WP2.

Murray, T. 1990. Review of research and of recent developments in South Pacific albacore fisheries with emphasis on large-scale pelagic driftnet fishing. *FAO Fish.Rep.*, 434:52-77.

Murray, T., and K. Bailey. 1989. Preliminary report on age determination of South Pacific albacore using caudal vertebrae. Paper presented at the Second South Pacific Albacore Research (SPAR) Workshop, 14-16 June 1989, Suva, Fiji. *S.Pac.Comm.*, WP20.

Richards, A. 1993. Historical perspectives and trends of South Pacific albacore fleets. Paper presented at the Fifth South Pacific Albacore Research (SPAR) Workshop, 29 March - 4 April 1993, Papeete, Tahiti, French Polynesia. *S.Pac.Comm.*, WP22.

Sharples, P., K. Bailey, P. Williams, and A. Allan. 1990. Report of observer activity on board JAMARC driftnet vessel R.V. Shin-hoyo Maru fishing for albacore in the South Pacific Ocean. 22 November - 23 December 1989 and 10 February - 3 March 1990. Paper presented at the Third South Pacific Albacore Research (SPAR) Workshop, 9-12 October 1990, Noumea, New Caledonia. *S.Pac.Comm.*, WP3.

Wang, C.-H. 1988. Seasonal changes of the distribution of South Pacific albacore based on Taiwan's tuna longline fisheries, 1971-1985. *Acta Oceanographica Taiwanica*, 20:13-40.

Watanabe, Y. 1990. Catch trends and length frequency of southern albacore caught by Japanese driftnet fishery. Paper presented at the Third South Pacific Albacore Research (SPAR) Workshop, 9-12 October 1990, Noumea, New Caledonia. *S.Pac.Comm.*, WP8.

Wetherall, J.A., and M.Y.Y. Yong. 1989. Use of longline catch rate statistics to monitor the abundance of South Pacific albacore. Paper presented at the Second South Pacific Albacore Research (SPAR) Workshop, 14-16 June 1989, Suva, Fiji. *S.Pac.Comm.*, WP11.

Wetherall, J.A. and M.Y.Y. Yong. 1990. South Pacific albacore longline CPUE monitoring. Paper presented at the Third South Paific Albacore Research (SPAR) Workshop, 9-12 October 1990, Noumea, New Caledonia. *S.Pac.Comm.*, WP9.

A BRIEF ANALYSIS OF FISHERY INTERACTION FOR BIGEYE TUNA IN THE PACIFIC OCEAN

Naozumi Miyabe
National Research Institute of Far Seas Fisheries
Shimizu, Japan

ABSTRACT

Fishery interaction involving bigeye tuna in the Pacific Ocean was briefly analyzed in terms of yield per recruit. Since fishing mortality rate on this species is not known precisely and catch is not known for all the fishery components, several assumptions of the level of fishing mortality and catch by fishery components were made to perform this study. The results indicate that in general, any increase of catch by the surface fishery will lead to a decrease of yield per recruit except when fishing intensity of the longline fishery is very small. There are also indications that at the current level of surface fishing an increase in longline catch will cause a significant increase yield per recruit.

1.   INTRODUCTION

Major fisheries which harvest tropical tunas are longline, purse-seine and baitboat fisheries. The longline fishery is conducted by three countries (Japan, Korea, and Taiwan) and covers the entire Pacific Ocean. Purse-seine and baitboat fisheries primarily occur in the eastern and western sides of the Pacific. In the western Pacific, the Japanese baitboat catch is the largest among the surface fisheries in terms of catch in weight probably followed by the surface fisheries of the Solomon Islands (see Table 10 in Miyabe,1993). The baitboat fishery in the eastern Pacific is relatively small in its magnitude compared to the purse-seine fishery. The major purse-seine fleets operating in the western Pacific include the USA, Japan, Taiwan, Korea, and the Philippines (Table 1). Major purse-seine fishing countries in the eastern Pacific include the USA, Mexico, Ecuador and Venezuela (IATTC,1989).

Bigeye tuna, which is one of the tropical tunas, is the main target species for the longline fishery but in other surface fisheries it forms smaller component of the catch compared to the other tunas (Miyabe,1993). In this sense, the problem of interaction among fisheries might be minimal for this species. There, however, do exist some interactions, although they haven't been previously analyzed. Interactions among fisheries (including ones within the same fishery) can occur among different areas, among different sizes of fish, and among the combination of these. It is considered that interaction on a geographically small scale, such as interaction among different fisheries operating both inside and outside the 200 mile EEZ of some countries, is hard to assess without having detailed information on catch, fishing effort, stock structure, migration and so on.

Keeping these in mind, interactions between two types of fishery, *i.e.*, the longline and surface fisheries which catch larger and smaller sizes of fish, respectively, are examined by means of yield-per-recruit analysis in this study, with the assumption that a single stock of bigeye tuna occurs throughout the Pacific Ocean.

Table 1. Number of fishing vessels operating in the SPC area, 1983-1990.

Baitboat

Country/Fleet	1983	1984	1985	1986	1987	1988	1989	1990
Australia	13	8	1	2	2	1	2	0
Fiji	0	0	7	6	8	8	8	9
Japan	103	94	84	83	77	63	59	61
Kiribati	0	0	0	4	4	6	5	5
New Caledonia	3	0	0	0	0	0	0	0
Solomon Is.	27	31	36	34	34	34	33	33
Tuvalu	1	1	1	1	1	1	1	1
Total	147	134	129	130	126	113	108	109

Purse-seiner

Country/Fleet	1983	1984	1985	1986	1987	1988	1989	1990
Indonesia	0	0	0	3	3	3	3	3
Japan, single	34	41	33	34	32	33	33	32
Japan, group	7	7	7	0	5	7	3	7
Korea	0	0	0	0	16	23	28	38
Mexico	0	2	0	0	0	0	0	0
New Zealand	7	5	5	4	3	4	0	0
Philippines	0	3	5	5	5	9	10	11
Solomon Is., group	1	1	1	1	1	1	1	1
Solomon Is., single	0	0	0	0	1	3	3	3
Soviet Union	0	0	5	7	0	0	0	0
Taiwan	3	6	7	10	13	19	25	32
U.S.A.	39	52	39	0	0	32	36	38
Total	91	117	102	64	79	134	142	165

Data source: SPC(1991).

2. MATERIAL AND METHOD

In order to perform yield-per-recruit analysis, it is necessary that fishing mortality rate (F) by age is given for each fishery in addition to average weight-at-age data. There are, however, no comprehensive analyses made to estimate F. Therefore, F was estimated here utilizing the available size and catch data under several assumptions.

The size data currently available are from the Japanese fisheries and the surface fisheries in the eastern Pacific. Due to this limited availability of data, fisheries for bigeye were categorized into four groups; *i.e.* (a) the longline fishery, (b) the Japanese baitboat fishery in the northwest Pacific, (c) the western tropical surface fishery, and (d) the eastern tropical surface fishery. Length data which represent these groups are taken from the Japanese longline fishery, the Japanese baitboat fishery, the Japanese tropical purse-seine fishery, and the eastern tropical purse-seine fishery, respectively. Data sources and years covered are shown in Table 2. Annual length frequencies were constructed for several recent years for these fisheries and weighted by annual catch in weight to get the average annual length frequency (Figure 1). Then they were aged using the growth equation of Suda and Kume (1967) assuming catches were made during mid-year. This estimated age composition was tabulated in Table 3.

Table 2. Data sources and years covered for length data used in this study.

Fishery group	Length Data	Year	Average weight (kg)	Data source
Longline fishery	Japanese longline	1985-87	51.7	Miyabe (1993)
Japanese baitboat northwest fishery	Japanese baitboat	1987-89	14.7	NRIFSF[*1]
Western tropical surface fishery	Japanese tropical purse seine	1986-88	10.3	NRIFSF
Eastern tropical surface fishery	Eastern tropical purse seine		21.1	
	Nearshore area	1983-86	27.5	IATTC[*2]
	Offshore area	1983-86	18.0	IATTC
	Baja California area	1985-86	19.3	IATTC

[*1] National Research Institute of Far Seas Fisheries.
[*2] Inter-American Tropical Tuna Commission.

Two sets of catch levels were assumed. One set is taken from the available statistics (Set 1), *e.g.*, FAO statistics or other published ones shown in Miyabe (1993). The second set includes potential catches of this species not reported elsewhere or wrongly reported as an other species (Set 2). Set 2 was created to focus on the effect of missing catch on stock assessment of this species. It is known that certain amounts of bigeye were reported as yellowfin in the western surface catch, especially in the purse-seine catch. In the Japanese tropical purse-seine catch there seems to be some unreported catch of age 0 fish despite almost no samples being observed in length frequency data. Our preliminary measurements at canneries on those catches indicated about 10% of small fish classified as yellowfin (age 0, less than 45 cm) were bigeye. Suzuki (1993) showed catch at size of yellowfin caught by the Japanese tropical purse-seine fishery during 1988 and 1989. From this data, the portion of yellowfin catch less than 45 cm was estimated by length-weight relationship to be about 10% in weight. That means approximately 1% (10% of 10%) of the total yellowfin catch may be age 0 bigeye. At the same time this mis-identification of species is highly likely for large-size fish as well. In addition, there must have been unreported catch of bigeye tuna for Indonesian and Philippine domestic fisheries in the Pacific Ocean; these fisheries do not report any bigeye catches regardless of large catches of yellowfin tuna. Taking this information into account, 10% of the reported yellowfin catches for surface fisheries in the SPC area (SPC,1991) was arbitrarily assumed to be bigeye, and additional catches of 1,000 metric tons (mt) each of age 0 and age 1 bigeye were added to the second set of catch levels to represent unknown catches. Although there is no information so far on this matter for the eastern Pacific surface fishery, the highest catch level (8,000 mt) was adopted.

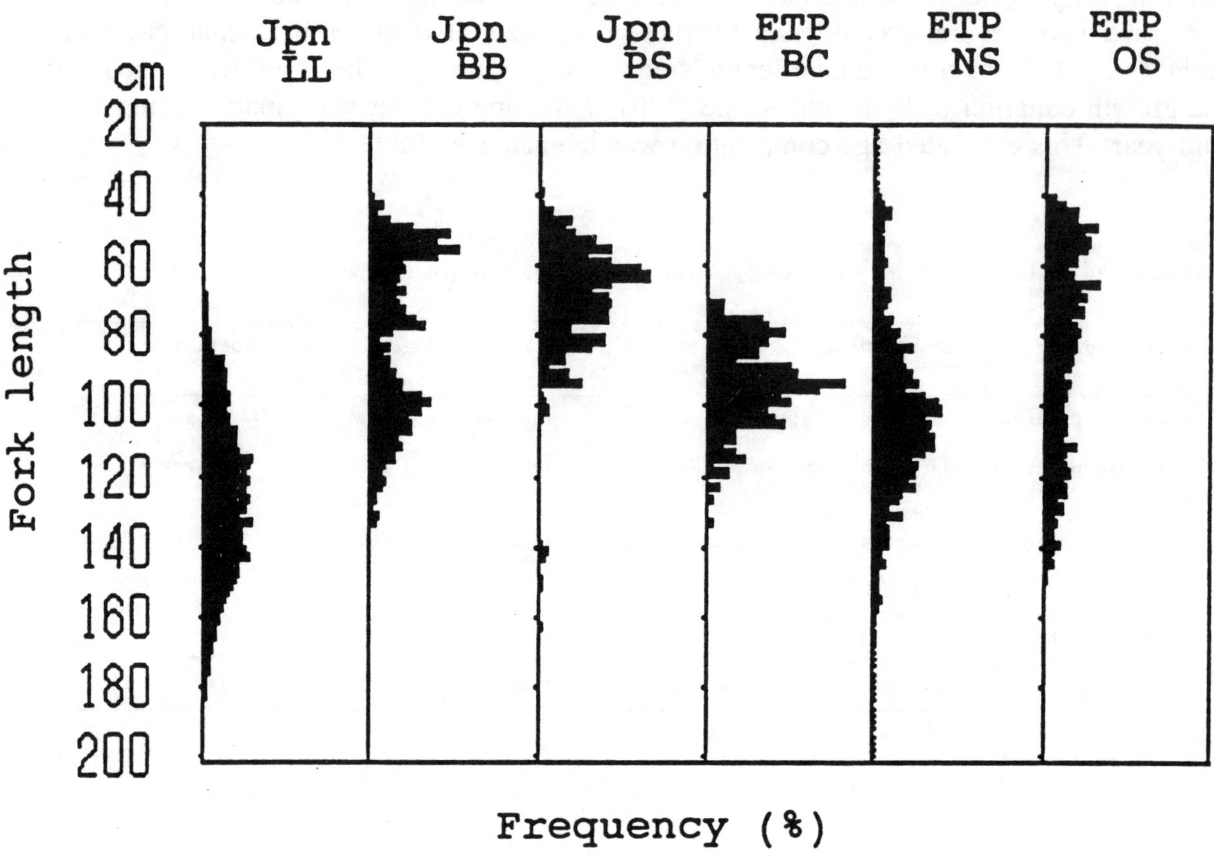

Figure 1.   Average annual length frequency by fishery. From left, Japanese longline, Japanese baitboat, Japanese purse seine, eastern tropical Pacific purse seine in Baja California area (north of 25○N), that in the nearshore area (10○S-10○N, east of 85○W), and that in the offshore area (15○N-20○S, west of 85○W).

Table 3.   Average age composition (%) of catch by fishery. BB, LL and PS stand for baitboat, longline and purse-seine fishery, respectively.

Age	Japan LL	Japan BB	Japan PS	IATTC PS
0	0.0	0.0	0.0	0.0
1	0.2	20.4	14.7	13.5
2	5.4	35.8	69.6	30.3
3	19.6	31.3	11.9	28.9
4	29.6	11.1	0.5	19.3
5	23.5	1.3	2.1	6.5
6	13.7	0.1	0.8	1.4
7	5.1	0.0	0.5	0.1
8+	2.8	0.0	0.0	0.0

These assumed catches were listed in Table 4. The 10% misreporting of yellowfin catch by surface fisheries in the western Pacific is thought to be maximum considering that misreporting tends to be smaller for larger sizes of fish. The catch of bigeye by the USA purse seiners also seems to be smaller than the Japanese catch because the sizes of yellowfin caught by the USA purse-seine fishery in the western Pacific are somewhat larger than that of the Japanese purse-seine fishery perhaps reflecting the difference in mode of operation. The Japanese seiners target mainly log- or FAD-associated schools, whereas the USA fleets set more on free-swimming schools.

Table 4. Assumed catch level by fishery for Pacific bigeye tuna.

Fishery group	Fishery	Catch (MT)
(Set 1) Longline fishery	Japan + Korea + Taiwan	132,700
Japanese baitboat northwest fishery	Japan	2,400
Western tropical surface fishery	Japan + Solomon Is.	2,200
Eastern tropical surface fishery	IATTC area	3,400
Total		140,700
(Set 2) Longline fishery	Japan + Korea + Taiwan	132,700
Japanese baitboat northwest fishery	Japan	2,400
Western tropical surface fishery	Japan + Solomon Is. Potential catch in SPC area	2,200 19,500
Eastern tropical surface fishery	IATTC area	8,000
Other unknown catch	Age 0 Age 1	1,000 1,000
Total		166,800

Catch-at-age in number was calculated by dividing catch in weight by the average weight, which is shown in Table 2, for each fishery. Average weight was obtained from length frequency data applying a length-weight relationship by Morita (1973). For age 0 and age 1 fish, the average weight was calculated assuming they were 0.5 and 1.0 years old, respectively. Catch-at-age for two sets of catch level is shown in Table 5 and Figure 2.

Table 5. Estimated catch-at-age by fishery group.

Set 1

Age	Long-line	Bait-boat	Western Surface	Eastern Surface	Surface Total	Grand Total
0	0	0	0	0	0	0
1	4297	33356	31334	21781	86470	90768
2	139791	58439	148614	48891	255944	395734
3	502894	51083	25377	46502	122962	625856
4	760477	18094	1009	31104	50207	810684
5	604135	2153	4399	10449	17002	621137
6	351435	142	1774	2253	4169	355604
7	130600	0	1086	163	1249	131849
8+	73101	0	0	0	0	73101

Set 2

Age	Long-line	Bait-boat	West. Surface	East. Surface	Western un-reported	Surface Total	Grand Total
0	0	0	0	0	2857143	2857143	2857143
1	4297	33356	309063	51249	526316	919983	924281
2	139791	58439	1465871	115039	0	1639348	1779139
3	502894	51083	250308	109417	0	410808	913703
4	760477	18094	9955	73186	0	101234	861711
5	604135	2153	43394	24587	0	70134	674269
6	351435	142	17497	5302	0	22941	374375
7	130600	0	10713	384	0	11097	141697
8+	73101	0	0	0	0	0	73101

A natural mortality rate of 0.4 was used since this value is close to the past estimate by Suda and Kume (1967) and seems a reasonable value considering the growth and life span of this species.

The next step is to estimate F at age, but this is difficult to do without making some assumptions. Here, it is assumed that age 5 and older fish are subjected to constant F. The constant decreasing trend in catch in log scale (Figure 2) appears to support this assumption and F for these ages appears to be the highest among all ages. Linear regression for ages 4 to 7 suggests the total mortality rate (Z) to be 0.6-0.8. The F values of 0.4 and 0.8 for age 5 and older (Ft) are arbitrarily selected, although Ft of 0.8 is possibly too high. The F values for longline and surface fisheries were obtained by applying the ratio of catch in number at age. The exploitation patterns for two sets of data and two Ft values are listed in Table 6.

Then a yield-per-recruit analysis was performed on the two fisheries using the programme written by Nagai (1990).

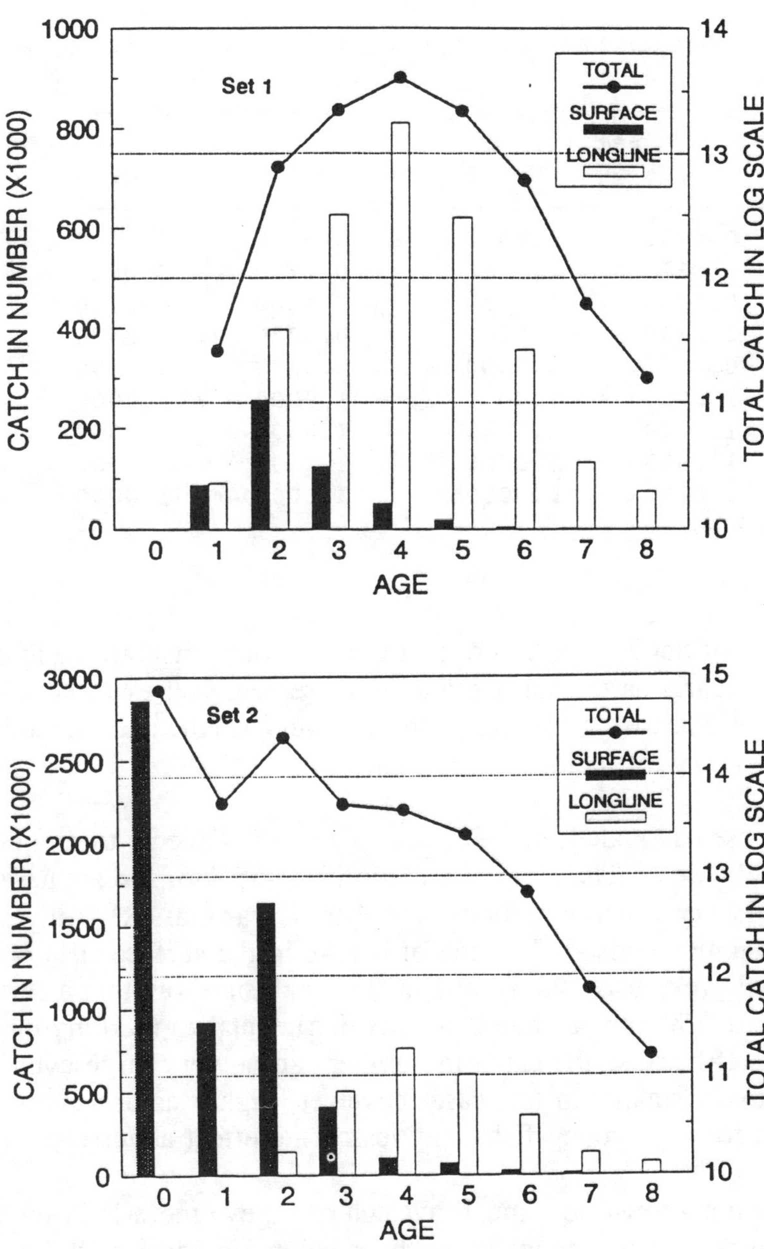

Fig. 2. Estimated catch at age in number for the longline and the surface fisheries, and total catch at age on a log scale.

3. RESULTS AND DISCUSSION

Trajectories of yield per recruit in a relative scale are shown in Figure 3. Thick lines indicate 25%, 50% and 75% levels of unexploited stock. Any increase of surface catch is harmful in terms of yield per recruit in all cases. For Set 1 data, yield per recruit does not change much within the range (0.1 - 2.0) of fishing intensity even if the fishing intensity of surface fisheries is increased. In contrast, intensifying longline fishing leads to the linear increase of yield per recruit especially when Ft is 0.4. It appears that the stock level of Set 1 data is at more than 50% of the unexploited level.

Table 6. Exploitation pattern for Pacific bigeye tuna.

	Set 1		Set 2	
Age	F=0.4	F=0.8	F=0.4	F=0.8
0	0.0000	0.0000	0.2750	0.1500
1	0.0175	0.0112	0.1250	0.0810
2	0.1000	0.0750	0.4000	0.2500
3	0.2750	0.2000	0.3750	0.2500
4	0.6250	0.5000	0.6250	0.5000
5	1.0000	1.0000	1.0000	1.0000
6	1.0000	1.0000	1.0000	1.0000
7	1.0000	1.0000	1.0000	1.0000
8+	1.0000	1.0000	1.0000	1.0000

In the case of Set 2 data, where many more young fish are caught, yield per recruit decreases clearly as the surface fishing increases. When Ft is 0.8, the gain of yield per recruit by increasing longline fishing is very small, and the stock is at a very low level (close to 25% of its initial condition).

Among these cases analyzed, Set 2 with $Ft = 0.4$ seems to model the current situation reasonably well. The $Ft = 0.8$ is too high (55% of the annual exploitation rate) for the longline fishery which mainly catches 4 to 5 year classes. Although it may not be appropriate to assume a constant bycatch of bigeye in the surface fishery in western Pacific as is stated previously, the results of this analysis show that a further increase of catch of very small fish, such as age 0 fish, is detrimental to the longline fishery as well as to the stock itself because the catch in number can be very large even though the catch in weight is relatively small. In this case, however, higher natural mortality, which seems reasonable for very young fish, may lessen the effect adversely.

Since it is not known how much bycatch of bigeye there is in the surface fishery, especially for southeast Asian countries such as Indonesia and the Philippines, and the number of purse seiners has been increasing (Table 1) particularly in most recent years, it is strongly recommended that precise catch data on this species in those fisheries be acquired as soon as possible.

4.   FUTURE STUDY

To adequately address the subject of interaction problems and to improve stock assessment of bigeye tuna, the following studies should be encouraged for future study:

a.   Study of stock identification; this should include genetic analysis.

b.   Collection of complete catch and effort statistics, especially statistics from surface fisheries which are still developing and from which no data are currently available. These statistics should correct the problems of underreporting and misidentification of species.

c.   Establishment of a growth equation which is applicable to the whole Pacific Ocean.

d.   Investigation of the movement (migration) of fish.

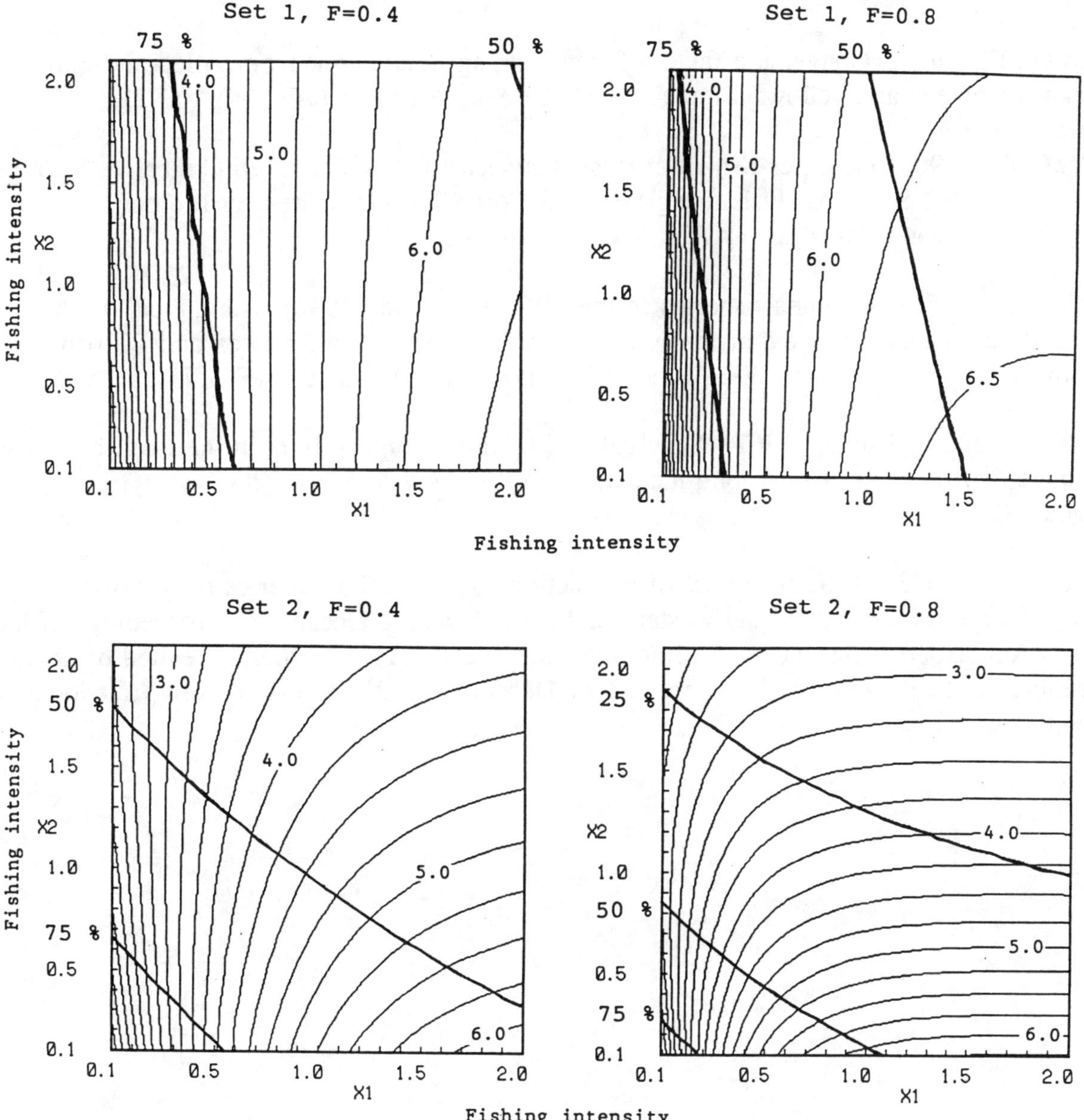

Fig. 3. Results of two-fishery yield per recruit analysis on bigeye tuna in the Pacific Ocean. X1 and X2 stand for long-line and surface fishery, respectively. Thick lines indicate stock levels of 25 %, 50 % and 75 % of unexploited stock. Fine lines show yield in kg per 10^6 fish (recruits).

5.   REFERENCES CITED

IATTC. 1989. Annual report of the Inter-American Tropical Tuna Commission, 1988. *Annu.Rep.I-ATTC*, (1978):288 p.

Miyabe, N. 1993. A review of the biology and fisheries for bigeye tuna, *Thunnus obesus*, in the Pacific Ocean. *In* Proceedings of the First FAO Expert Consultation on Interactions of Pacific Tuna Fisheries, edited by R.S. Shomura, J. Majkowski, and S. Langi, 3-11 December 1991, New Caledonia. [See this document.]

Morita, Y. 1973. Conversion factors for estimating live weight from gilled-and-gutted weight of bigeye and yellowfin tunas. *Bull.Far Seas Fish.Res.Lab.*, (9):109-21.

Nagai, T. 1990. Yield per recruit analysis between two fisheries. *In* Collective Volume of Programs on Stock Analysis using Personal Computer (II), edited by Biological Ecology Division. *Nat.Res.Inst.Fish.Sci.*, pp. 204-11.

SPC. 1991. Status of tuna fisheries in the SPC area during 1990, with annual catches since 1952. Paper presented at the Fourth Standing Committee on Tuna and Billfish, South Pacific Commission, 17-21 June, 1991, Port Vila, Vanuatu, WF/3:75 p.

Suda, A., and S. Kume. 1967. Survival and recruit of bigeye tuna in the Pacific Ocean, estimated by the data of tuna longline catch. *Rep.Nankai Reg.Fish.Res.Lab.*, (25):91-104.

Suzuki, Z., 1993. A brief review of interactions between the fisheries on yellowfin tuna, *Thunnus albacares*, in the western and central Pacific Ocean. *In* Proceedings of the First FAO Expert Consultation on Interactions of Pacific Tuna Fisheries, edited by R.S. Shomura, J. Majkowski, and S. Langi, 3-11 December 1991, Noumea, New Caledonia. [See this document.]

INTERACTIONS AMONG FISHERIES FOR NORTHERN BLUEFIN TUNA, *THUNNUS THYNNUS*, IN THE PACIFIC OCEAN

William H. Bayliff
Inter-American Tropical Tuna Commission
La Jolla, California, USA

ABSTRACT

The most important interactions which take place among fisheries for northern bluefin tuna are those between the fisheries of the western and the eastern Pacific Oceans and those among the various fisheries of the western Pacific Ocean. The yields per recruit for both the western and eastern Pacific Ocean would increase if the age at entry for all fisheries were increased to about 1 year. The yield per recruit for the western Pacific Ocean would increase even more if the age at entry for all fisheries were increased to about 2 1/2 years, but this would nearly eliminate the eastern Pacific fishery. Only limited information is available on the interactions among the fisheries of the western Pacific.

1. INTRODUCTION

Parts of this report are similar to parts of a section entitled NORTHERN BLUEFIN in Bayliff (1993b).

2. INTERACTION ISSUES

There are two basic types of interactions among the fisheries which take northern bluefin tuna, interactions between ocean areas (eastern, central, and western Pacific) and interactions within ocean areas.

2.1 <u>Interactions Between Ocean Areas</u>

Northern bluefin are recruited to the troll fishery for small fish in the western Pacific Ocean (WPO). Some of them remain in the WPO and others, after exposure to the troll fishery and other fisheries in the WPO, migrate to the eastern Pacific Ocean (EPO). In the EPO they are exploited by the purse-seine fishery and, to a much lesser extent, by other types of commercial gear and by sport gear. After a sojourn in the EPO, the fish which are not caught and do not succumb to natural mortality return to the WPO, where they are exposed to further exploitation. Accordingly, there is important interaction between the fisheries of the WPO and the EPO.

It is known that bluefin have been caught by the gillnet and baitboat fisheries of the central Pacific Ocean (CPO) (Bayliff *et al.*, 1991: Table 6). The data presented by Bayliff (1993c: Table 9) include catches by Japanese vessels in the EPO (longline) and the CPO (longline, gillnet, and baitboat), as well as the WPO, as data in which these catches are stratified by area are not available. Bayliff (1993c) used unstratified data to obtain a rough estimate of 1,000 bluefin caught by the Japanese squid gillnet fishery in the north Pacific Ocean during 1990. Unless this estimate is greatly in error, it appears that the squid gillnet fishery has had little effect on the fisheries of the EPO and the WPO.

Substantial amounts of bluefin are caught by longlining (Bayliff, 1993c: Figure 2 and Table 9). Since longline-caught fish are usually larger than fish caught by other gears, the fisheries by other gears presumably have a considerable effect on the longline fishery, but the longline fishery has a much lesser effect on the other fisheries. The longline fishery, if it reduces the recruitment, could reduce the catches by the other fisheries, but there have been no attempts to determine whether this is the case.

2.2 Interactions Within the Eastern Pacific Ocean

Northern bluefin are caught almost exclusively by purse seining in the EPO, so the principal interaction is that between Mexican- and U.S.-flag vessels. The fish usually appear first off Mexico in about May or June, and are taken in waters off both Mexico and the United States during July, August, September, and October (Bayliff, 1993c: Figure 6).

2.3 Interactions Within the Western Pacific Ocean

Northern bluefin are taken by trolling vessels, traps, purse seines, baitboats, gillnets, handlines, and longlines in the WPO (Bayliff, 1993c: Table 9), and the activities of each type of gear affect the catches of those which take fish of the same size or larger fish.

2.4 Interactions Within the Central Pacific Ocean

Northern bluefin have been taken by longlines, gillnets, and baitboats in the CPO. The gillnets and baitboats take smaller fish than do the longlines.

3. DATA CURRENTLY AVAILABLE WHICH ARE RELEVANT TO STUDIES OF INTERACTIONS

Data on catches, standardized fishing effort, growth of the fish, fishing and natural mortality, age composition of the fish in the catches, and their movements and migrations are needed for studying the interactions of the fisheries which exploit northern bluefin. Information on these is given by Bayliff (1993c).

Good data for the catches by the surface fishery of the EPO are available (Bayliff, 1993c: Tables 7 and 8). Fairly good data on the annual catches by the surface fisheries of the WPO and the CPO are available (Bayliff, 1993c: Tables 8 and 9), although the catches of small bluefin are estimated by proration from the reported catches of *meji* (small bluefin, bigeye, and yellowfin). Also, the data are not stratified by area, month, or size groups.

Bayliff (1993c: Figure 8) gives estimates of the purse-seine effort directed toward bluefin and the catches per unit of effort of bluefin in the EPO during the 1960-1991 period. Because there were insufficient data to compare the catch rates of vessels of different sizes, he made no attempt to standardize the data by size classes of vessels. Except for data on the numbers of Japanese purse-seine vessels and traps (Skillman and Shingu, 1980), no information is available on fishing effort for northern bluefin for the WPO.

The growth of northern bluefin for the first few years of life has been estimated from tagging data (Bayliff *et al.*, 1991: Table 5) and from length-frequency data (Yokota *et al.* 1961: 217; Bayliff, 1993a). The length-frequency data show seasonal variation in the growth, but the tagging data do not, probably because the numbers of returns of tagged fish are insufficient for this purpose. The growth of bluefin has also been estimated from hard-part data (Bayliff *et al.*, 1991: Table 3). The estimates of growth from hard-part data for fish up to about 150 cm in length agree fairly well with those from tagging and length-frequency data, but there is no way to verify the estimates for the larger fish.

No estimates of the rates of fishing mortality of northern bluefin in the Pacific Ocean have been made. The natural mortality has been estimated only by the method of Pauly (1980), in which estimates of K and L_∞ in the von Bertalanffy growth equation and data for the average temperature at which bluefin occur are compared with corresponding data for other stocks for which estimates of the natural mortality are available. The estimate derived by this method, 0.276 on an annual basis (Bayliff, 1993c), is fairly close to those made by other methods for northern bluefin in the Atlantic Ocean (Clay, 1991) and for southern bluefin (Hearn, 1991).

Estimates of the age compositions of northern bluefin in the catches of both the EPO and the WPO have been made (Bayliff, 1993c: Tables 11 and 12). The estimates are probably reasonably accurate for the younger fish (0 to about 3 or 4 years old), but they may be inaccurate for the older fish. (In this report fish in their first year of life are referred to as 0-year olds, age-0 fish, or fish 0 years of age, and so on.)

The principal features of the movements and migrations of northern bluefin are known, but the information is not sufficiently quantitative for many purposes. It has been established (Bayliff *et al.*, 1991; Bayliff, 1993a) that the proportion of young fish which migrate from the WPO to the EPO varies considerably from year to year and that the timing of this migration also varies from year to year. Information on the timing of the migration from the EPO to the WPO is sparse.

4. CURRENT ASSESSMENTS OF INTERACTIONS

4.1 Interactions Between the Western and Eastern Pacific

4.1.1 Migration between the two ocean areas

There has been a decline in the catches of northern bluefin in the EPO in recent years. This decline could be due to (1) a decrease in recruitment, (2) a decrease in the overall abundance of fish greater than about 60 cm in length caused by heavy exploitation of age-0 fish in the WPO, (3) reduced fishing effort in the EPO, (4) a decrease in vulnerability to capture of the fish which have migrated to the EPO, and/or (5) a decrease in the availability of bluefin in the EPO (*i.e.* a decrease in the proportion of the population which has migrated to the EPO or a shorter average sojourn in the EPO of the fish which have made that migration).

In regard to Points 1 and 2 above, data for the age composition of the catch of bluefin during 1966-1986 by Japanese vessels (Bayliff, 1993c: Table 12) indicate that there has not been a decline in the catches of age-0 fish, which seems to rule out the first possibility, nor an increase in the proportion of age-0 to older fish, which seems to rule out the second possibility.

Proceeding now to Point 3, the numbers of smaller purse seiners, which previous to the late 1970s had been responsible for most of the bluefin catches in the EPO, declined during the late 1970s and the 1980s. Bayliff (1993c: Section 11.1.3) concluded that reduced effort is partly, but not entirely, responsible for the reduced catches in the EPO.

In regard to Point 4, the distribution of bluefin in the EPO seems to have changed during this century. Prior to 1930 they were caught only off California, although they probably occurred off Baja California as well. During the 1930-1947 period they were caught off both California and Baja California, but greater catches were made off California in most years. From 1948 to the present most of the catch has been made off Baja California. Fishermen based in California seem to direct more of their effort toward bluefin than do those based in Mexico, so the shift in distribution may have decreased the vulnerability of bluefin to capture. This shift took place well before the 1980s, however, so it does not appear that it is the cause of the poor catches during that decade.

The tagging data (Bayliff *et al.*, 1991) and age-composition data (Bayliff, 1993a) provide some useful information in regard to Point 5. It can be seen in Table 1 that only the 1981 year class contributed significant numbers of returns to the EPO fishery during Year 1. It can also be seen that for Year 2 nearly half the returns for the 1979 year class and more than half of those for the 1983 and 1984 year classes were from fish caught in the EPO. This information suggests that the proportion of age-1 fish which migrated to the EPO was greatest for the 1981 year class and that the proportions of age-2 fish which migrated to the EPO were greatest for the 1983 and 1984 year classes, intermediate for the 1979 year class, and least for the 1980, 1981, 1982, and 1985 year classes.

A large proportion of the catch of bluefin in the EPO in 1982 consisted of age-1 fish (Bayliff, 1993c: Table 11). This is consistent with the evidence from tagging (Table 1) that a large proportion of the fish of the 1981 year class appeared in the EPO as age-1 fish in 1982. The catch of northern bluefin in the EPO was poor in 1983, however, perhaps because the fish of the 1981 year class experienced heavy mortalities in the EPO in 1982 or mostly began their return trip to the WPO before the start of the 1983 season.

The greatest catches of northern bluefin in the EPO in recent years were those of 1985 and 1986 (Bayliff, 1993c: Table 7), and the catches in those years consisted mostly of age-2 fish (Bayliff, 1993c: Table 11), *i.e.* 1983-year-class fish in 1985 and 1984-year-class fish in 1986. This is consistent with the evidence from tagging (Table 1) that large proportions of the fish of the 1983 and 1984 year classes appeared in the EPO as age-2 fish in 1985 and 1986.

If it were certain that the fish which were tagged in the WPO were selected randomly it would be concluded that greater proportions of age-2 fish of the 1983 and 1984 year classes migrated to the EPO, and that this resulted in greater catches of northern bluefin in the EPO in 1985 and 1986. It is possible, however, that there are separate non-migrant and migrant subpopulations, and that greater proportions of the migrant subpopulation were selected for tagging during the first year of life of the 1979, 1981, 1983, and 1984 year classes. Thus the relatively high proportion of EPO returns for the 1979 year class, even though the catch in the EPO in 1981 was poor, might be the result of heavy concentration of tagging effort on a

TABLE 1. Releases and recaptures of bluefin tagged off Japan during 1980-1988 (from Bayliff *et al.*, 1991). The abbreviations are as follows: W, western Pacific; C, central Pacific (between 160°E and 130°W); E, eastern Pacific; IATTC, Inter-American Tropical Tuna Commission; FSFRL, Far Seas Fisheries Research Laboratory.

Release			Recapture																	
			Year 0			Year 1			Year 2			Year 3			Year 4			Total		
Year class	Organisation	Number	W	C	E	W	C	E	W	C	E	W	C	E	W	C	E	W	C	E
1979	IATTC	739	-	-	-	157	0	0	33	0	24	9	1	0	3	0	0	202	1	24
1980	IATTC	106	10	0	0	1	0	0	1	0	0	0	0	0	0	0	0	12	0	0
	FSFRL	802	64	0	0	46	0	0	8	0	4	2	0	0	0	0	0	120	0	4
1981	IATTC	3,297	264	0	0	48	2	24	5	0	1	0	0	0	0	0	0	317	2	25
	FSFRL	1,653	127	0	0	67	3	21	17	0	3	2	0	0	0	0	0	213	3	24
1982	IATTC	237	24	0	0	5	0	0	0	0	0	0	0	0	0	0	0	29	0	0
	FSFRL	614	2	0	0	25	0	0	3	0	1	2	0	1	0	0	0	32	0	2
1983	FSFRL	788	8	0	0	111	0	1	9	0	19	1	0	0	0	0	0	129	0	20
1984	FSFRL	1,944	109	0	0	54	0	2	3	0	26	0	0	0	0	0	0	166	0	28
1985	FSFRL	993	1	0	0	84	0	0	10	0	4	1	0	0	-	-	-	96	0	4
1986	FSFRL	863	45	0	0	37	0	0	0	0	1	-	-	-	-	-	-	82	0	1
1987	FSFRL	729	35	0	0	10	0	1	-	-	-	-	-	-	-	-	-	45	0	1
1988	FSFRL	588	14	0	0	-	-	-	-	-	-	-	-	-	-	-	-	14	0	0
Total		13,353	703	0	0	645	5	49	89	0	83	17	1	1	3	0	0	1,457	6	133

253

relatively small subpopulation of migrants. This possibility can be evaluated by examining the data in Table 2 for both age-1 and age-2 fish. For the age-1 fish it appears that the proportions of migrants were high for the 1981 year class and low for the other year classes, regardless of the areas or months of release of the fish. For the age-2 fish it can be seen that the fish of the 1980-1982 and 1985 year classes released during December and January tended to be non-migrants and those of the 1983 and 1984 year classes released during December and January tended to be migrants. Fish of all year classes released during August-November tended to be non-migrants, but there are only 10 returns of these from the 1983 and 1984 year classes, and all of these fish were released at Shimane in the Sea of Japan, so these fish would seem to be less likely than any others, because of the physical barriers, to migrate to the EPO. These data do not offer much support for the subpopulation hypothesis. They indicate that for the age-1 fish the tendency to migrate to the EPO was strong for the 1981 year class and much weaker for all the others, and that for the age-2 fish the tendency to migrate was strongest for the fish of the 1983 and 1984 year classes, intermediate for those of the 1979 year class, and weakest for those of the 1980-1982 and 1985 year classes.

Age-composition data (Table 3) provide further information pertinent to Point 5. Correlation coefficients were calculated for 12 pairs of data (Table 4, upper panel) from Table 3. Five of the 12 tests were significant at the 5-percent level. The numbers of age-2 fish in the EPO and the WPO are negatively correlated (Test 6), indicating that the poor catches in the EPO could be due at least partly to less-than-normal proportions of the total population migrating from the WPO to the EPO. The catches of age-1 and -2 fish in the EPO are positively correlated (Test 7), indicating that the catch of age-2 fish in the EPO can be predicted, albeit poorly, from the catch of age-1 fish in the EPO one year previously. The catches of age-0 and -1 fish in the WPO are highly correlated (Test 8), indicating that the catch of age-1 fish in the WPO can be predicted from the catch of age-0 fish in the WPO one year previously.

Test 11 gave an r value which was significant at the 1-percent level, which is not surprising in view of the fact that an even higher r value was obtained from Test 8. An r value which was significant at the 5-percent level was obtained for Test 12; this, also, is not surprising in view of the fact that a nearly-significant value was obtained from Test 2.

The coefficient of correlation for the catches of age-0 fish in the WPO and the catches of age-2 fish in the EPO two years later is 0.442 (Table 4, Test 2). Although this relationship is not significant at the 5-percent level, it may indicate that the catch of age-2 fish in the EPO is related to recruitment two years previously, assuming that the catch of age-0 fish in the WPO is a valid index of recruitment. Since the catch of age-2 fish in the EPO appears to be related to the recruitment two years previously (Test 2) and the catch of age-2 fish in the WPO in the same year (Test 6), a multiple correlation coefficient was calculated (Test 13). The resulting coefficient of multiple determination was highly significant, indicating that 55.6 percent (0.746^2 x 100) of the variation of the catches of age-2 fish in the EPO is explained by (1) a positive relationship to recruitment in the WPO and (2) a negative relationship to the catch of age-2 fish in the WPO. Since the catches of age-2 fish make up the majority of the catch by weight in the EPO in most years, recruitment two years previously and emigration from the WPO appear to have major influences on the total catches in the EPO.

TABLE 2. Data for tagged bluefin released in the western Pacific and recaptured during Year 1 (upper panel) and Year 2 (lower panel) in the western (W) and eastern (E) Pacific (from Bayliff *et al.*, 1991).

Year classes	Areas released	Month of release																								Total	
		7		8		9		10		11		12		1		2		3		4		5		6			
		W	E	W	E	W	E	W	E	W	E	W	E	W	E	W	E	W	E	W	E	W	E	W	E	W	E
1979-1980	Kochi	2	1	18	0																					20	1
1982-1987	Shizuoka	1	0	2	0	7	0	3	0																	13	0
	Nagasaki									20	0	108	1	61	1	20	0									209	2
	Kagoshima											9	1	45	0	65	0	67	0	8	0	3	0			197	1
	Hokkaido							2	0															6	0	8	0
	Total	3	1	20	0	7	0	5	0	20	0	117	2	106	1	85	0	67	0	8	0	3	0	6	0	447	4
1981	Kochi	1	0	40	24	7	0																			48	24
	Shizuoka			4	8	2	0																			6	8
	Nagasaki									19	6	38	7													57	13
	Hokkaido									4	0															4	0
	Total	1	0	44	32	9	0			23	6	38	7													115	45
1979	Kagoshima													10	10			17	12	6	2					33	24
1980-1982	Kochi			6	1																					6	1
1985	Shizuoka			2	1																					2	1
	Shimane									1	1															1	1
	Nagasaki									10	2	13	5	2	0											25	7
	Kagoshima																	6	3							6	3
	Toyama Bay									1	0															1	0
	Hokkaido							3	0																	3	0
	Total			8	2			3	0	12	3	13	5	2	0			6	3							44	13
1983-1984	Shimane									8	2															8	2
	Nagasaki											1	16	2	15											3	31
	Kagoshima											1	11													1	11
	Hokkaido																							0	1	0	1
	Total									8	2	2	27	2	15									0	1	12	45

TABLE 3. Estimated numbers of bluefin, in thousands, caught in the western and eastern Pacific Oceans (after Bayliff (1993a).

| | Western Pacific Ocean | | | | | Eastern Pacific Ocean | | | |
| | 0 | 1 | 2 | 3 | 4 | 1 | 2 | 3 | 4 |
Year class									
1958	-	-	-	-	-	12	66	6	1
1959	-	-	-	-	-	58	347	9	5
1960	-	-	-	-	-	524	608	80	4
1961	-	-	-	-	-	829	521	29	1
1962	-	-	-	-	30	526	256	16	0
1963	-	-	-	62	8	972	305	35	1
1964	-	-	91	5	17	327	715	34	0
1965	-	266	3	52	1	622	308	5	1
1966	1270	461	1	13	17	97	155	32	1
1967	3607	964	78	23	24	416	311	35	0
1968	2300	371	48	14	18	290	258	7	-
1969	2970	378	2	18	16	14	711	-	0
1970	1938	443	15	4	3	467	·	2	0
1971	3316	682	20	12	29	-	589	1	>0
1972	498	124	28	115	26	609	440	2	24
1973	4875	1403	46	61	95	51	722	172	21
1974	3953	676	96	44	10	260	150	21	8
1975	1277	222	61	14	28	488	250	7	>0
1976	1784	698	151	38	7	55	81	>0	0
1977	2542	478	98	76	84	508	539	1	>0
1978	5091	1452	119	584	99	78	284	11	1
1979	2088	611	180	64	56	48	70	2	>0
1980	2810	605	200	54	15	4	120	6	1
1981	1975	785	139	21	75	249	62	6	0
1982	665	213	44	86	30	12	44	>0	>0
1983	1362	421	49	123	-	33	315	5	>0
1984	2417	757	61	-	-	76	388	6	>0
1985	2046	760	-	-	-	22	74	21	>0
1986	1470	-	-	-	-	7	45	6	1

In general, the results of the analysis of the catch-at-age data tend to support the conclusion from the studies of the tagging data that variations in the proportions of fish which migrate from the WPO to the EPO are at least partly responsible for the variations in the catches in the EPO.

4.1.2 Yield-per-recruit studies

Yield-per-recruit analyses (Bayliff, 1993c) indicate that increasing the age at entry into the fishery to about 2 1/2 years (about 90-100 cm) would maximize the overall yield per recruit of northern bluefin which migrate to the EPO. Information on the yields per recruit to the fisheries of the EPO and the WPO is given in Figure 1. Because the values of natural and

TABLE 4. Correlations for various combinations of catches of bluefin. WPO and EPO stand for western and eastern Pacific Ocean, respectively (from Bayliff, 1993b).

	Correlation	Degrees of freedom	r
1.	WPO, age 0, *versus* EPO, age 1	18	-0.157
2.	WPO, age 0, *versus* EPO, age 2	18	0.442
3.	WPO, age 1, *versus* EPO, age 1	18	-0.390
4.	WPO, age 1, *versus* EPO, age 2	18	0.175
5.	WPO, age 2, *versus* EPO, age 1	18	-0.322
6.	WPO, age 2, *versus* EPO, age 2	18	-0.460*
7.	EPO, age 1, *versus* EPO, age 2	35	0.458*
8.	WPO, age 0, *versus* WPO, age 1	18	0.843**
9.	WPO, age 0, *versus* WPO, age 2	17	0.174
10.	WPO, age 1, *versus* WPO, age 2	18	0.344
11.	WPO, age 0, *versus* WPO + EPO, age 1	17	0.757**
12.	WPO, age 0, *versus* WPO + EPO, age 2	16	0.511*
13.	EPO, age 2, *versus* WPO, age 0, and WPO, age 2	15	0.746**

* significant at the 5-percent level
** significant at the 1-percent level

fishing mortality, especially the latter, are little more than guesses, the results of these analyses should not be taken literally. Nevertheless, it appears that the yield per recruit for the EPO would be maximized by setting the age at entry at about 1 year and that the yield per recruit for the WPO would be greatest if the age at entry were set at about 2 1/2 years.

4.2 Interactions Within the Eastern Pacific Ocean

The principal interaction which takes place within the EPO is briefly described in Section 2.2. This does not appear to be of sufficient interest to warrant study.

4.3 Interactions Within the Western Pacific Ocean

No information, other than that given in Section 2.3, is available on this subject.

4.4 Interactions Within the Central Pacific Ocean

No information, other than that given in Section 2.4, is available on this subject.

5. FUTURE STUDIES

5.1 Work Contemplated

A workshop on Pacific northern bluefin involving staff members of the Inter-American Tropical Tuna Commission (IATTC) and the National Research Institute of Far Seas Fisheries (NRIFSF) of Japan was held at the headquarters of the IATTC in La Jolla,

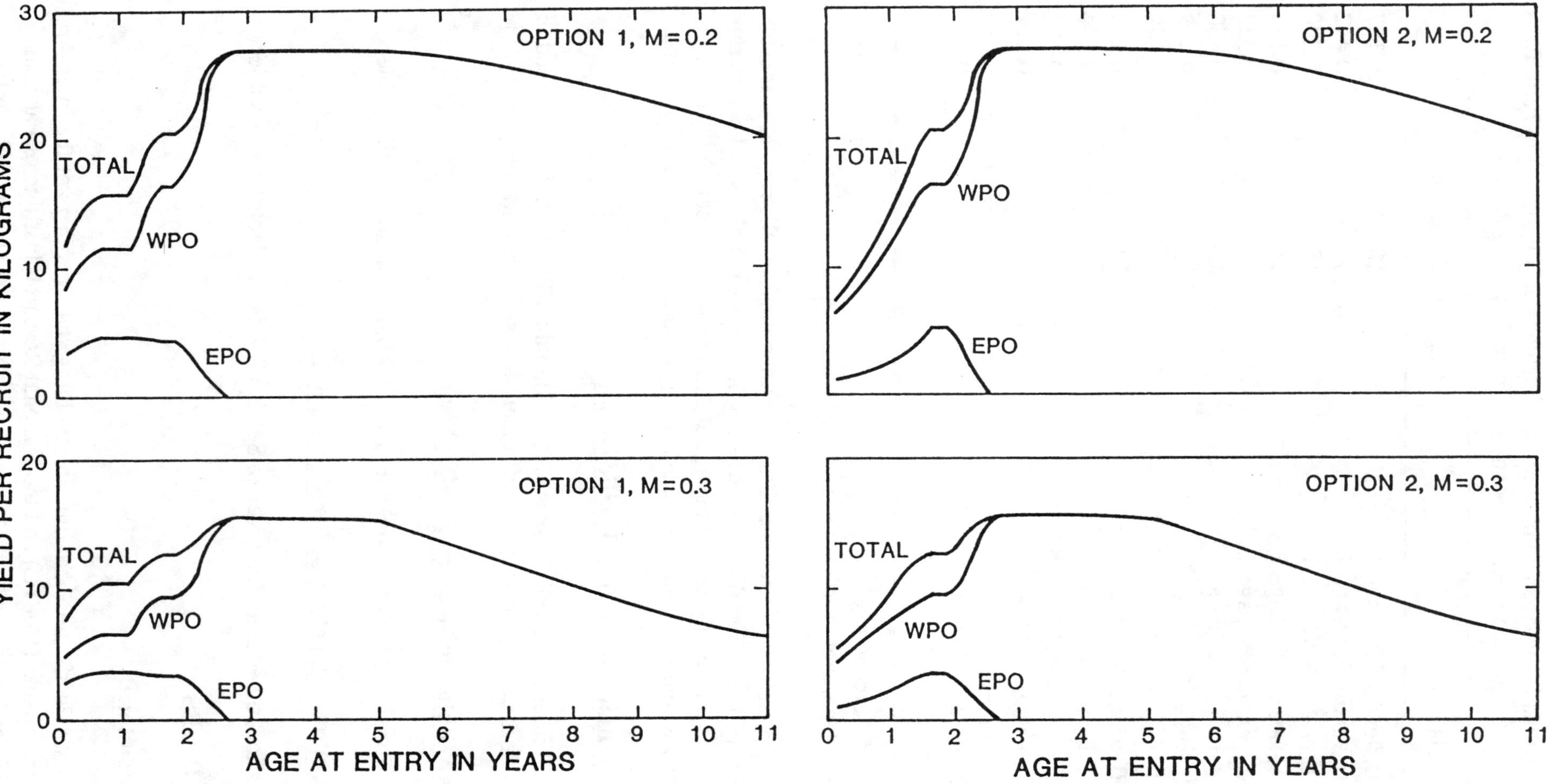

FIGURE 1. Yields per recruit of northern bluefin to the fisheries of the eastern and western Pacific Ocean. Option 1 pertains to fish which begin a west-east migration during the first year of life and an east-west migration during the third year of life, and Option 2 refers to fish which begin a west-east migration during the second year of life and an east-west migration during the third year of life. M stands for the annual coefficient of natural mortality and WPO and EPO stand for western Pacific Ocean and eastern Pacific Ocean, respectively.

on April 12-13, 1993. It was agreed that the following tasks would be completed before the next workshop, which will be held in 1994:

1. prepare an inventory of data on bluefin taken by the fisheries of the WPO, including catch and effort statistics by area and time period and size frequencies of fish caught, which are filed at regional and prefectural fisheries laboratories and other locations in Japan;
2. using the above data, if available, and data at the NRIFSF laboratory, prepare descriptions of the fisheries of the WPO which take bluefin, including catch and effort statistics by area and time period and size frequencies of fish caught;
3. review data on the total catches of bluefin during 1987-1991 by Japanese vessels to correct, if necessary, the data published in the FAO yearbooks;
4. update estimates of age composition of bluefin caught by Japanese vessels for the years subsequent to 1986;
5. prepare a summary of catches of bluefin by gillnets in offshore waters of the Pacific Ocean, by area and season;
6. prepare a summary of catches of bluefin by Japanese longline vessels in the Pacific Ocean, by area and season, for 1952-1987 (and subsequent years, if data available);
7. assemble data on the offshore purse-seine fishery of the WPO for the purpose of calculating an index of abundance of bluefin in that area;
8. redo the cohort analyses of bluefin in the Pacific Ocean, using different estimates of age composition and a wider range of estimates of the rates of natural mortality;
9. redo the yield-per-recruit analyses of bluefin in the Pacific Ocean, using different estimates of age composition and a wider range of estimates of the rates of natural mortality;
10. review data on attrition of tagged bluefin in the WPO to attempt to estimate the period(s) during which fish bound for the EPO (or the CPO) leave the WPO;
11. assemble data on distribution and abundance of bluefin larvae in the WPO for the purpose of developing indices of abundance of the larvae;
12. carry out analyses of spawning biomass per recruit for bluefin in the WPO, using wide ranges of fishing and natural mortalities, fecundities, *etc.*;
13. carry out further studies of the feasibility of using micro-constituents of otoliths for the purpose of determining more about the movements of bluefin;
14. investigate ways of funding a study of micro-constituents of otoliths of bluefin for the purpose of determining more about the movements of bluefin.

5.2 <u>Data Needed</u>

Catch data for vessels of the various countries which catch northern bluefin in the WPO and the CPO, stratified by area, month, and gear, are needed, as are length-frequency data corresponding to the catch data. Standardized effort data should be available, so that estimates of the relative abundances of the various sizes of fish can be calculated.

Tagging data do not offer much support for the hypothesis that there are separate migrant and non-migrant subpopulations of northern bluefin in the Pacific Ocean (Section 4.1.1), so it is tentatively concluded that there is only one population. This should be investigated by other means, however, as lack of knowledge of the stock structure can lead to incorrect conclusions regarding interactions between the fish of the WPO and the EPO. The biochemical genetic studies of Smith and Clemens (1973), based upon samples taken off California and Baja California, indicated that that "fishery is supported by a single population." However, because they did not have any samples from the WPO, this finding is of limited value. If samples for such studies were taken simultaneously in the WPO and the EPO they might answer the question as to whether there are separate migrant and non-migrant subpopulations of bluefin in the Pacific Ocean. Another approach which might prove useful is analysis of the micro-constituents of the otoliths. Such studies are currently being conducted with skipjack in the Pacific Ocean and with southern bluefin. Because northern bluefin of the same age reside in two distinct areas, the WPO and the EPO, this species is ideal for such a study.

Northern bluefin should be tagged with tags which (1) do not cause immediate or delayed mortalities of the fish, (2) do not affect their growth, (3) are retained throughout the life of the fish, and (4) are always detected and returned to the agency which tagged and released the fish. The best source of such fish would probably be the troll fishery of Japan. The data from this tagging programme could be used to produce accurate estimates of the growth, mortality, and abundance of the fish for various age-structured analyses. In addition, more could be learned about the movements and migrations of the fish. A tag which is invisible to fishermen, fish buyers, *etc.*, but which can be seen or detected by employees of fisheries research organizations, would be ideal, because there would be no errors due to failure to find tags, failure to report tags that were found, *etc.* Three types of tags are listed in Table 5. The visual implant tag (Haw *et al.*, 1990) is an externally-readable tag which is implanted into living transparent tissue, most commonly on the head. According to the manufacturer, the "expected retention varies with fish size and species but tag loss is generally below five percent." A tagging programme with this type of tag would cost about the same as a programme with dart tags, but it is not certain that visual implant tags would be superior to dart tags. The coded-wire tag (Jefferts *et al.*, 1963) can be detected externally (Morrison, 1990), but it must be removed from the fish to read the code. The detector for the PIT (passive integrated transponder) tag (Prentice *et al.*, 1990) can read its code without removing it from the fish. The costs of a tagging programme with either the coded-wire or PIT tags would be considerable, especially since there are many locations where fish are landed in Japan. It is more important to obtain tag returns for larger than for smaller fish, however, and if large fish are landed in relatively few locations most of the sampling effort could be concentrated there. If other species of fish were being tagged with similar tags during the same period the costs of the detectors could be shared among the various programmes. The estimated numbers of tagged fish which would be recaptured and returned each year are shown in Table 6.

Archival tags, each of which would record data on temperature, light, and pressure at frequent intervals (Smith and Goodman, 1986), are likely to become commercially available within the next few years. The kinds of tags discussed in the previous paragraph give information on the locations of the fish at two times, the date of release and the date of recapture. Archival tags would give data on where the fish were at all times between the dates of release and recapture. It is possible that some fish start to cross the ocean, and then turn back. Data from an archival tagging programme would reveal whether that is the case and, if so, where and at what time the fish turn back. This information, combined with oceanographic data, might be used to predict the availability of bluefin of various ages to the fisheries of the WPO and the EPO.

TABLE 5. Information on internal tags which might be used for bluefin. The tag costs are for 5,000 tags, and the applicator and detector costs are for one unit each.

Type	Size	Cost			Manufacturer
		Tags	Applicator	Detector	
visual implant	2.5 x 1.0 x 0.1 mm	$1,250	$1,500	not needed	Northwest Marine Technology
	3.5 x 1.5 x 0.1 mm	$3,500			Shaw Island, Washington, USA
coded wire	0.53 x 0.25 mm	$365	$7,700-12,000	$2,700-9,000	Northwest Marine Technology
	1.07 x 0.25 mm	$365			Shaw Island, Washington, USA
	1.60 x 0.36 mm	$402			
	2.13 x 0.36 mm	$402			
PIT	10.0 x 2.1 mm	$23,750	$24-$41	$1,666	Biosonics
					3670 Stone Way North
					Seattle, Washington, USA

TABLE 6. Estimated numbers of tagged fish recaptured and returned from 5,000 northern bluefin released at 3 months of age. M and F stand for the annual coefficients of natural and fishing mortality, respectively.

M	F	Age												
		0	1	2	3	4	5	6	7	8	9	10	11	12
0.2	0.5	1,459	1,064	528	262	130	65	32	16	8	4	2	1	<1
0.2	1.0	2,473	1,184	356	107	33	10	3	1	<1				
0.2	2.0	3,673	776	86	10	1	<1							
0.3	0.5	1,410	944	424	191	86	39	17	8	3	2	1	<1	
0.3	1.0	2,395	1,055	288	78	21	6	2	1	<1				
0.3	2.0	3,573	697	70	7	1	<1							

5.3 Institutions Involved

The principal institutions involved would be the IATTC and the NRIFSF. In addition, catch statistics and size data should be obtained from the Taiwan Fisheries Bureau, Republic of China. Analyses of micro-constituents of the otoliths of bluefin would probably involve scientists or technicians from the University of California at San Diego.

5.4 Assistance Required

The IATTC and the NRIFSF would need considerably more money if the research described above, especially the tagging, were to be performed. In addition, catch statistics and size data for fish caught by vessels of the Republic of China would be required.

6. ACKNOWLEDGEMENTS

Suggestions for improvement of drafts of this report were received from Drs. Richard B. Deriso, Ziro Suzuki, and Alexander Wild, and Messrs. Doyle A. Hanan and Richard S. Shomura.

7. REFERENCES CITED

Bayliff, W.H. 1993a. Growth and age composition of northern bluefin tuna, *Thunnus thynnus*, caught in the eastern Pacific Ocean, as estimated from length-frequency data, with comments on trans-Pacific migrations. *Bull.I-ATTC*, 20(9):501-40.

Bayliff, W.H. (ed.) 1993b. Annual report of the Inter-American Tropical Tuna Commission, 1992. *Annu.Rep.I-ATTC*, (1992):(in press).

Bayliff, W.H. 1993c. A review of the biology and fisheries for northern bluefin tuna, *Thunnus thynnus*, in the Pacific Ocean. *In* Proceedings of the FAO Expert Consultation on Interactions of Pacific Tuna Fisheries, edited by R.S. Shomura, J. Majkowski, and S. Langi, 3-11 December 1991, Noumea, New Caledonia. [See this document.]

Bayliff, W.H., Y. Ishizuka, and R.B. Deriso. 1991. Growth, movements, and attrition of northern bluefin, *Thunnus thynnus*, in the Pacific Ocean, as determined from tagging experiments. *Bull.I-ATTC*, 20(1):1-94.

Clay, D. (ed.) 1991. Atlantic bluefin tuna (*Thunnus thynnus thynnus* (L.)): a review. *Spec.Rep.I-ATTC*, (7):89-179.

Haw, F., P.K. Bergman, R.D. Fralick, R.B. Buckley, and H.L. Blankenship. 1990. Visible implanted fish tag. *In* Fish-marking Techniques, edited by N. Parker, A.E. Giorgi, R.C. Heidinger, D.B. Jester, Jr., E.D. Prince, and G.A. Winans, *Symp.Ser.Am.Fish.Soc.*, (7):311-5.

Hearn, W. 1991. Natural mortality. *In* Review of aspects of southern bluefin tuna biology, population and fisheries, edited by A.E. Caton. *Spec.Rep.I-ATTC*, (7):243-6.

Jefferts, K.B., P.K. Bergman, and H.F. Fiscus. 1963. A coded wire identification system for macro-organisms. *Nature, Lond.*, 198(4879):460-2.

Morrison, J.A. 1990. Insertion and detection of magnetic micro-wire tags in Atlantic herring. *In* Fish-marking Techniques, edited by N. Parker, A.E. Giorgi, R.C. Heidinger, D.B. Jester, Jr., E.D. Prince, and G.A. Winans, *Symp.Ser.Am.Fish.Soc.*, (7):272-80.

Pauly, D. 1980. On the interrelationships between natural mortality, growth parameters, and mean environmental temperatures in 175 fish. *J.Cons.CIEM*, 39(2):175-92.

Prentice, E.F., T.A. Flagg, C.S. McCutcheon, D.F. Brastow, and D.C. Cross. 1990. Equipment, methods, and an automated data-entry station for PIT tagging. *In* Fish-marking Techniques, edited by N. Parker, A.E. Giorgi, R.C. Heidinger, D.B. Jester, Jr., E.D. Prince, and G.A. Winans, *Symp.Ser.Am.Fish.Soc.*, (7):335-40.

Skillman, R.A., and C. Shingu (rapporteurs) 1980. Pacific northern bluefin tuna. *FAO Fish.Tech.Pap.*, (200):24-30.

Smith, A.C., and H.B. Clemens. 1973. A population study by proteins from the nucleus of bluefin tuna eye lens. *Trans.Am.Fish.Soc.*, 102(3):578-83.

Smith, P., and D. Goodman. 1986. Determining fish movements from an "archival" tag: precision of geographical positions made from a time series of swimming temperature and depth. *NOAA Tech.Mem.NMFS-SWFC, La Jolla*, (60):13 p.

Yokota, T., M. Toriyama, F. Kanai, and S. Nomura. 1961. Studies on the feeding habit of fishes. *Rep.Nankai Reg.Fish.Res.Lab.*, (14):1-234.

AN OVERVIEW OF INTERACTION ISSUES AMONG THE FISHERIES FOR SOUTHERN BLUEFIN TUNA

Tom Polacheck
CSIRO Division of Fisheries
GPO Box 1538
Hobart, Tasmania 7001
Australia

ABSTRACT

The history of fisheries for southern bluefin tuna has involved a complex set of interaction issues. These include interactions between surface and longline vessels operating in the same area and season, interactions between Australian surface catches and the subsequent catches in the longline fishery from the same cohorts, interactions among longline vessels fishing in different areas, interactions among longline vessels of different nationalities, and interactions between longline and surface fisheries with respect to recruitment and future catches. The present paper reviews briefly these interaction issues and suggests future areas of research. Emphasis for the latter is directed toward better understanding of movement, migration routes, mixing rates, and site fidelity of the southern bluefin tuna.

1.   INTRODUCTION

Southern bluefin tuna (SBT) have a circumpolar distribution in the southern hemisphere with complex, but poorly understood, movement and dispersal pathways. Spawning takes place in the tropical Indian Ocean with major adult feeding grounds off New Zealand, South Africa and southern Australia. The species is managed as a single population and represents the most widely dispersed stock of any tuna. A large, but unknown fraction of the juveniles (*i.e.* < age 4) migrates progressively from the spawning grounds in the tropical Indian Ocean along the western and southern Australian coast line. These juveniles are often found in large surface aggregations and are vulnerable to purse-seine and pole-and-line gear. Pre-adults (ages 6 & 7) and adult fish (8 and above) are rarely found at the surface and are primarily harvested with longline gear, while older juvenile fish (4-5) are commonly caught by both surface and longline gear (see Caton *et al.*, 1990 and Caton, 1991 for reviews of the biology and fisheries for SBT).

Apart from the central and eastern South Pacific, fisheries for SBT occur in all southern temperate oceans plus the tropical Indian Ocean. Surface and longline fisheries for SBT were developed in the 1950s, principally by Australia and Japan respectively. The longline fishery developed rapidly, achieving peak catches in 1961. Since this peak of 77,000 mt, longline catches have steadily declined despite continuous increases in effort. The Australian surface fishery developed more slowly with maximum catches occurring in 1982. Subsequently, both surface and longline catches have dramatically declined, in part as the result of management measures aimed at conserving and rebuilding the stock. All recent assessments of the SBT stock indicate that the combined catches from the surface and longline fisheries have resulted in a continuous decline in spawning stock biomass since 1960. Current (1989) levels are estimated to have declined to at least 16 to 25% of pre-exploitation levels (Anon., 1991).

In addition to the Australian and Japanese fisheries, a locally important troll and handline fishery for SBT has existed in New Zealand since 1980 and domestic longline fisheries have been initiated in New Zealand and Australia. Significant SBT by-catches are thought to be taken by Taiwanese gillnetters and Indonesian and Taiwanese longline vessels, and were previously taken by Korean longliners. Furthermore, the Australian surface and the Japanese longline fisheries are made up of numerous spatially and seasonally distinct components which harvest different size ranges of fish (see Caton, 1991, for a comprehensive review of the different SBT fishery components).

The large area, the complex spatial and movement patterns and the numerous components comprising the global fishery provide the potential for a wide range of interactions at different spatial and temporal scales which need to be considered in the management of this stock. In addition, the importance of different interactions has changed in response to changes in population size and to developments in the fishery. In this paper, the major interaction issues relevant to SBT fisheries are described, previous analyses of interactions are reviewed, sources of relevant data are identified and future research needs are discussed.

2. INTERACTION ISSUES

Significant interaction issues among fisheries develop when the operations in one or more fisheries has an effect on the catch rate or yield (biological and/or economic) in one or more other fisheries. A fishery in this case is defined as a set of vessels which can be classified as distinct by common characteristics affecting their operations (*e.g.* gear, size, area, season, target species, nationality, *etc.*). Among the vessels that harvest SBT, the potential number of different fisheries which can be defined is large and the classification of vessels into the various major fisheries components does not result in a small number of mutually exclusive subsets. This paper will focus only on major interactions among SBT fisheries.

Six main interaction issues can be identified among the SBT fisheries:

a. Interactions among the three main components of the surface fishery operating within Australia's exclusive economic zone (AFZ)[1].

b. Interaction between surface and longline vessels operating in the same geographic area and season of the year.

c. Interaction between Australian surface catches and the subsequent catches in the longline fishery from the same cohorts.

---

[1] One major component of the Australian surface fishery (*i.e.* that which occurred off New South Wales) collapsed in the 1980s and more recently a new trolling component has been developing off eastern Tasmania.

d.   Interactions among longline vessels fishing in different areas.

e.   Interactions among longline vessels of different nationalities.

f.   Interactions between longline and surface fisheries with respect to recruitment and future catches.

3.   DATA AVAILABLE FOR STUDYING INTERACTIONS

The primary data available for studying interaction issues in SBT include:

a.   total catch (weight) by fishing area, and size (length) of fish for the three principal Australian surface fisheries,

b.   total catch in number, and effort (number of hooks) by month and by five-degree geographic square for the Japanese longline fishery,

c.   length-frequency estimates of the Japanese longline catch by quarter and statistical reporting area (Figure 1),

d.   information from SBT tagged and released in the different components of the Australian fishery between 1959 and 1991 (Table 1), and

e.   daily total catch by length from the New Zealand domestic fishery.

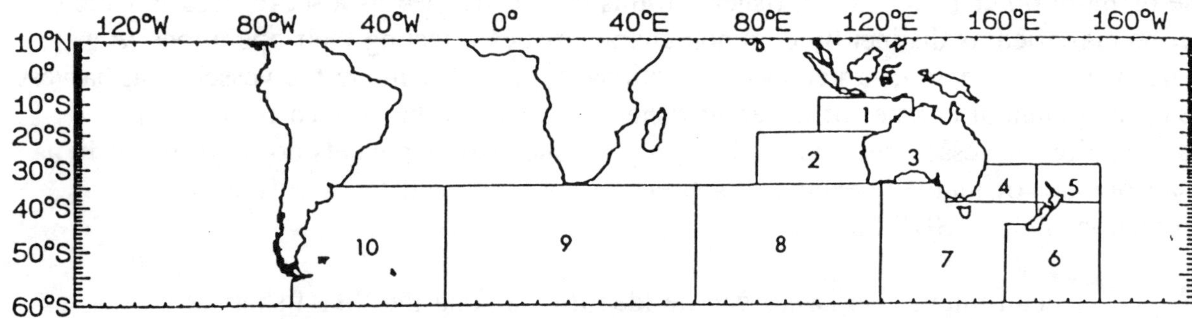

Figure 1.   Large statistical areas used for reporting catch by length for SBT caught by Japanese longline vessels.

In addition to these primary data sets, vessel reports of set-by-set data from longliners operating within Australian and New Zealand EEZs are available for more recent years as well as data collected by observers aboard a limited number of vessels. In addition, set-by-set, catch-and-effort data including the size distribution of the catch are also available for 1991 for 12 vessels participating in a special real time monitoring programme (RTMP). Finally, daily set-by-set data for most of the Japanese longline fleet which includes catch, effort, and location (but not size information) are stored in Japan but are unavailable for analysis (also see Caton, 1991 and Majkowski and Morris, 1986 for more information on available data). Korea and Taiwan also publish longline catch-

and-effort statistics by five-degree square, but there are indications that these statistics are not complete with respect to SBT (Caton, 1991).

Table 1a. Summary of available tagging data from SBT tagging experiment conducted between 1960 and 1980. Determination of cohorts was based on the age estimated at time of release from the measured length and the Kirkwood (1983) growth curve. Cohorts with fewer than 25 releases have been omitted from the table. Key: WA = Western Australia; SA = South Australia; NSW = New South Wales; LL = longline (returns from longline vessels from all oceans)

Cohort	Number of releases			Number of recaptures from release in WA				Number of recaptures from release in SA				Number of recaptures from release in NSW			
	WA	SA	NSW	WA	SA	NSW	LL	WA	SA	NSW	LL	WA	SA	NSW	LL
1956	0	0	28	0	0	0	0	0	0	0	0	0	0	2	0
1957	0	0	66	0	0	0	0	0	0	0	0	0	2	3	1
1958	0	0	0	0	0	0	0	0	0	0	0	0	0	0	0
1959	543	831	520	19	10	2	0	0	8	0	8	0	2	44	2
1960	4479	966	164	84	61	34	9	0	16	1	13	0	0	5	1
1961	5263	2446	910	84	105	23	39	0	87	9	60	0	4	59	8
1962	4427	1215	1382	21	63	62	27	0	27	10	40	0	19	257	14
1963	2037	346	262	2	27	40	19	0	20	2	11	0	3	26	6
1964	778	0	686	0	3	3	1	0	0	0	0	0	12	87	7
1965	4326	107	3729	65	74	20	31	0	19	1	1	0	42	1384	22
1966	915	619	3881	15	6	30	3	0	160	29	0	0	18	2144	10
1967	0	2112	1217	0	0	0	0	7	577	151	18	0	13	358	14
1968	355	0	356	37	3	3	0	0	0	0	0	0	8	68	3
1969	963	0	0	35	2	0	0	0	0	0	0	0	0	0	0
1970	117	0	0	0	1	0	0	0	0	0	0	0	0	0	0
1971	638	0	0	31	9	0	1	0	0	0	0	0	0	0	0
1972	730	0	0	50	8	4	1	0	0	0	0	0	0	0	0
1973	681	0	0	26	20	8	4	0	0	0	0	0	0	0	0
1974	567	66	0	1	41	5	1	0	21	2	1	0	0	0	0
1975	163	838	0	7	4	0	0	3	178	20	1	0	0	0	0
1976	56	0	0	3	1	1	0	0	0	0	0	0	0	0	0
1977	1085	0	0	55	24	2	0	0	0	0	0	0	0	0	0
1978	457	0	0	82	38	5	2	0	0	0	0	0	0	0	0

Table 1b. Summary of available tagging data from SBT tagging experiment conducted in 1983/84.

Cohort	Location of tagging	Number tagged	Number of Recaptures			
			Esperance	Albany	South Australia	Other[1]
1979	Esperance	1079	227	10	190	2
1980	Esperance	3470	554	219	500	7
	Albany	1478	45	111	513	7
	South Aus.	669	0	0	219	10
1981	Albany	885	72	35	180	0
	South Aus.	2484	4	19	1124	10

1) caught by either Japanese longline vessels or fishermen from NSW, Australia

4. INTERACTIONS WITHIN THE JUVENILE SURFACE FISHERIES OFF AUSTRALIA

The commercial surface fishery for SBT has historically included three main components. The New South Wales (NSW) and the South Australia (SA) fisheries date back to the 1950s while the Western Australia (WA) component did not begin until 1969. Both the absolute and relative magnitude, as well as the size compositions, of the catch have changed markedly over time (Tables 2-5). Major concerns about interactions between these different components arose in response to a twofold increase in the annual number of juvenile fish caught between 1976 and 1983 and a total collapse of the surface NSW fishery in 1983 (*e.g.* Majkowski *et al.*, 1988).

Table 2. Southern bluefin tuna catch by state and by quota year for 1951/52 to 1989/90 (from Caton, 1991).

Quota year	Western Australia		South Australia		New South Wales	
	Tonnes	Number	Tonnes	Number	Tonnes	Number
1951/52			20	1 344	49	4 132
1952/53			30	2 030	244	19 643
1953/54			5	316	479	37 858
1954/55			24	1 620	419	33 019
1955/56			199	13 637	298	24 243
1956/57			387	26 548	765	60 763
1957/58			554	38 017	877	70 715
1958/59			700	48 043	1 768	140 121
1959/60			1 396	95 843	1 786	142 023
1960/61			2 255	154 835	2 149	169 290
1961/62			3 377	231 888	1 423	112 530
1962/63			3 589	246 447	1 259	101 626
1963/64			5 517	378 883	2 610	251 282
1964/65			4 730	288 659	2 261	227 602
1965/66			5 994	416 813	2 246	162 451
1966/67			3 385	245 253	2 144	166 149
1967/68			2 926	263 376	3 672	362 347
1968/69	299	69 219	3 255	427 716	5 129	665 188
1969/70	708	189 015	3 123	333 705	5 885	628 736
1970/71	600	121 405	2 817	343 550	3 611	537 385
1971/72	757	128 537	4 374	454 015	5 033	371 471
1972/73	308	63 946	6 835	506 172	6 133	288 436
1973/74	273	59 799	6 988	756 126	1 811	83 481
1974/75	1 142	202 828	4 842	599 045	5 276	310 630
1975/76	395	43 033	6 938	865 455	2 466	195 544
1976/77	841	103 716	8 789	1 159 693	308	37 067
1977/78	1 846	528 157	4 934	548 020	4 814	243 398
1978/79	2 311	450 000	4 338	631 782	4 332	223 555
1979/80	2 358	366 055	6 855	1 082 576	3 611	159 157
1980/81	2 822	516 116	9 877	819 400	3 427	137 519
1981/82	3 816	651 964	12 748	1 184 435	3 267	117 172
1982/83	5 478	1 113 144	13 831	1 244 140	1 648	122 121
1983/84	4 516	774 782	10 419	831 794	899	20 521
1984/85	2 097	321 189	11 271	727 230	118	2 480
1985/86	1 146	186 074	12 088	887 054	4	89
1986/87	1 234	212 592	10 029	640 665	45	1 095
1987/88	1 104	207 069	9 849	871 514	24	790
1988/89	426	55 220	4 872	412 805	2	100
1989/90	200 (est.)	35 000 (est.)	4 199	285 022	19*	

*Includes predominantly the catches of small trolling vessels operating off eastern Tasmania; a small proportion represents New South Wales longline catch.

Prior to 1969, the only potential for major, direct interactions among these fisheries would have been an affect of the NSW fishery on the SA fishery. The NSW fishing season took place in the latter half of the year and sometimes during January of the following year, while the SA fishery occurred in the first half of the year. Before 1969, the same cohort tended to dominate catches from both fisheries during a season, with the NSW fishery usually taking predominately 2 year old fish just prior to their becoming 3 years of age, while the SA fishery was dependent on 3 year old fish (Figure

2 and Tables 3-4). In both the NSW and SA fisheries, the fish became unavailable during the off-season, which suggests the possibility of significant interchange given the timing and their close proximity. However, the total catches in the years prior to 1969 were small (relative to subsequent removals - Table 2) and the two fisheries were likely to have had only minimal effects on each other. Results from tagging studies (discussed below) indicate little direct interaction during this period.

Table 3. Estimates of the catch by age in numbers (thousands of fish) for New South Wales surface fishery by calendar year. (Based on data held by CSIRO; from various sources.)

			Age		
Year	1	2	3	4	>4
1952	0.3	8.1	5.2	1.6	0.4
1953	0.5	15.8	11.0	4.1	1.9
1954	0.5	13.9	11.0	5.5	3.5
1955	0.3	9.9	8.3	4.5	3.0
1956	0.8	25.3	17.0	5.9	2.3
1957	0.9	29.0	21.8	9.9	5.6
1958	1.9	58.4	40.4	15.1	6.7
1959	1.9	59.1	45.6	21.7	12.9
1960	2.3	71.1	53.0	23.6	13.1
1961	1.5	47.1	40.3	22.6	15.4
1962	1.4	41.7	33.1	16.6	10.3
1963	2.8	86.3	59.5	22.1	9.6
1964	1.3	230.7	40.6	23.0	18.8
1965	5.4	19.6	100.2	26.1	8.0
1966	11.9	56.0	71.0	24.5	7.7
1967	0.0	262.6	26.4	9.6	2.2
1968	23.5	465.7	118.2	36.0	10.6
1969	5.8	545.7	73.1	17.0	2.0
1970	13.6	319.4	64.4	23.9	22.4
1971	0.7	249.6	124.8	5.6	0.0
1972	0.0	106.1	115.7	143.6	32.4
1973	0.1	5.5	1.5	75.3	29.2
1974	5.0	122.6	34.5	42.1	81.2
1975	9.9	163.9	17.9	11.8	13.8
1976	0.0	16.0	6.4	5.1	9.5
1977	1.5	31.0	55.7	61.5	32.3
1978	2.6	66.2	32.5	66.0	106.0
1979	0.2	8.8	35.7	76.8	48.3
1980	0.3	12.9	44.7	46.8	46.2
1981	0.8	36.4	9.4	8.7	68.9
1982	4.7	59.2	30.2	14.9	13.1
1983	0.0	0.2	0.3	1.3	18.8
1984	0.0	0.0	0.0	0.1	2.3
1985	0.0	0.0	0.0	0.0	0.0
1986	0.0	0.0	0.0	0.0	1.0
1987	0.0	0.0	0.4	0.2	0.2
1988	0.0	0.0	0.0	0.0	0.0
1989	0.0	0.0	0.0	0.0	0.0

Beginning in 1969, the SA fishery began to change. The fishery expanded to the west and began to take significant numbers of 2 year old fish (Figure 3, Table 4). During the same period, the WA fishery developed (Table 5) and its catch was initially dominated by 2 year old fish (in some years 1 year old fish dominated the catch in the last quarter of the year and 2 year old fish in the rest of the year). As such, all three fisheries were sequentially harvesting the same cohort of fish. However, the NSW fishery was no longer the first fishery harvesting a recruiting cohort. The SA fishery also continued to take significant numbers of 3 year old fish. These changes meant that the potential interactions between the SA and NSW fisheries had shifted and significant two-way as well as one-way interactions were possible.

Table 4. Estimates of the catch by age in numbers (thousands of fish) for South Australia surface fishery by calendar year. (Based on data held by CSIRO; from various sources.)

			Age		
Year	1	2	3	4	>4
1952	0.0	0.0	0.7	0.5	0.2
1953	0.0	0.0	1.0	0.8	0.3
1954	0.0	0.0	0.2	0.1	0.0
1955	0.0	0.0	0.8	0.6	0.2
1956	0.0	0.0	6.7	5.0	1.9
1957	0.0	0.1	13.1	9.8	3.6
1958	0.0	0.1	18.7	14.0	5.2
1959	0.0	0.1	23.7	17.7	6.5
1960	0.0	0.3	47.2	35.3	13.1
1961	0.0	0.4	76.3	57.0	21.1
1962	0.0	0.6	114.2	85.4	31.6
1963	0.0	0.7	121.4	90.8	33.6
1964	0.0	1.0	186.6	139.6	51.6
1965	0.0	1.6	69.5	162.7	54.8
1966	0.0	0.8	150.4	238.8	26.8
1967	0.0	8.0	97.4	119.2	20.7
1968	0.0	10.7	182.4	59.3	11.0
1969	7.5	198.5	179.9	51.6	0.4
1970	1.4	236.2	76.8	8.8	0.2
1971	0.0	116.8	171.8	51.7	3.3
1972	14.6	67.5	302.8	98.0	0.9
1973	0.1	44.3	111.0	299.8	21.2
1974	88.1	302.8	277.2	148.2	31.8
1975	147.3	230.5	236.9	64.0	18.0
1976	139.5	357.1	267.7	162.2	7.1
1977	68.5	396.5	393.1	161.8	13.9
1978	66.1	94.7	253.5	100.4	8.5
1979	213.5	254.5	234.6	61.3	0.0
1980	68.9	508.9	395.5	33.2	10.1
1981	94.1	152.8	249.7	244.3	59.8
1982	119.3	390.7	762.4	148.6	90.9
1983	15.4	299.6	473.9	160.3	95.3
1984	15.1	156.3	355.7	119.1	90.4
1985	4.5	115.5	223.3	149.8	171.6
1986	0.3	93.1	389.7	181.1	159.6
1987	0.3	31.5	175.1	287.9	109.3
1988	0.3	153.2	526.5	129.0	55.6
1989	0.0	15.7	242.6	138.3	17.8

Table 5. Estimates of the catch by age in numbers (thousands of fish) for Western Australia surface fishery by calendar year. (Based on data held by CSIRO; from various sources.)

			Age		
Year	1	2	3	4	>4
1969	18.8	50.3	0.1	0.0	0.0
1970	45.5	142.6	0.6	0.0	0.0
1971	2.4	117.1	1.9	0.0	0.0
1972	2.6	101.1	24.6	0.2	0.0
1973	3.3	62.5	2.7	0.1	0.0
1974	1.4	54.4	3.3	0.5	0.0
1975	3.6	169.9	43.4	0.7	0.0
1976	0.4	17.3	5.7	0.1	0.0
1977	32.0	97.5	0.0	0.0	0.0
1978	154.9	334.9	5.5	0.0	0.0
1979	198.2	247.9	3.9	0.0	0.0
1980	50.9	324.2	67.6	6.1	2.5
1981	44.9	394.5	105.9	8.6	3.5
1982	71.4	565.3	91.0	4.3	1.6
1983	79.9	856.1	60.5	1.3	0.1
1984	12.4	602.4	89.2	2.9	0.0
1985	8.4	261.0	38.0	3.8	1.9
1986	1.4	127.8	39.0	5.8	0.5
1987	8.3	179.8	13.2	8.0	2.5
1988	56.9	137.7	12.5	3.1	2.2
1989	1.2	40.6	5.5	6.0	1.5

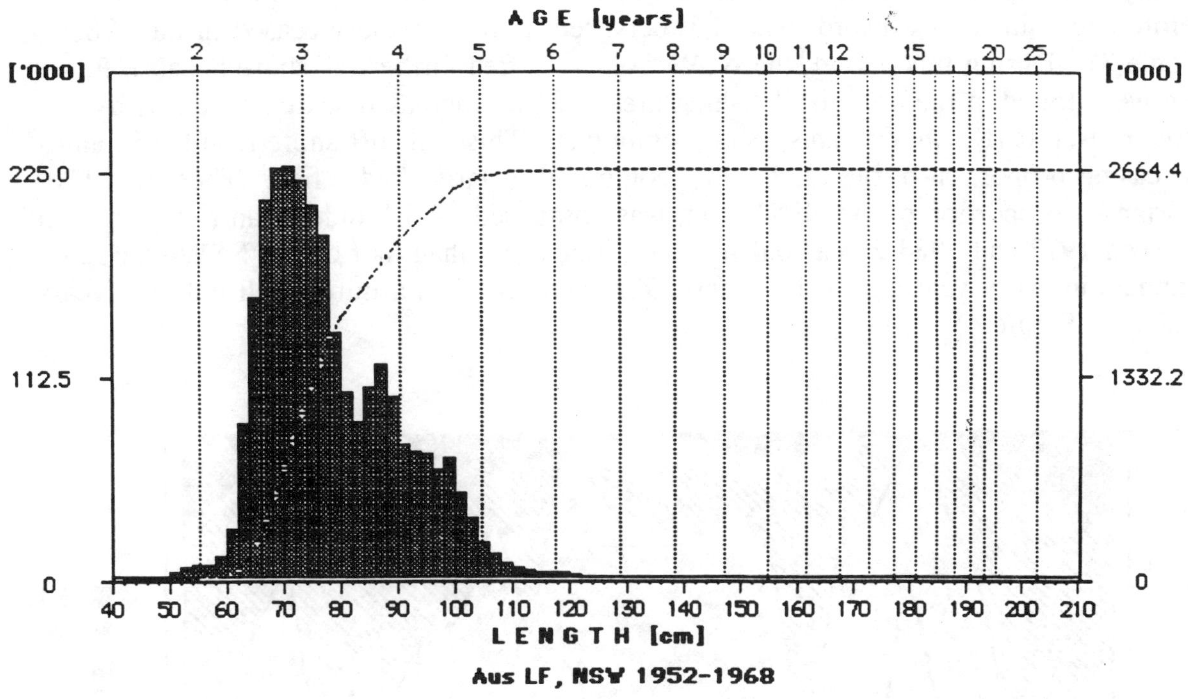

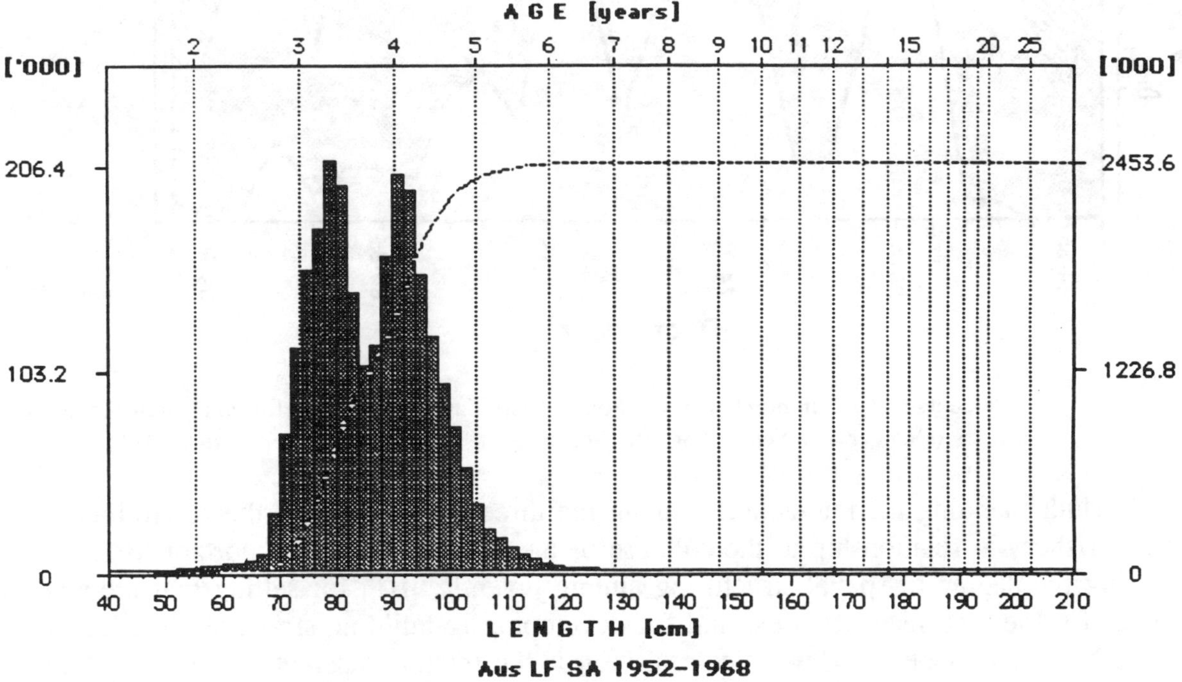

Figure 2.   Comparison of the length-frequency distributions of the surface catches in NSW and SA for 1952 to 1968 (data from all years combined). The vertical lines indicating age are the estimated length at age of a fish in January 1 based on the Kirkwood (1983) growth curve. The left axis is the number of fish in thousands in each size category; the right axis is for the cumulative number. (Based on data held at CSIRO from various sources.)

During the same period, the NSW fishery expanded with catches (in numbers) doubling between 1966 and 1967 and nearly doubling again between 1967 and 1968 to around 650,000 fish. This was the maximum number of fish ever harvested by this fishery, although catches (in terms of numbers) remained high for another two years before beginning a downward trend that persisted until the fishery ceased in the 1980s (Table 3). During this period, the NSW fishery shifted farther off-shore because fewer fish were found in the traditional fishing areas despite increased searching effort by spotter aircraft (Kevin Williams, pers. commun.). This shift off-shore resulted in an increasing proportion of the catch being composed of larger/older fish. Thus, in 1971 a substantial proportion of the NSW catch was composed of fish older than two years, and between 1972 and 1984 2 year old fish constituted less than half of the NSW catch in numbers in every year except 1975 and 1982, and less than a third in all but four years (Table 3, Figure 4).

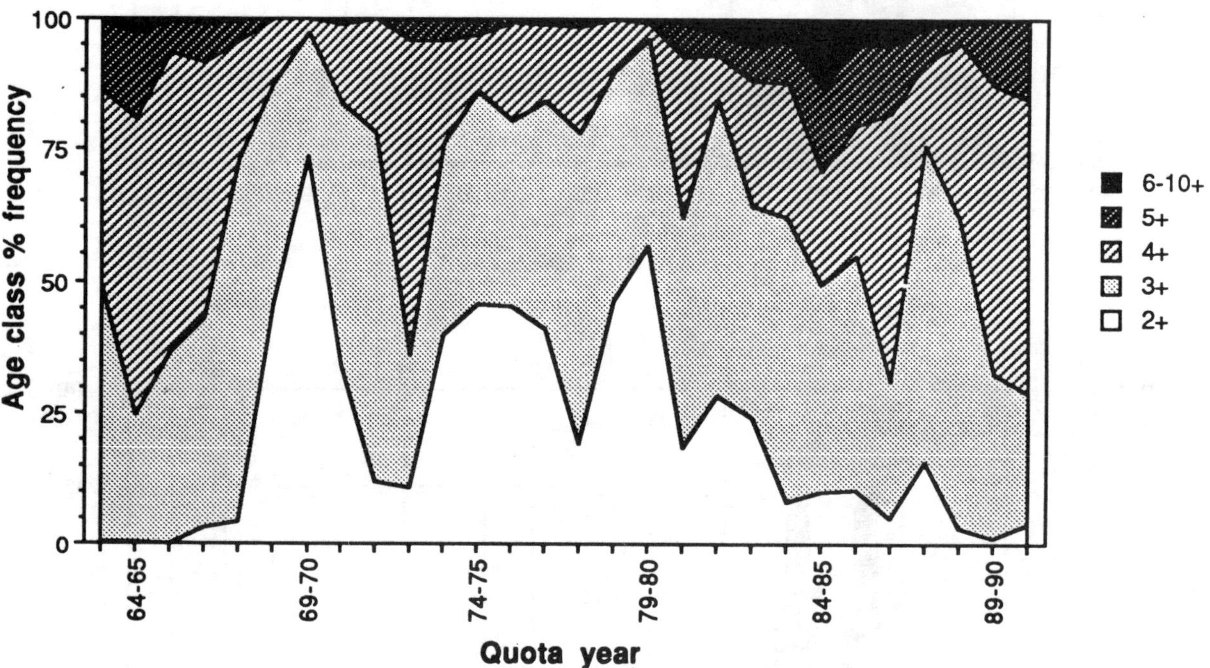

Figure 3.   Percentage catch in number in the South Australian southern bluefin tuna catch for quota years (October 1 to September 30) 1963-64 to 1990-91 (from Caton *et al.*, 1991).

Understanding the interactions among the three components of the Australian surface fishery in relationship to the collapse of the NSW fishery is important for assessing the degree of spatial structuring among juvenile SBT, for setting management policies for these surface fisheries, and for developing re-building strategies, particularly for the NSW component. The critical question is the relative discreetness of the NSW fishery (*i.e.* was its demise the result of over fishing a non-interacting component of the stock and/or changing local environmental conditions or was the collapse due to the lack of escapement from the other two Australian surface fisheries which were removing substantial numbers from each cohort prior to their potential exploitation by the NSW fishery).

In this regard it is worth noting that the substantial reductions in catches in western and south Australia which have taken place since the mid-1980s as the result of quota restrictions, did not result in an immediate recovery of the NSW fishery and only recently have some sightings of SBT been reported (K. Williams, pers. commun.). However, any interpretation of this delay and lack of recovery off NSW in terms of spatial discreetness would require analyses that also take into account declines in recruitment that were occurring (Polacheck and Klaer, 1991).

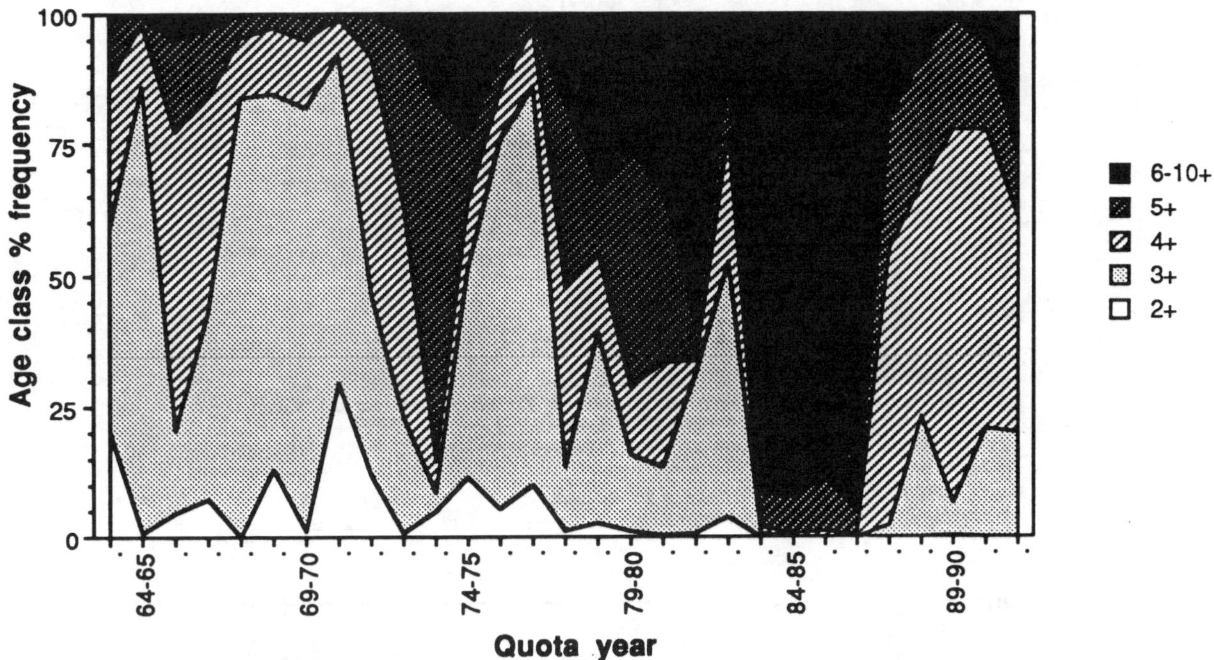

Figure 4.   Percentage catch in number of 2+, 3+, 4+, and 6+ to 10+ fish in the New South Wales southern bluefin tuna catch for quota years (October 1 to September 30) 1959-60 to 1989-90 (from Caton, 1991).

Another large interaction issue with respect to the Australian surface fisheries has been the potential interaction between surface catches in WA and SA. In general, the fishery in WA caught younger fish than in SA (Figures 2 and 5), although this was not always true in the early years of the WA fishery. WA tended to be the first fishery that harvested a recruiting cohort, although some overlap with SA in age composition occurred. As such most of the potential interaction would be a one-way effect of the catches in WA on the subsequent yield in SA. This is supported by tagging results in which substantial numbers of fish tagged in WA have been recaptured in SA with the reverse not being true for the same age group (Table 6).

In addition, analytical analyses of tagging data from the 1960s suggest that the effect of the WA fishery on SA was minimal probably due to the relatively low levels of catch (see below and Majkowski *et al.*, 1988). However, when the WA fishery expanded in the late 1970s with its total catch expanding to over 400 thousand fish per year in 1978, the effect of these large catches on the other Australian surface fisheries was an issue of concern, and results from tagging experiments conducted during 1983/84 indicated a large negative effect (Hearn and Majkowski, 1987).

Interactions within the Australian SBT surface fishery have been important historically and can provide insights into the spatial/temporal dynamics of juvenile SBT. Historically, these interaction issues were a source of conflict and had major political implications in relation to fishing rights, particularly in NSW where vessels from both SA and NSW would compete. However, these interactions are not currently a major management issue because the NSW fishery collapsed in the 1980s and is non-existent at

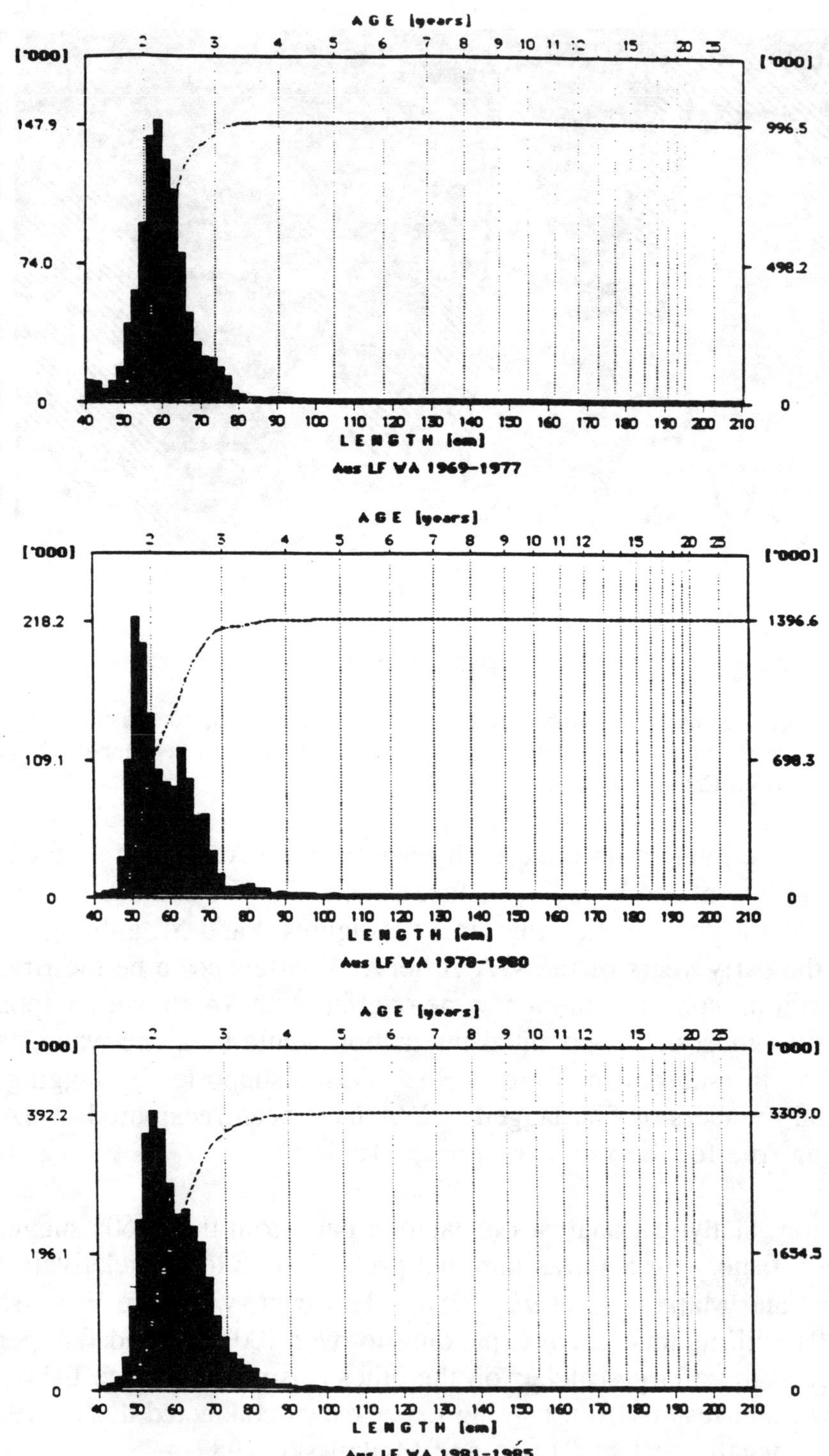

Figure 5. Length-frequency distributions of the surface catches in Western Australia. The vertical lines indicating ages are the estimated length at age of a fish on January 1 based on the Kirkwood (1983) growth curve. The left axis is the number of fish in thousands in each size category; the right axis is for the cumulative number. (Based on data held at CSIRO from various sources.)

the present time. Also the WA fishery has been reduced to a very low level (due to quota restrictions and economic factors). In the foreseeable future, economic considerations, combined with low stock levels are likely to prevent the re-development of a major WA fishery. However, if SBT re-appear off NSW in substantial numbers then interactions between fisheries in NSW and SA may again be an issue. Because the Australian SBT fishery has been restructured with the introduction of individual transferrable quotas, management issues are not likely to be concerned with access rights but with questions of the optimal strategies for recovery and utilization.

Tagging Experiment	Location of Tagging	Number of Releases	Number of Recaptures	
			Western Australia	South Australia
1960–1980	Western Aus.	36,127	1,904	1,899
	South Aus.	12,941	34	2,505
1983/1984	Western Aus.	6,913	1,273	1,383
	South Aus.	3,153	23	1,342

Table 6. Summary of tag returns in Western Australia and South Australia surface fishery from releases in each fishery. (Based on data collected and stored at CSIRO, Division of Fisheries, Hobart, Tasmania.)

5. OVERLAPPING SURFACE AND LONGLINE FISHERIES INTERACTIONS

Longline and surface fisheries overlap in their operations within the AFZ and to a lesser extent within the EEZ of New Zealand. Within AFZ, the potential for overlap has been reduced by voluntary seasonal closures adopted by the Japanese longline fleet since 1971 within the Great Australian Bight and Victoria/NSW coast (Figures 6-7) and restrictions placed on the longline vessels under access arrangements.

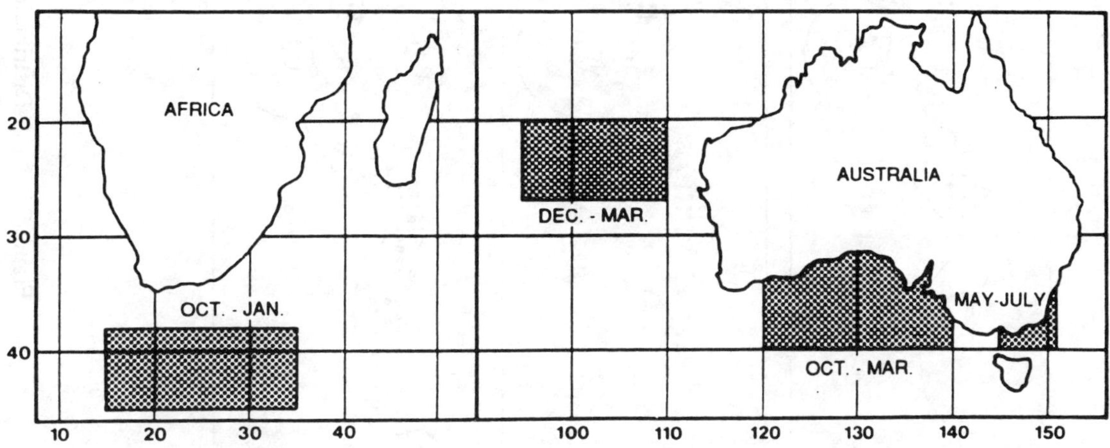

Figure 6. Areas and seasons of voluntary closures for SBT adopted by Japanese longline vessels (from Caton, 1991).

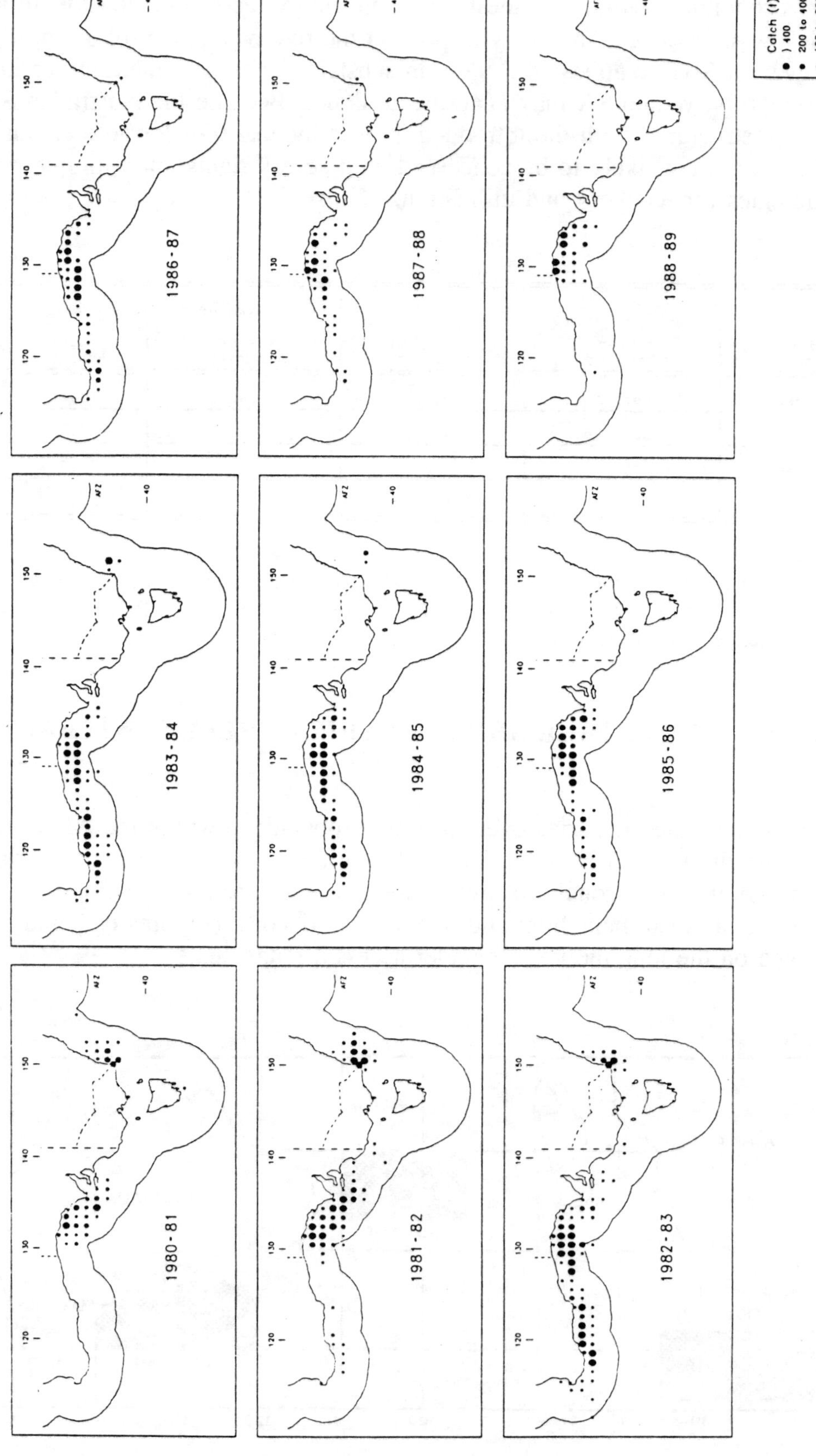

Figure 7.   Distribution of Australian surface fishery catches of southern bluefin tuna 1980-81 to 1988-89 (from Caton, 1991).

In general, direct interactions (*i.e.* competition for the same cohorts of fish during the same fishing season) between the Japanese longline and Australian surface fisheries are expected to be small. The relative magnitude of the global longline catches within the dominant age range harvested by the surface fishery (*i.e.* 2 to 4 years) has traditionally been small[2]. As such, the immediate impact of the longline fishery on surfaces catches is likely to be minimal in most situations. Some possible exceptions would be in NSW during the 1970s and in SA from 1968-79. As noted above, when the NSW fishery began to decline, it shifted its area of operation off-shore and became more dependent upon older fish. Comparison of the length-frequency distributions of the Japanese longline catches in statistical area 4[3] and those in the NSW surface fishery indicates that substantial proportions of the total catch in Area 4 for fish over 85 cm were caught by each fishery (Figure 8) during the 1970s.

Majkowski *et al.* (1981) made a "back-of-the-envelope" estimate of the reduction in the surface fishery catch off NSW due to the Japanese longline fishery in the same area for 1974-1978. Their analysis was based on the high correlations they found between the NSW catch by age and the Japanese longline catch rates for the same ages during these years. (However, no high correlations exist when the data for either earlier or all years are examined). Their estimate of 400 mt per year would suggest that the interactions were significant given that NSW catches during the same time period ranged from 2,465 to 4,350 mt [excluding 1976 when catch fell to 308 mt due to a large, persistent, warm-core eddy off Eden in that year]. However, the estimates are based on numerous untested or weakly-supported assumptions about the relationship between catches, effort, and catch rates in the two fisheries and their relationship to the global SBT fishery. More direct information and analyses on the actual movement of fish, and interaction between these two fisheries would be needed in order to evaluate the reliability of this estimate of 400 mt per year.

The reverse situation of a minimal direct impact of the surface fishery on adjacent and nearby longline fisheries is not necessarily true. In general, the main target of the Japanese longline fishery has been large SBT and to this extent direct competition for the same fish has not been not of major concern. However, this does not mean that interactions do not exist. The extent that subsequent recruitment of larger fish in an area near the surface fishery is directly affected by local removals by the surface fishery is unknown. This is a complex question related to migration rates and site fidelity for which little information is available.

It is worth noting that historically in longline statistical Area 3 (*i.e.* the Great Australian Bight), the size range of fish harvested by longliners and the SA surface

---

[2] However, the possibility exists that a substantially greater number of 2 to 4 year old fish are caught, but not retained and thus not included in the official catch statistics. This is more likely in recent years since a large differential exists in the price paid for large *versus* small longline-caught SBT and the fishery has been operating under restrictive quota management. Substantial discarding of small fish has been noted by Australian observers aboard Japanese longliners in the AFZ prior to 1991. Since 1991, all SBT caught within the AFZ are required to be retained as part of the bi-lateral access agreement.

[3] Statistical Area 4 (Figure 1) is the Japanese statistical reporting area which includes the waters adjacent to the NSW fishery. Finer spatial resolution of the length-frequency distributions of the Japanese longline catches are not available.

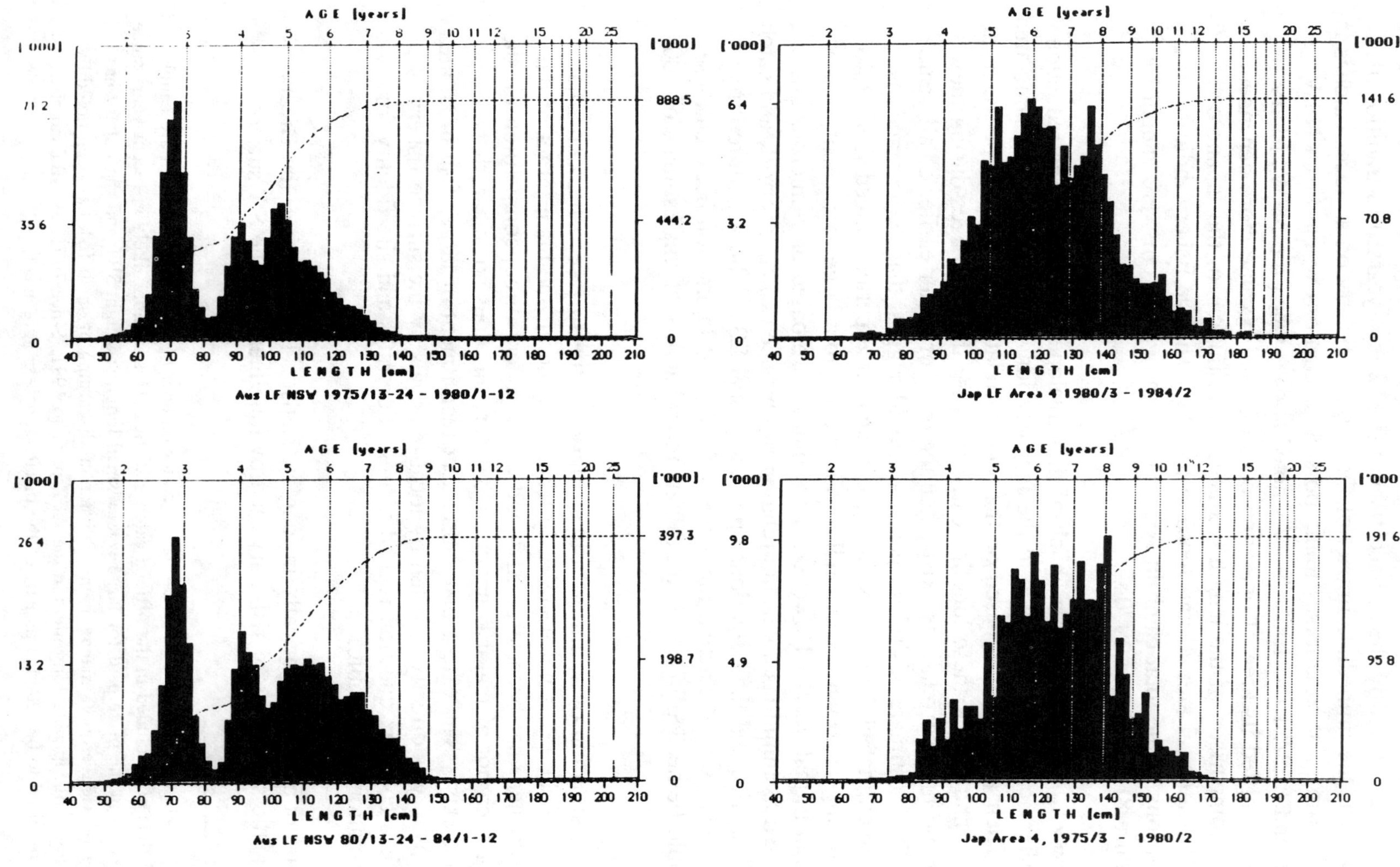

Figure 8. Comparison of the length-frequency distribution of the catch by Japanese longliners operating in statistical area 4 and the catch by the NSW surface fishery during the 1970s and 1980s. The vertical lines indicating age are the estimated length at age of a fish on January 1 based on the Kirkwood (1983) growth curve. The left axis is the number of fish in thousands in each size category and the right axis is for the cumulative number. (Based on data held at CSIRO from various sources.)

fishery largely overlapped. While Area 3 was never a major longline area, it was fished regularly between 1966 and 1979 and in some years substantial effort (*e.g.* > 1.2 million hooks) occurred. In those years, the possibility exists that the adjacent and overlapping surface fisheries in SA and WA may have been having a significant negative effect on the longline catch rates in Area 3. However, there is no direct evidence to indicate this. The number of reported fish tagged and released in SA and WA, and recovered by longliners operating in Area 3 within two years has been small.

In the most recent years with the imposition of quota restrictions, the focus of concern about local interactions between longline and surface fisheries has shifted from concerns about yield to concerns about escapement to the parental biomass. Of particular concern is the effect that the recent concentration of longline effort around eastern Tasmania may have on the escapement from the surface fisheries as the result of quota restrictions.

The catch rates of 4 year old fish by Japanese longliners operating in Australia's EEZ based on both radio and observer reports indicate a marked increase between 1990 and 1991 (Sainsbury *et al.*, 1991). Part of the increase is likely to be due to a change in fishing regulations which required vessels to retain all hooked fish in the AFZ in order to avoid high grading and the discarding of damaged or dead small fish. Prior to 1991, no such requirement existed and observers reported substantial discarding of smaller fish. Catch rates of 4 year olds in 1991 based on observer reports were over twice those based on radio reports provided by the vessels. While the available data are incomplete, extrapolation of the reported catch rate to the expected total catch off eastern Tasmania results in an estimate of catch of either 37,000 or 91,000 4 year old fish (depending upon whether vessel-radio or observer-reported catch rates are used). A catch of this magnitude would represent a substantial number of 4 year olds and raises the possibility of a potential interaction between the surface and longline fisheries with respect to escapement.

A domestic handline and troll fishery for SBT has been operating in New Zealand waters since 1980 off the west coast of the South Island primarily between 41° and 44°35'S latitude. This fishery catches primarily large SBT (>145 cm - Figure 9). Also operating within the New Zealand EEZ are licensed longliners and more recently joint-venture longliners. The SBT catches from these longliners also tends to be primarily large fish (Figure 9) and there is a large degree of overlap in the size range of fish caught by both fisheries. However, the spatial distribution of the handline and trolling vessels does not overlap the areas of greatest longline effort (Figure 10). There is no direct evidence with which to evaluate whether the spatial separation between these fisheries is sufficient or not to prevent any direct interaction between them, although the distances separating them are within the range that an SBT could traverse within a short time period. The short distance separating these fisheries combined with the considerable overlap in the size composition of the catch suggests the potential for significant interactions.

Results from tagging data can provide direct indications of the potential for interaction between surface and longline fisheries. However, since almost all SBT tagging has been on surface-caught fish, tagging results only provide an indication of one-way interactions of surface fisheries on longliners and can lead to a biased focus which

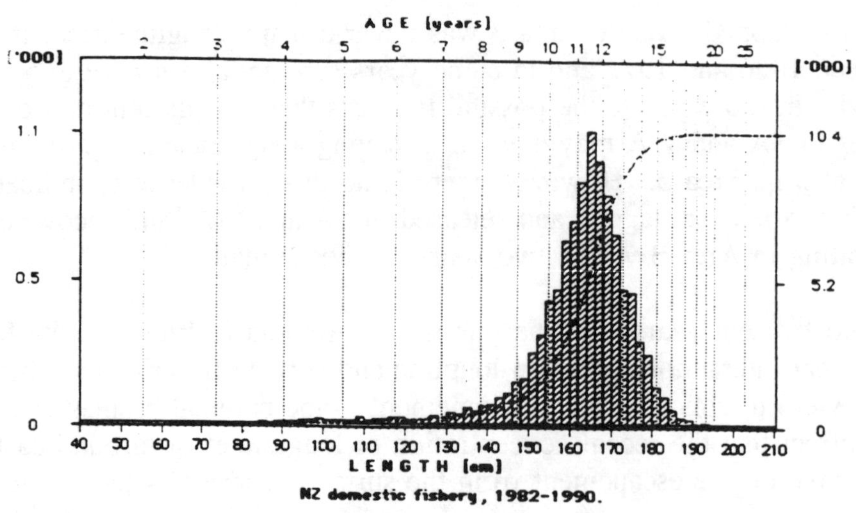

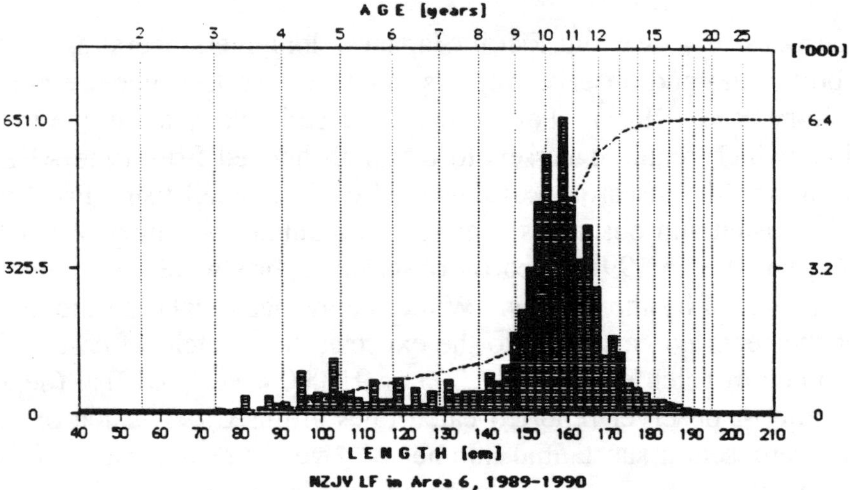

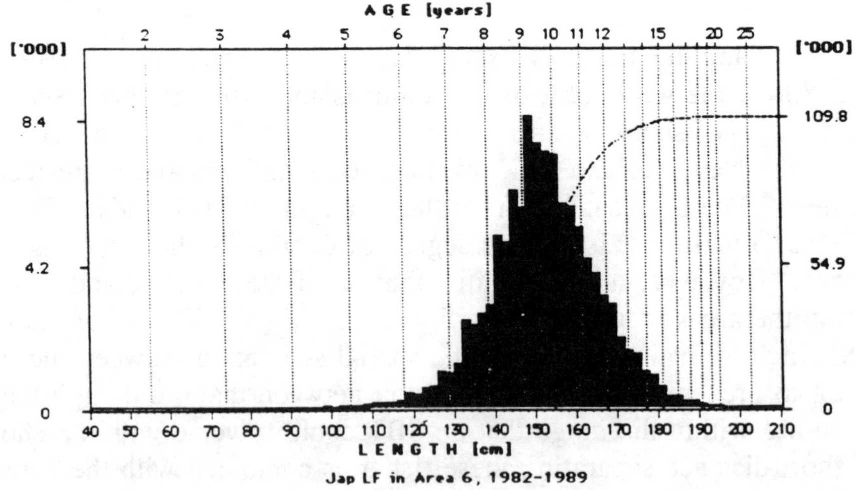

Figure 9. Comparison of the length-frequency distribution of the SBT catch in the New Zealand domestic handline and troll fishery, the New Zealand joint-venture longline fishery, and the Japanese longline fishery operating in Area 6. Upper panel - New Zealand handline and troll fishery, 1982-1990; middle panel - New Zealand joint-venture fishery, 1989-90; and lower panel - Japanese longline fishery in Area 6, 1982-89. (Based on data supplied by MAF of New Zealand and NRIFSF and stored at CSIRO, Division of Fisheries, Tasmania.)

disregards the potential for two-way interactions. Nevertheless, tagging results are useful for addressing the question of surface/longlining interactions and recent tagging results clearly demonstrate the potential for interaction between these two components.

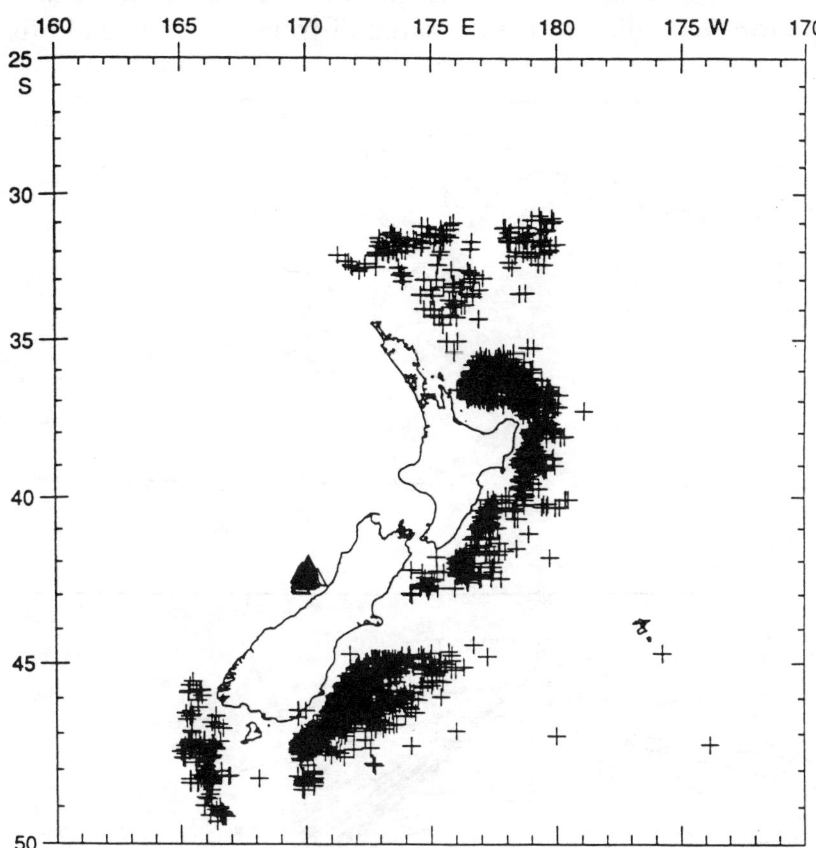

Figure 10.   Location of catch of southern bluefin tuna by New Zealand domestic (triangles) and foreign (crosses) vessels. Positions represent the location of daily-position reports for 1980-1988 (from Caton, 1991).

For example, tag returns from a small-scale, pilot tagging project off eastern Tasmania conducted during June-July 1991, suggest high rates of direct interaction between surface and longline fisheries for juvenile fish. In this programme, 89 SBT (primarily in the 85 to 110 cm size range) were tagged from trolling vessels operating in inshore waters. Over ten percent of these tagged fish were recovered within two months by longline vessels operating in waters off eastern Tasmania (Figure 11). This is the highest reported tag-recovery rate from a tuna longline fishery anywhere in the world. While the sample size is small, this high recovery rate combined with the short time and distance intervals between tagging and recaptures indicates a high degree of rapid mixing between fish that are available and vulnerable to both surface and longline gear.

On a larger geographical scale, results from the 1990/91 CSIRO tagging programme in SA, also suggest a high degree of relatively rapid mixing of fish in >80 cm size range between the surface fishery in SA and nearby longline fisheries, principally off eastern Tasmania (Table 7 and Figure 11). Thus, a total of 32 releases from SA were recaptured by longline vessels within the first ten months of tagging. This compares to a total of 135 recaptures in the SA surface fishery, 39 of which occurred after the first month of release (Table 7). The return rate from longliners after short times at liberty from the 1990/91 tagging experiment in SA are substantially greater than those reported previously (Table 8). The source of the difference may be due to increased reporting rates[4], changes in movement rates, and/or changes in the distribution and intensity of longline fishing effort in relation to the off-shore distribution of juvenile fish.

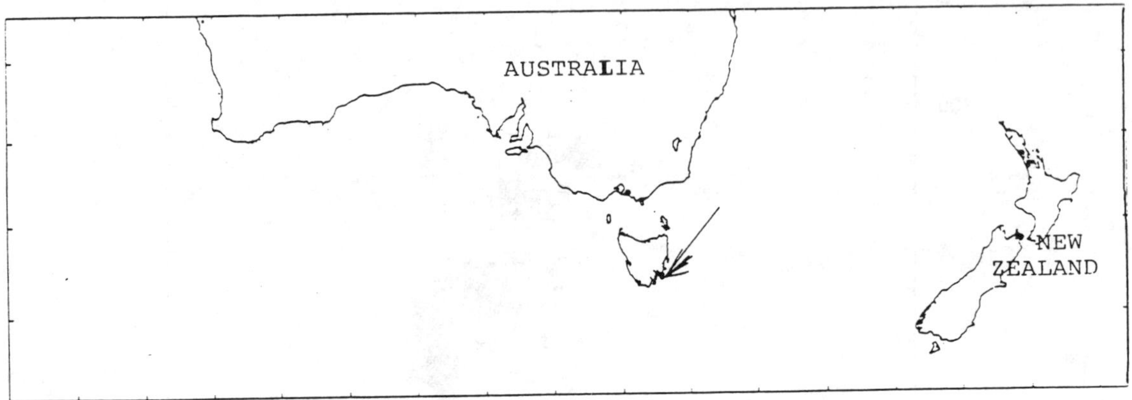

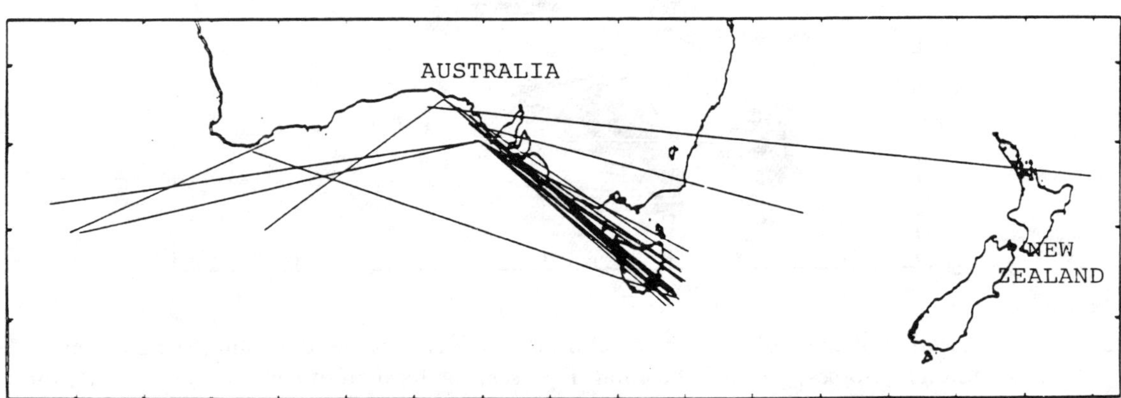

Figure 11.   Location of recaptures of tagged fish released in 1991 and recaptured by longline vessels before November 1991. Upper panel - returns from fish tagged in inshore waters of eastern Tasmania in June and July 1991. All tagged fish were initially caught by trolling vessels; all recaptures occurred within three months and were caught by Japanese longline vessels. Lower panel - recaptures of tagged fish released in South Australia during January to February from pole-and-line caught fish.

The difference in the recapture rate by longliners in 1991 off Tasmania for similar sized fish tagged and released in SA and Tasmania was around a factor of 8. While the sample size for Tasmania releases is small, the large difference in recapture rates suggests that a substantial portion of the 4 year old fish in SA during January and February did not

---

[4] A greater effort to retrieve tags from longliners in terms of publicity and personal contact with Japanese fishing masters visiting Australian ports was expended as part of the 1990/91 tagging programme than in previous experiments. In addition in 1991, Australian observer coverage was relatively high aboard Japanese longliners operating in the AFZ. These observers were instructed to look for tagged fish and may have contributed to a high reporting rate.

become vulnerable to capture by the longline fishery off Tasmania during May to August of the same year.

Finally, in August/September 1977 fifty-two tagged small SBT were released from a longliner operating southwest of Australia. Only two of these tagged fish were ever recaptured; both recaptures were in the SA surface fishery during the next fishing season and within 6 months of their time of release (Shingu, 1978). While no quantitative assessment can be made based on such small sample sizes, these two recoveries further indicate the potential for rapid mixing of fish that are available to the longline and surface fisheries.

Table 7.   Summary of tag release and return data from the 1991 CSIRO tagging experiment in South Australia. (Based on data collected and stored at CSIRO, Division of Fisheries, Hobart, Tasmania.)

Size Range (at release)	Number of releases	Time at Liberty < 1 Month		Time at Liberty >1 Month	
		South Australia	Longline Fishery	South Australia	Longline Fishery
< 71 cm	1000	0	0	7	1
71-84	2108	44	0	17	9
>84	1244	52	0	15	22

Table 8.   The percentage of the total annual Japanese longline catch of southern bluefin tuna (in numbers) that was caught in each of the ten statistical areas in Figure 1. (Based on data supplied by NRIFSF and stored at CSIRO, Division of Fisheries, Hobart, Tasmania.)

Year	1	2	3	4	5	6	7	8	9	10
1952	100	0	0	0	0	0	0	0	0	0
1953	100	0	0	0	0	0	0	0	0	0
1954	97	3	0	0	0	0	0	0	0	0
1955	99	1	0	0	0	0	0	0	0	0
1956	99	0	0	0	1	0	0	0	0	0
1957	47	0	0	0	53	0	0	0	0	0
1958	44	9	0	1	46	0	0	0	0	0
1959	18	51	0	0	31	0	0	0	0	0
1960	16	74	0	0	10	0	0	0	0	0
1961	16	58	0	5	20	0	0	0	0	0
1962	9	52	0	21	1	3	15	0	0	0
1963	5	53	0	28	6	0	6	0	0	0
1964	3	59	0	23	6	1	8	0	0	0
1965	3	53	0	23	11	1	9	0	0	0
1966	3	33	0	27	9	1	22	5	0	0
1967	1	23	0	11	4	1	10	51	0	0
1968	1	2	4	10	4	2	33	32	12	0
1969	0	1	3	4	4	8	14	24	41	0
1970	0	2	0	5	2	9	29	15	36	1
1971	0	1	0	12	6	15	38	9	19	0
1972	0	0	0	12	9	10	29	6	34	0
1973	0	0	0	14	6	5	19	10	45	0
1974	0	1	0	10	6	5	25	9	44	0
1975	0	0	1	5	7	11	21	20	35	0
1976	0	0	1	3	10	15	22	24	27	0
1977	0	0	1	10	2	5	18	32	32	0
1978	0	0	3	13	0	2	23	10	50	0
1979	0	0	0	8	4	9	12	9	57	0
1980	0	1	0	9	9	12	19	13	37	0
1981	0	0	0	10	8	12	15	8	46	2
1982	0	0	0	8	6	9	8	14	54	1
1983	0	0	0	5	3	3	13	14	61	0
1984	0	0	0	3	3	4	17	25	47	0
1985	0	0	0	1	3	6	20	31	39	0
1986	0	0	0	3	4	6	9	41	37	0
1987	0	1	0	1	5	6	19	23	44	0
1988	0	0	0	3	4	3	19	19	51	0
1989	0	0	0	6	2	3	25	23	42	0

6. INTERACTION AMONG AUSTRALIAN SURFACE CATCHES AND THE SUBSEQUENT CATCHES IN THE LONGLINE FISHERY FROM THE SAME COHORTS

The effects of the large increase in surface catches that occurred in the late 1970s and early 1980s raised concerns about the effect these catches may have had on future longline catches. Prior to this time, the effect of Australian surface fisheries on the global SBT longline catches was thought to be small, if not negligible (Murphy, 1977). Japanese longline catch rates declined from the 1960s through the 1980s. The VPA estimates of stock size and age-one recruitment also exhibit declining trends since 1960 (Polacheck and Klaer, 1991; Ishizuka and Tsuji, 1991), with the declines in stock size generally being steeper than the decline in recruitment. The extent to which these long-term declines in catch rates are due to reduced escapement from surface fisheries or due to decreased recruitment at age one (as the result of reduced spawning biomass) is not clear.

Hampton (1989) constructed yield-per-recruit surfaces for the SBT fisheries which provide an indication of the effects on the yield from a cohort for different relative amounts of effort in the Australian surface and Japanese longline fisheries (Figure 12). The results suggest that the yield per recruit in the longline fishery declines rather steeply as a function of increasing effort in the Australian surface fishery (Figure 13). The finer detailed shapes of these surfaces are dependent upon assumptions about the age-specific selectivity estimates used for each gear. In a stock like SBT which exhibits a high degree of spatial structure, age-specific selectivities are likely to vary considerably with changes in the spatial distribution of effort, targeting and retention practices. However, these changes in selectivity are only likely to effect the steepness of the relationship. The general shape of the yield surfaces is likely to be relatively robust as long as the age of most of the surface fishery catches is less than the age being caught in the longline fishery.

Hampton's yield-per-recruit surfaces assume that all recruiting juvenile SBT pass through the Australian surface fishery. To the extent that a proportion of the juveniles are not vulnerable to the Australian surface fisheries, then the actual yield surface would tend to flatten out and approach a non-zero constant as a function of an increasing fishing mortality rate in the Australian surface fishery. The migration and dispersal of juveniles is not well understood. In the traditional model of juvenile migration, young fish migrate down the west coast of Australia from the spawning grounds and first appear in the WA surface fishery at age one and two. They then remain off SA and NSW for two to three years with dispersal into the longline fishery occurring throughout this period.

However, some juveniles appear in the longline fishery off South Africa at an early age (as young as age 2) indicating that not all portions of the juvenile stock are equally vulnerable to the Australian surface fishery. Murphy (1977) suggested that a large portion of the juveniles moving down the west coast of Australia turn westward and never enter the Australian fishery and Harden-Jones (1984) suggested that some SBT larvae may be carried westward in the South Equatorial Current to African waters directly from the spawning ground. Analyses of tagging data suggests incomplete mixing of 2 year old fish from WA and SA in SA at ages 3 and 4. In contrast, 3 year old fish appear more evenly mixed between the two areas. This suggest that not all of the 2 year old fish which occur in WA pass through SA. In addition, a higher proportion of longline tag returns from fish released off WA have been from the waters around South Africa than around either SA or NSW (*e.g.* Murphy 1977; Ishizuka, 1987; Hearn and Polacheck, 1991).

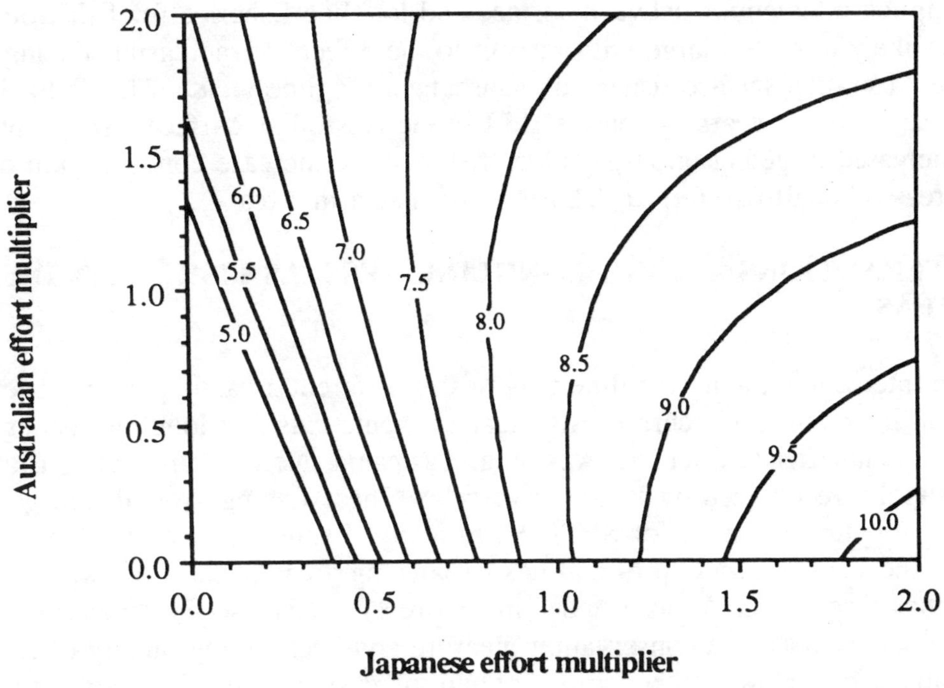

Figure 12.   Southern bluefin tuna yield-per-recruit contours for different levels of Australian and Japanese relative fishing power (from Hampton, 1989).

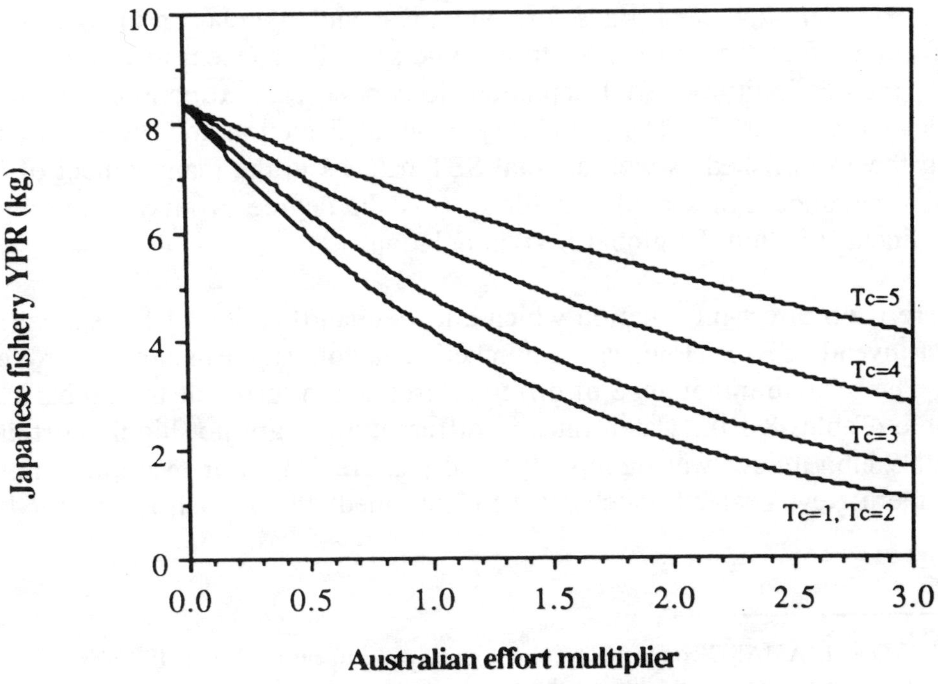

Figure 13.   The relationship between Japanese fishery yield-per-recruit and Australian fishing effort for various age-at-entry to the surface fishery (from Hampton, 1989).

Thus, longline catch-and-effort statistics and the tagging data suggest a significant proportion of 2 year old fish may not enter the SA and NSW fishery[5]. Such movement or dispersal of young fish away from the Australian coast means that a refuge from the surface fisheries exists for part of the juvenile stock which will eventually become recruited into the longline fishery. Consequently, Hampton's yield-per-recruit surfaces (and the implicit interactions between surface and longline fisheries for fish from the same cohort) are likely to be too large with respect to the effects of increasing fishing-mortality rates in the Australian surface fishery on subsequent longline yield. This is likely to be even more so in recent years, given the shift in the Australian surface fishery away from WA and increased targeting on larger/older fish and the increase concentration of longline effort in areas of small fish (*e.g.* off Eastern Tasmania in Area 7).

7.   INTERACTIONS AMONG LONGLINE VESSELS FISHING IN DIFFERENT AREAS

The interaction among longline fisheries in different areas is possibly one of the more critical interactions for current SBT management. Distinct longline fishing grounds for SBT off South Africa, Southern Australia, Tasmania, New South Wales, Brazil, and New Zealand are recognized by Japanese vessels. These fishing grounds are generally considered to be feeding areas for SBT. Historically, fishing grounds for SBT also occurred in the spawning, and pre- and post- spawning areas in the tropical and sub-tropical Indian Ocean (*i.e.* Areas 1 and 2 in Figure 1). Japanese fishermen voluntarily closed these latter areas as a conservation measure and because the quality of fish caught was generally poor. However, a significant and increasing by-catch of SBT is being taken in these two areas by Japanese, Taiwanese and Indonesian longliners targeting other tuna species (Caton, 1991).

Relative catch rates among different age classes have varied among the feeding areas, with extremely low catch rates for some areas and age classes in recent years. Longline fishing effort has also tended to become spatially concentrated and concerns exist about "the possibility of serial depletion (to commercial extinction)" of one area after another (Anon., 1991). The possibility of serial depletion and the importance of considering the areas fished as well as total SBT catches in the management of the stock depend upon the amount of spatial structuring (*i.e.* the degree of mixing and interchange versus site fidelity) within the global SBT population.

There is no direct information which allows quantification of the spatial structuring among post juvenile SBT. However, a number of factors suggests a complex spatial structure. The relative importance of different grounds in terms of the global catch has been variable (Tables 8-10). Catch rates in different areas are not highly correlated although all exhibit a downward temporal trend (Figure 14). For example, in area 5 around northern New Zealand, catch rates for intermediate size fish in the 1964-81 period

---

[5] The CSIRO and JAMARC conducted a joint tagging experiment in WA in 1990/91 and CSIRO conducted a tagging experiment in SA during the same fishing season. A set of similar tagging experiments for the 1991/92 season is currently underway. Part of the objective of these tagging experiments is to tag the same cohort in multiple years and areas. Results from these tagging experiments should provide further information on the proportion of recruits in WA that enter different fisheries.

were among the highest of any statistical area, but since that time catch rates for these intermediate sizes have been the lowest of any area and in fact declined to near zero.

Table 9. The percentage of the annual Japanese longline catch of 8 to 12 year old southern bluefin tuna (in numbers) that were caught in each of the ten statistical areas in Figure 1. Estimates of the catch-at-age in each area were based on quarterly length-frequency distributions of the catch supplied by the National Research Institute of Japan and a knife-edge partition into age classes based on the Kirkwood (1983) growth curve. (Based on data supplied by NRIFSF and stored at CSIRO, Division of Fisheries, Hobart, Tasmania.)

Year	Statistical Area 1	2	3	4	5	6	7	8	9	10
1952	100	0	0	0	0	0	0	0	0	0
1953	100	0	0	0	0	0	0	0	0	0
1954	96	4	0	0	0	0	0	0	0	0
1955	99	1	0	0	0	0	0	0	0	0
1956	69	0	0	0	31	0	0	0	0	0
1957	3	0	0	0	97	0	0	0	0	0
1958	6	7	0	1	86	0	0	0	0	0
1959	4	30	0	0	66	0	0	0	0	0
1960	2	65	0	0	32	0	0	0	0	0
1961	4	17	0	17	62	0	0	0	0	0
1962	2	18	0	50	2	8	21	0	0	0
1963	1	13	0	61	14	1	9	0	0	0
1964	1	20	0	50	12	2	14	0	0	0
1965	2	21	0	45	22	1	9	0	0	0
1966	0	6	0	56	14	1	17	5	0	0
1967	0	6	0	30	8	1	12	43	0	0
1968	0	0	9	20	4	2	24	23	16	0
1969	0	0	6	7	3	6	8	7	61	0
1970	0	0	0	9	3	6	21	10	49	1
1971	0	0	1	21	7	12	27	5	28	0
1972	0	0	0	19	8	8	22	4	38	0
1973	0	0	0	19	7	3	11	6	54	0
1974	0	0	0	15	7	4	16	7	51	0
1975	0	0	1	10	7	7	20	18	37	0
1976	0	0	1	4	10	10	14	18	44	0
1977	0	0	2	18	3	4	14	27	31	0
1978	0	0	4	22	0	1	15	6	51	0
1979	0	0	0	13	4	8	7	4	63	0
1980	0	0	0	18	2	6	14	11	49	0
1981	0	0	0	20	4	6	16	7	48	0
1982	0	0	0	17	3	6	5	6	63	0
1983	0	0	0	9	1	1	4	5	80	0
1984	0	0	0	3	0	0	15	18	64	0
1985	0	0	0	1	0	1	10	27	60	0
1986	0	0	0	3	0	1	3	30	63	0
1987	0	0	0	2	0	1	12	11	74	0
1988	0	0	0	2	0	0	29	13	56	0
1989	0	0	0	8	0	0	26	17	49	0

In addition, there has been a differential in the location of tagged fish recaptured by longliners depending upon the initial location of tagging. For example, about equal numbers of all longline recaptures in the 1960-70 tagging experiments from fish tagged in WA come from east and west of 70°E, while for releases from SA about one third of recaptures occurred west of 70°E and only a small percentage of recaptures this far west occurred for the releases from NSW. However, interpretation of these differential recapture rates is confounded by the differences in age and year of tagging. A more detailed analysis which considers both the temporal, spatial and size/age distribution of the longline catch and effort is needed. Finally, genetic studies from samples of SBT collected around New Zealand suggest heterogeneity both between localities and seasons (Smith *et al.*, 1990; Dr. P. J. Smith, pers. commun.), but the amount of information on genetic variability among SBT is very limited. The tagging and limited genetic information both suggest complex spatial structuring with correspondingly complex interactions among longline fisheries in different areas.

Table 10.   The percentage of the annual Japanese longline catch of 7 to 8 year old southern bluefin tuna (in numbers) that were caught in each of the ten statistical areas in Figure 1. Estimates of the catch-at-age in each area were based on quarterly length-frequency distributions of the catch supplied by the National Research Institute of Japan and a knife-edge partition into age classes based on the Kirkwood (1983) growth curve. (Based on data supplied by the NRIFSF and stored at CSIRO, Division of Fisheries, Hobart, Tasmania.)

Year	1	2	3	4	5	6	7	8	9	10
1952	100	0	0	0	0	0	0	0	0	0
1953	100	0	0	0	0	0	0	0	0	0
1954	96	4	0	0	0	0	0	0	0	0
1955	99	1	0	0	0	0	0	0	0	0
1956	69	0	0	0	31	0	0	0	0	0
1957	3	0	0	0	97	0	0	0	0	0
1958	6	7	0	1	86	0	0	0	0	0
1959	4	30	0	0	66	0	0	0	0	0
1960	2	65	0	0	32	0	0	0	0	0
1961	4	17	0	17	62	0	0	0	0	0
1962	2	18	0	50	2	8	21	0	0	0
1963	1	13	0	61	14	1	9	0	0	0
1964	1	20	0	50	12	2	14	0	0	0
1965	2	21	0	45	22	1	9	0	0	0
1966	0	6	0	56	14	1	17	5	0	0
1967	0	6	0	30	8	1	12	43	0	0
1968	0	0	9	20	4	2	24	23	16	0
1969	0	0	6	7	3	6	8	7	61	0
1970	0	0	0	9	3	6	21	10	49	1
1971	0	0	1	21	7	12	27	5	28	0
1972	0	0	0	19	8	8	22	4	38	0
1973	0	0	0	19	7	3	11	6	54	0
1974	0	0	0	15	7	4	16	7	51	0
1975	0	0	1	10	7	7	20	18	37	0
1976	0	0	1	4	10	10	14	18	44	0
1977	0	0	2	18	3	4	14	27	31	0
1978	0	0	4	22	0	1	15	6	51	0
1979	0	0	0	13	4	8	7	4	63	0
1980	0	0	0	18	2	6	14	11	49	0
1981	0	0	0	20	4	6	16	7	48	0
1982	0	0	0	17	3	6	5	6	63	0
1983	0	0	0	9	1	1	4	5	80	0
1984	0	0	0	3	0	0	15	18	64	0
1985	0	0	0	1	0	1	10	27	60	0
1986	0	0	0	3	0	1	3	30	63	0
1987	0	0	0	2	0	1	12	11	74	0
1988	0	0	0	2	0	0	29	13	56	0
1989	0	0	0	8	0	0	26	17	49	0

8.   INTERACTION AMONG LONGLINE VESSELS OF DIFFERENT NATIONALITIES

The longline fishery for SBT has traditionally been dominated by Japanese vessels. However, SBT have been caught as by-catch by Taiwanese and Japanese drift gillnetters and Taiwanese, Korean and Indonesian longliners. The magnitude of the by-catch in these fisheries is not well documented. However, the catch appears to have been increasing and in some cases the catch may not represent by-catch but direct effort for SBT. This trend will likely continue to increase (partially in response to the high landed value of SBT). These increases in by-catches and the reduced quota under which the Japanese longline fishery has been operating means that the magnitude of the by-catches now represents a substantial and increasing fraction of the total global longline catch. The importance of interactions between catches from the directed Japanese SBT longline fishery and catches by other fleets both with respect to rebuilding of the stock and economic competition is likely to increase.

In addition, direct competition between longline vessels targeting SBT is also likely to increase. Some domestic Australian and Australian-flagged chartered longliners catch SBT and some competition must exist between these vessels and Japanese vessels

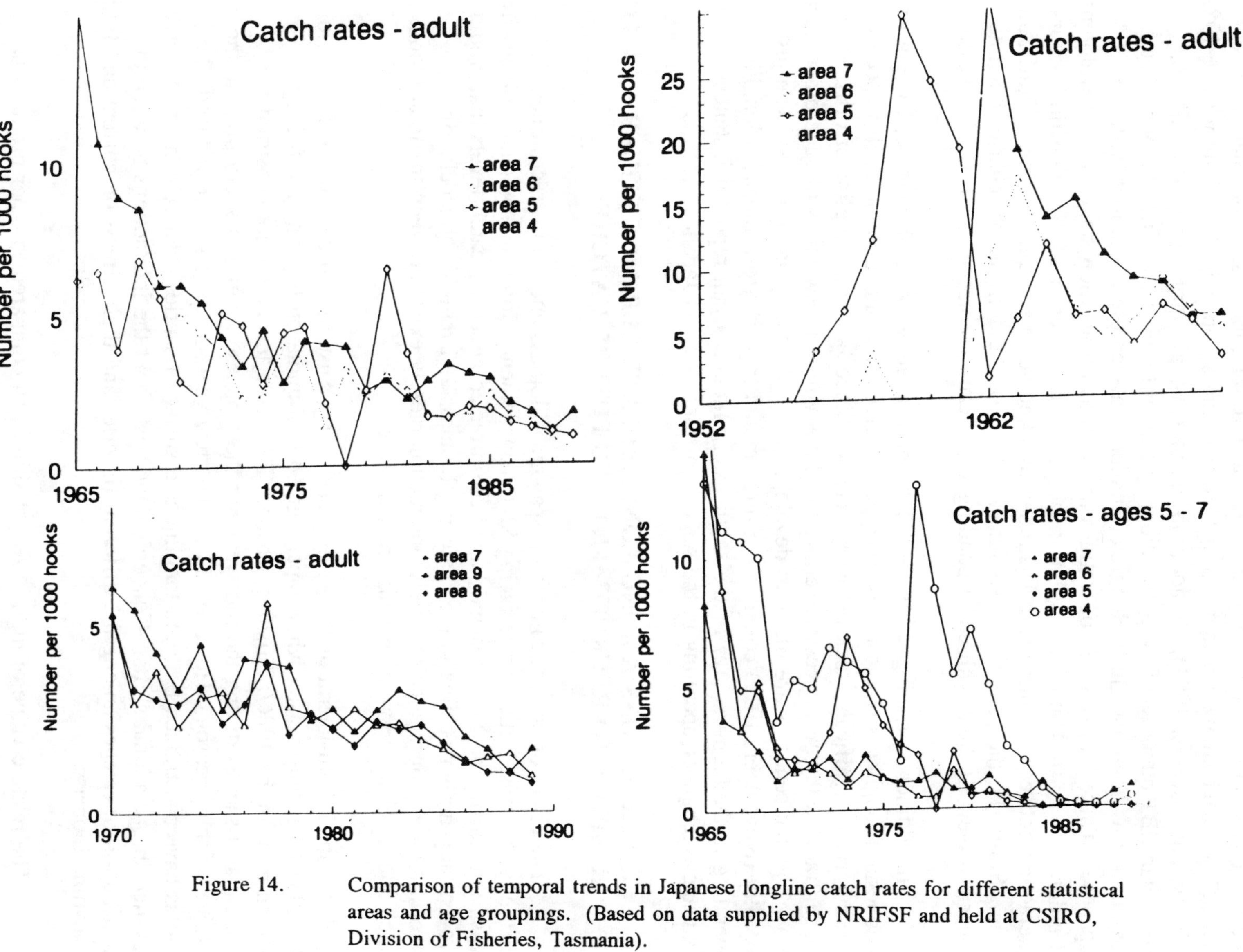

Figure 14.   Comparison of temporal trends in Japanese longline catch rates for different statistical areas and age groupings. (Based on data supplied by NRIFSF and held at CSIRO, Division of Fisheries, Tasmania).

licensed to fish in the AFZ. Similarly, four to five longline vessels on charter to New Zealand companies have participated in New Zealand's domestic SBT fishery since 1989 (Murray and Burgess, 1991). In addition, an increasingly larger proportion of the Australian SBT quota which was traditionally caught by pole-and-line and purse-seine vessels is now being caught by longliners under joint-venture operations with the Japanese. These joint-venture operations are seen as a transition stage in the development of a domestic Australian SBT longline fishery. To the extent that this transition occurs, interaction between Japanese and Australian longline vessels could become a significant issue. Similarly, the potential exists for expansion of the domestic or charter longline fishery in New Zealand with corresponding significant interactions with foreign vessels.

In addition, targeting shifts to SBT by non-Japanese (*e.g.* Taiwanese, Korean or Indonesian) longliners would not be unexpected given the landed value of this species, particularly if significant stock recovery occurs. Since Japanese longline effort is often concentrated with several vessels setting longlines within a few miles of each other, local as well global concerns are likely to develop. Thus, an Australian observer who just recently (November, 1991) completed a three-month trip aboard a Japanese longline operating in the high-sea area of the southeastern Indian Ocean reported instances of a Taiwanese longliner operating in the same immediate area as Japanese vessels.

9. INTERACTIONS AMONG LONGLINE AND SURFACE FISHERIES WITH RESPECT TO RECRUITMENT AND FUTURE CATCHES

The effect of current catches on spawning biomass and future recruitment is currently the most critical issue facing SBT management. Joint management policies by Australia, Japan and New Zealand have been directed at setting catch levels that would prevent any further decline in the spawning biomass and rebuild the stock. In determining appropriate catch levels, interaction clearly exists between the total global catch and its distribution among different fisheries.

Initially, uni-lateral quota levels were set by Australia on the surface fishery for the 1983/84 fishing season while Japan agreed to a quota on the longline fishery beginning with the 1986/87 fishing season. The quota levels did not constrain the Australian fisheries during the first year nor the Japanese longline fishery prior to the 1988/89 fishing season. However, in the last few years as quotas were reduced, the quotas have constrained both the longline and surface fisheries. These constraints have resulted in operational shifts. Concerns now exist about the interactions between the distribution of catches among the different fisheries and their effects on current and future spawning biomass.

The relative value of high-quality sashimi fish compared to other products has been and continues to result in a shift in targeting to large fish and in a shift in gear towards longlining. Thus, the WA SBT fishery is almost non-existent (*e.g.* the 1990/91 season's catch was around 200 mt); the SA surface fishery has been attempting to target larger fish with only partial success (Caton *et al.*, 1991); and an increasing proportion of the Australian quota is being caught by Japanese longline vessels under joint-venture arrangements. These shifts towards targeting larger fish have increased the necessity of understanding the interaction between current spawning biomass, future recruitment and the distribution of the catch among different fisheries.

From a yield-per-recruit perspective, a shift in selectivity towards larger/older fish will result in an increased contribution to the spawning stock over a wide range of size/age classes for a given quota. This is because the estimated annual gain in weight due to growth is greater than estimates of the loss in numbers due to natural mortality. Thus, from an equilibrium perspective, maximum gain in spawning biomass from a cohort is achieved when removals are taken out at later ages. However, during a transition phase, this will not necessarily be the case since the shifts in targeting may result in the same cohorts being tracked as they pass through the different fisheries. Tracking of a cohort is of particular concern if these shifts are occurring during a period of declining recruitment.

More critical, perhaps, in the case of SBT is the fact that increasing the proportion of the total catch by longliners could result in further reductions in the spawning biomass. Stock projections based on the results from virtual population analyses suggest that such shifts towards older fish would enhance the long term recovery rate of the stock but would decrease the recovery in the short term (Klaer and Kirkwood, 1990; Klaer et al., 1991). The estimated effects of shifting the size selectivity in the global fishery tend to be small if recent recruitment has been sufficient to allow enough escapement from the surface fishery for the spawning stock to have begun to recover. However, great uncertainty exists about the most recent recruitment levels. If recent recruitment has been declining and is insufficient to allow for recovery given current catch levels, then shifting the catch to longlining and older ages could not only retard short-term recovery, but could reduce the chances of it occurring. However, the general conclusions from these projection analyses suggest that the total magnitude of the catch appears to be a more critical factor for recovery than the distributions of the catch among the various fisheries.

10. INTERACTION, MOVEMENT, AND MIXING RATES BASED ON TAGGING STUDIES

A quantitative understanding of movements, mixing and migration rates is needed to fully assess and model the interactions among different fisheries, particularly ones that are spatially dispersed. Tagging data in combination with catch-and-effort data potentially provide the best source of information on interactions and movement. As noted above, extensive tagging programmes for SBT have been conducted using surface-caught fish in Australia (Table 1).

An examination of recapture by fishery and time of release from the three components of the Australian surface fishery demonstrate the potential for extensive and rapid movement (Figures 11 and 15). These recaptures confirm the general movement of young fish away from the WA surface fishery within the year of tagging (i.e. only a small fraction of the recaptures of WA releases recaptured within WA had times of liberty greater than one year) and suggest that a substantial number of these fish recruited into the NSW and SA fishery. The recapture data also indicate movement between the NSW and SA surface fisheries but tend to suggest unequal mixing with a greater proportion of recaptures from the area of release in the second year after tagging.

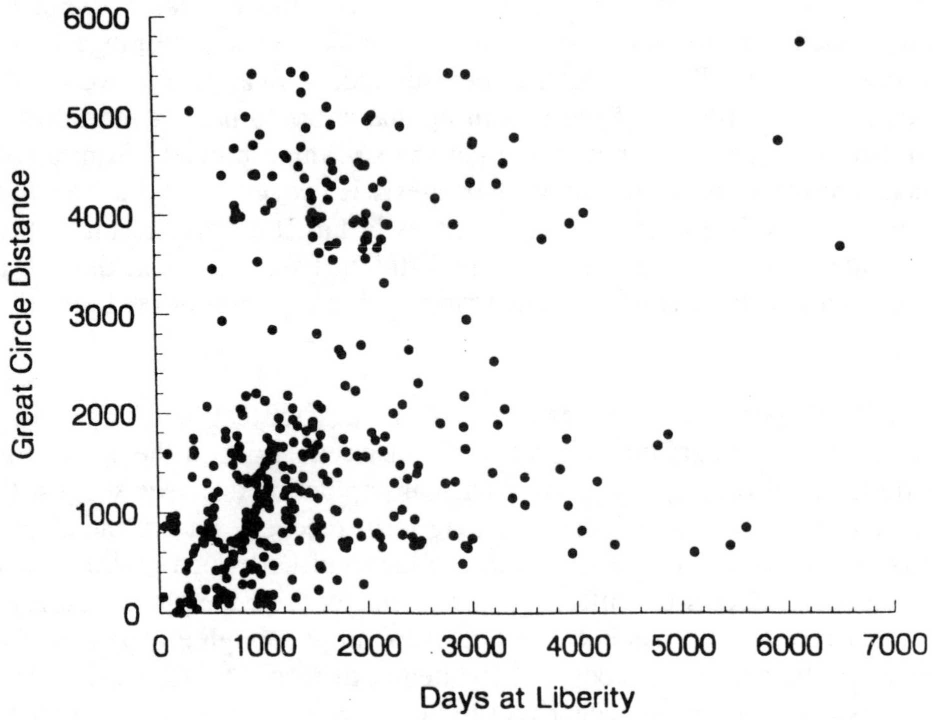

Figure 15.   Great circle distance between point of release and point of recapture *versus* time at liberty
for all longline recaptures from tagging experiments conducted between 1959-1980.
(Based on data collected and stored by CSIRO, Division of Fisheries, Tasmania).

The recapture data also indicate some interaction between the surface and longline
fisheries within the first year after release. However, the number of returns in the
longline fishery has generally been low and concerns have been expressed about the
possibility of a high rate of non-reporting of recaptured fish. Short-term tag returns from
the most recent (1990/1991) tagging experiment suggest a higher rate of interchange
between the surface and longline fisheries than previous experiments (Tables 7 and 8).

The above discussion of tag-return data is rather heuristic and any interpretation is
likely to be confounded because of a failure to account for the relative catch and effort in
the different fisheries, for changes over time, and changes in the age structure of the SBT
population. Quantitative assessment of interactions and movements requires analyses of
well developed models using appropriate statistical tests. Two different
movement/interaction models have been developed and applied to SBT tagging data.

The first approach seeks to quantify directly the effect that a change in the catch
or fishing effort in one fishery would have on the yield in another fishery (Majkowski *et
al.*, 1988; Hearn and Mazanov, 1991). In this approach, the fish tagged need to
constitute a random sample of the actual catch in one area (both with respect to size and
timing of the fishery) and a fraction of those tagged fish caught in that area need to be re-
released as a means of simulating the effect of a reduction in catch or effort in that
fishery. If this experimental protocol is followed and equilibrium conditions are assumed,
estimates of the change in yield can be derived from comparisons of the biomass of the
fish initially tagged with the biomass of those recaptured in the two fisheries (see
Majkowski *et al.*, 1988, and Hearn and Mazanov, 1991 for details). Note that while the
requirement for the re-release of tagged fish is unrealistic in most fisheries including

those for SBT, alternative field and analytical approaches can be used, at least partially, to overcome this requirement.

Majkowski *et al.* (1988) applied this method to pooled tagging data from the 1960s and Hearn and Majkowski (1987) applied this method to data from a tagging experiment conducted during the 1983/84 fishing season. Results based on the 1960s tagging returns indicate that changing the catch of any component of the Australian SBT fishery (*i.e.* WA, SA or NSW) would have little effect on the catches in any other of the surface components[6]. The maximum estimated effect of increasing the catch in any one component by one unit on the catches in other components would have been a decrease of 0.06 units. The results from the 1983/84 tagging experiment suggest larger interactions between the WA and SA fisheries (note that the NSW fishery had collapsed by this time). Thus, an increase by 1 unit of catch in WA was estimated to result in a decrease in the yield in the SA fishery of between 0.64 and 0.83 units[7]. However, the effect of increasing catches in the SA fishery on the catches in WA was still estimated to be minimal, reflecting the fact that the amount of movement from SA to WA must be low since the fishing-mortality intensity in both fisheries was high.

The second analytical approach based on tagging data used to investigate interactions among SBT fisheries is a parametric model that estimates transfer rates between fisheries, natural mortality rates and fishery-specific catchability coefficients (Hampton, 1989; 1991). These parameters are estimated using the rates of tag returns for tags released and recovered in the different fisheries. In contrast to the first approach, this second approach assumes that a random sample of the population has been tagged. Models of these types have been developed by Beverton and Holt (1957) and Sibert (1984) and the one used with the SBT data is an extension of these previous models to a three-fishery situation (Hampton, 1989; 1991).

Hampton applied his model to two sub-sets of the SBT tagging data from the 1960-1970 period and derived estimates for the Japanese longline fishery and the NSW and SA/WA combined surface fisheries. The resulting estimates of the transfer rates between the different components suggested relatively high rates of mixing and movement. The estimates of the transfer rate between the NSW and SA/WA fisheries and vice-versa were of the order of 0.1 to 0.2 per year while transfer rates into the longline fishery ranged from 0.6 to 1.8 (only one-directional movement from the surface fisheries into the longline fishery was allowed in the model). The model as developed and applied assumes that both transfer rates and catchability coefficients are independent of age. The one-directional transfer rate from the surface fishery to the longline fishery seems unrealistically high for juveniles and is a likely consequence of the assumption of a constant rate over all ages and the drop off to zero recaptures in the surface fishery for older fish.

---

[6] High incidence of undetected tag shedding or under-reporting of recaptures could mask the true level of interaction (*e.g.* Hearn *et al.*, 1991).

[7] These estimates did not account for differences in the size distributions of the fish when tagged and the actual size distribution of the catch in the fishery. If these differences were taken into account, the estimated effects would be slightly greater (Dr. W. Hearn, pers. commun.).

Both analytical approaches required assumptions in their development and application to the SBT tagging data. Some of these assumptions appear unrealistic and the robustness of the results to deviation from these assumptions has not been well explored. Thus, the tag/re-release approach of Majkowski *et al.* (1988) entails assumptions about equilibrium conditions and requires a linear relationship between catch and effort in the surface fishery (see Hearn and Mazanov, 1991 for additional discussion). For the 1960s period, the validity of pooling data from a number of different tagging programmes over several years is of concern during a period in which both the surface fishery was expanding (Table 2) and stock sizes were variable[8]. Nevertheless, the approach appears to provide a reasonable measure of the level of interaction between fisheries at the time of the tagging programme, but appears to have limited predictive use when either the magnitude of the fishery and/or stock sizes change markedly. Thus, the large difference in the degree of interaction between the 1960s and early 1980s as estimated by the two different sets of tagging experiments is not predictable based soley on the results from one of the experiments.

The results from Hampton's parametric approach provide a direct framework for predicting interaction as fisheries change. However, Hampton's analysis needed to pool tagging data from different years (and thus ignore cohort and year effects both with respect to the stock and the fishery). The analysis also discounts the possibility that once fish become vulnerable to longliners that they remain vulnerable to surface fisheries. Finally, Hampton's model assumes age-independent catchability coefficient and transfer rates. All of these factors are important with respect to SBT. Consequently, using the derived parameter estimates for assessment and predictive purposes could be misleading. The problems in this case are not a limitation of the approach, but limitations in the catch, effort and tagging data which would allow construction and estimation of a sufficiently detailed parametric model. However, overcoming these problems is difficult because of the need to simultaneously tag in all of the fisheries combined with the need for large sample sizes and replication in order to get meaningful parameter estimates (*i.e.* low coefficients of variation and low correlations among the parameter estimates).

11. RESEARCH AND DATA NEEDS

The most important interaction issues for SBT differ from those for most other tuna due to the current low stock levels and the potential for over-exploitation. Given the current low biomass, the stock/recruitment dynamics for SBT are the most critical management and assessment issue. While traditional approaches for examining the stock/recruitment dynamics do not directly involve interaction issues, concerns about spatial structure and serial depletion mean that an understanding of the interaction among the spatially-dispersed SBT fisheries is of fundamental importance for understanding the recruitment dynamics. As such the most important research and data needs with respect to interactions for SBT continue to centre around questions of movement, migration routes, mixing rates and site fidelity. Two areas of future research can be identified - those which can be done with already existing data and those requiring new data collection schemes.

---

[8] VPA estimates vary by about a factor of two in the number of either 1, 2 ,3 or 4 year old fish during the period 1960-1970 (Polacheck and Klaer, 1991).

There is a need for improved models and analyses of existing tagging data that account for both the age-specific and temporal dynamics of the stock, changes in the fishery and environmental factors. There is also a need for assessment models which take into account not only the temporal dynamics of the catch-and-effort data but their spatial dynamics. These types of modelling activities are likely to need fine scale catch, effort and size-composition data from the longline fishery. Historically, sufficient fine-scale resolution catch-and-effort data have been collected but have not been available for research. The fine-scale information on the size distribution of the catch that has been collected is also unavailable. The overall level of data coverage also needs to be improved.

Improvement to the analyses of existing catch, effort, and tagging data (including the fine-scale catch-and-effort data when available) may provide further insights into the spatial structure and movements of SBT. However, by themselves, the existing data are not likely to be sufficient for addressing the most fundamental interaction questions. The fundamental shortcoming of the present data is that all tagging experiments (with the exception of 50 longline-caught fish - see above) have been done on juvenile fish captured in the Australian surface fishery. This precludes any direct information on movement and mixture rates of juvenile and adult fish, and on juvenile fish in areas outside the Australian surface fishery. Consequently, in terms of the current interaction issues in the fisheries for SBT, there is a need for additional data. There are several complementary approaches that could be undertaken to overcome the fundamental shortcoming in the present data. These include:

1) tagging in fisheries and areas other than the Australian surface fishery;

2) developing methods for tracking the real-time movements of fish over long distances and time periods (*e.g.* "archival tags", see Hunter *et al.*, 1986);

3) genetic analyses of different age classes of fish caught over a range of geographic locations;

4) micro-analyses of the isotope ratios of hard parts from fish that have been tagged and released in various locations; and

5) collection of behavioural data on feeding, temperature preference, *etc.* in the context of developing models to explain why SBT move.

There are problems and difficulties with all of these approaches. Significant progress is most likely to be made through a multi-faceted strategy involving a combination of at least several, if not all, of these approaches.

12.   REFERENCES CITED

Anonymous. 1991. Report of the tenth meeting of Australian, Japanese and New Zealand Scientists on Southern Bluefin Tuna. Wellington, New Zealand, 23-27 September 1991.

Beverton, R.J.H. and S.J. Holt. 1957. On the dynamics of exploited fish populations. Fishery Investigations (Series II) 19:1-533.

Caton, A.E. 1993. Commercial and recreational components of the Southern Bluefin tuna (*Thunnus maccoyii*) fishery. *In* Interactions of Pacific Tuna Fisheries: Proceedings of the First FAO Expert Consultation on Interactions of Pacific Tuna Fisheries, edited by R.S. Shomura, J. Majkowski, and S. Langi, 3-11 December 1991, Noumea, New Caledonia. [See this document].

Caton, A.E. (editor). 1991. Review of aspects of southern bluefin tuna biology, population and fisheries. *Spec.Rep.I-ATTC*, (7):181-350.

Caton, A., K. Williams, P. Ward, and C. Ramirez. 1991. The 1990-91 Australian southern bluefin tuna season. Tenth Scientific Meeting on Southern Bluefin Tuna SBFWS/91/1.

Caton, A., K. McLoughlin, and M.J. Williams. 1990. Southern bluefin tuna: scientific background to the debate. *Bull.Aust.Bur. Rural Resources*, (3):41 p.

Hampton, W.J. 1989. Population Dynamics, Stock Assessment and Fishery Management of the Southern Bluefin Tuna (*Thunnus maccoyii*). Ph.D. Thesis. University of New South Wales, Kensington, Australia, 273 p.

Hampton, W.J. 1991. Estimation of southern bluefin tuna (*Thunnus maccoyii*) natural mortality and movement rates from tagging experiments. *Fish.Bull.NOAA-NMFS*, (89):591-610.

Harden-Jones, F.R. 1984. A view from the ocean. *In* Mechanisms of migration in fishes, edited by J.D. McCleave, G.P.Arnold, J.J. Dodson, and W.H. Neil. New York. Plenum Press, pp. 1-26.

Hearn, W.S., G.M. Leigh, and R.J.H. Beverton. 1991. An examination of a tag-shedding assumption, with application to southern bluefin tuna. *ICES J.Mar.Sci.*, 48:41-51.

Hearn, W.S. and J. Majkowski. 1987. Interactions among the Australian southern bluefin tuna fisheries: a preliminary analysis based on tag release-recapture experiment. Paper presented at Expert Consultation on Stock Assessment of Tunas in the Indian Ocean, 4-8 December 1986, Colombo, Sri Lanka.

297

Hearn, W.S. and A. Mazanov. 1991. Interaction between two fisheries: Tag-recapture methods for estimating the effect on catches of changing fishing intensity. Paper presented at FAO Expert Consultation on Interactions of Pacific Ocean Tuna Fisheries, 3-11 December 1991, Noumea, New Caledonia, 21 p.

Hearn, W. and T. Polacheck. 1991. Estimation of the average total mortality rate of southern bluefin tuna from tagging data. Tenth Scientific Meeting on Southern Bluefin Tuna SBFWS/91/5.

Hunter, J.R., A.W. Argue, W.H. Bayliff, A.E. Dizon, A. Fonteneau, D. Goodman, and G.R. Seckel. 1986. The dynamics of tuna movements: an evaluation of past and future research. *FAO Fish.Tech.Pap.*, (227):78 p.

Ishizuka, Y. 1987. Migration and growth of southern bluefin tuna based on Australian tagging data. Ninth Trilateral Scientific Meeting on Southern Bluefin Tuna, Hobart, Australia.

Ishizukua, Y. and S. Tsuji. 1991. Assessment of the southern bluefin tuna stock. Tenth Scientific Meeting on Southern Bluefin Tuna SBFWS/91/8.

Klaer, N. and G.P. Kirkwood. 1990. Future projections of southern bluefin tuna biomass incorporating stochastic variation in the stock-recruitment relationship. Ninth Scientific Meeting on Southern Bluefin Tuna SBFWS/90/7.

Klaer, N., T. Polacheck and K. Sainsbury. 1991. Future projections of southern bluefin tuna biomass incorporating stochastic variation in the stock-recruitment relationship. Tenth Scientific Meeting on Southern Bluefin Tuna SBFWS/91/7.

Kirkwood, G.P. 1983. Estimation of von Bertalanffy growth curve parameters using both length-increment and length-at-age data. *Can.J.Fish.Aquat.Sci.*, 40:1405-11.

Majkowski, J., W.S. Hearn, and R.L. Sandland. 1988. A tag-release/recovery method for predicting the effect of changing the catch of one component of a fishery upon the remaining components. *Can.J.Fish.Aquat.Sci.*, 45:675-84.

Majkowski, J, and G. Morris (editors). 1986. Data on southern bluefin tuna (*Thunnus maccoyii* (Castlenau)). *Rep.CSIRO Mar.Lab.*, 179: 95 p.

Majkowski, J.K. Williams, and G.I. Murphy. 1981. Research identifies changing patterns in Australian tuna fishery. *Aust.Fish.*, 40(2):5-10.

Murray, T. and D. Burgess. 1991. Trends in southern bluefin tuna fisheries operating in New Zealand waters, 1980-1990. Tenth Scientific Meeting on Southern Bluefin Tuna SBFWS/91/3.

Murphy, G.I. 1977. New understanding of southern bluefin tuna. *Aust.Fish.*, 36(1):2-6.

Polacheck,T. and N. Klaer. 1991. Assessment of the status of the southern bluefin tuna stock using virtual population analysis - 1991. Tenth Scientific Meeting on Southern Bluefin Tuna SBFWS/91/6.

Sainsbury, K., T. Polacheck, and N. Klaer. 1991. Interpretation of recent changes in the longline catch rate of 4+ southern bluefin tuna. Tenth Scientific Meeting on Southern Bluefin Tuna SBFWS/91/4.

Shingu, C. 1978. Ecology and stock of southern bluefin tuna. *Fish.Study Jap.Assoc.Fish.Resources Protection,* No. 31. [English translation available by M.A. Hintze as *Rep.CSIRO Div.Fish.Oceanogr.,* (1981), 131:79 p.]

Sibert, J.R. 1984. A two-fishery tag attrition model for the analysis of mortality, recruitment and fishery interaction. *Tech.Rep.Tuna Billfish Assess.Programme, S.Pac.Comm.,* (13):27 p.

Smith, P.J., A. Jamieson and A.J. Birley. 1990. Electrophoretic studies and the stock concept in marine teleosts. *J.Cons.CIEM,* 47:231-41.

INTERACTIONS BETWEEN FISHERIES FOR SMALL TUNAS OFF THE SOUTH CHINA SEA COAST OF THAILAND AND MALAYSIA

Mitsuo Yesaki
Indo-Pacific Tuna Development and Management Programme
Colombo, Sri Lanka

ABSTRACT

Data of small tunas caught by Thai and Malaysian pelagic fisheries in the South China Sea are examined for possible interactions between and among fisheries. The report provides brief descriptions of the several fisheries and changes in species composition. Although direct evidence, *e.g.*, tagging data are not available, the markedly increased exploitation of small tunas by the Thai purse-seine fishery since the early 1980s, accompanied by declines in catches of small tunas by the Thai gillnet fishery and the Malaysian troll, gillnet, and purse-seine fisheries during the same period, suggests possible interactions.

1. INTRODUCTION

Landings of small tunas from the South China Sea coast of Thailand and Malaysia increased from about 7,000 mt in 1970 to 141,000 mt in 1989, with much of this increase occurring since 1982. The species comprising this group are the longtail tuna (*Thunnus tonggol*), kawakawa (*Euthynnus affinis*) and frigate tuna (*Auxis thazard*). Longtail tuna is the most abundant and largest of these three species in this area.

2. FISHERIES

The rapid increase in landings of small tunas was primarily a result from the development of the Thai tuna purse-seine fishery, but also of increased landings by the traditional fisheries of Thailand and Malaysia. These include fisheries that aim specifically for small tunas as well as fisheries that capture small tunas only incidentally while fishing for other species. The objective of this analysis is to examine available information in order to determine interactions between these fisheries.

2.1 <u>Thailand</u>

Small tunas were captured only incidentally in the drift gillnet and purse seine fisheries prior to the eighties. The gillnet fishery traditionally focused on narrow-barred Spanish mackerel (*Scomberomorus commerson*) and the purse-seine fisheries sought a variety of small pelagic species. In the late seventies, a processing plant in Bangkok started canning small tunas for export, thereby creating a ready market for these species. To meet this demand, the gillnet fishery started seeking small tunas in the Gulf of Thailand in about 1981, followed about a year later by the purse-seine fisheries. The gillnets used for small tunas are virtually the same as those for narrow-barred Spanish mackerel, and have essentially not changed since the introduction of nylon webbing, except that the length has increased with adoption of mechanical haulers.

The two-boat purse seine was introduced into the Gulf of Thailand in 1925. The one-boat purse seine was developed in 1930 and has now virtually replaced the older type. A major innovation in purse seining occurred in 1973 when fish lures (Fish Aggregating Devices - FADs) were developed (Cheunpan, 1986). The FADs were initially deployed by one-boat purse seiners in coastal areas to capture small pelagic species. These FADs were deployed further and further offshore with the growth of the luring purse-seine fishery, resulting in larger incidental catches of small tunas. High tuna prices also encouraged purse-seine fishermen to aim for free-swimming schools of small tunas. Fishermen based in Rayong developed a special purse seine to capture small tunas, which is longer, deeper, and fabricated if webbing with larger meshes in the wings and the bunt. These purse seines are used for small tunas and other large pelagic fish, including scad (*Megalaspis cordyla*), yellowtail scad (*Caranx mate*), *etc.* (Supongpan and Saikliang, 1987). Although there are one-boat, luring and tuna purse seines for different species groups, there are no types of vessels specifically for small tunas. Larger vessels use these different types of purse seines depending upon fishing conditions. Details of Spanish mackerel gillnets and one-boat, luring, and tuna purse seines used in Thailand are given in Okawara *et al.* (1986) and general characteristics of vessels fishing these gears are shown in Table 1. Tuna purse seiners are the largest and highest-powered of the vessels fishing small tunas in the South China Sea.

Table 1. Number, size and horse-power ratings of vessels fishing small tunas off the South China Sea coast of Thailand and Malaysia.

Country	Vessel Type	Total	Number[1]				LOA(m)[2]	HP rating[2]
			By length class (m)					
			14	14-18	18-25	25		
Thailand	Purse seine							
	- luring	432	79	45	308	-	10.0-20.0	30-240
	- one-boat	522	28	120	394	10	19.5-20.0	185-320
	- tuna						28.0-32.0	350-520
	Spanish mackerel gillnet	401	45	172	175	9	17	120
Malaysia	Purse seine	312	-	-	-	-	-	-
	- luring	-	-	-	-	-	19.7-25.0	180-240
	- one-boat	-	-	-	-	-	15.3-18.9	33-195
	Gillnet	1387	-	-	-	-	-	-
	- Spanish mackerel	-	-	-	-	-	10.0-13.5	12-24
	Hook-and-line	1158	-	-	-	-	-	-
	- troll-line	-	-	-	-	-	15	45

1/ - IPTP, 1991
2/ - Okawara et al., 1986; Munprasit et al., 1989

2.2   <u>Malaysia</u>

Small tunas are captured primarily by the troll, gillnet, and purse-seine fisheries off the east coast of Peninsular Malaysia. A troll fishery for small tunas has existed off this coast at least since the thirties and possibly earlier. This fishery has traditionally aimed at small tunas for local fresh markets and katsuobushi factories. Small tunas are captured incidentally by the gillnet fishery and one-boat and luring purse-seine fisheries. Details of troll lines, gillnets, and purse seines used off the east coast of Peninsular Malaysia are given in Munprasit *et al.* (1989). Spanish mackerel gillnetters and tuna trollers are the smallest and lowest-powered of the vessels fishing small tunas in the South China Sea (Table 1).

There are at present two fisheries in the South China Sea that aim for small tunas; Thai tuna purse seine and Malaysian troll fisheries. The Thai gillnet, one-boat, and luring purse seine and the Malaysian gillnet, one-boat, and luring purse-seine fisheries capture small tunas only incidentally while fishing for other species.

3. DATA CURRENTLY AVAILABLE WHICH ARE RELEVANT TO STUDIES OF INTERACTIONS

3.1 <u>Sources of Data</u>

3.1.1 Thailand

Four sources of information were utilized in reconstructing the small tuna fisheries of Thailand. The first source was the catch statistics compiled by the Statistics Division of the Department of Fisheries (IPTP, 1991). The second was the catch and effort statistics collected by the Statistics Division from logbooks completed for selected vessels (IPTP database). Catch and effort of small tunas (not separated by species) are available from 1972 for drift gillnetters and from 1973 for purse seiners. Catches have been separated into longtail and other tunas (kawakawa and frigate) since 1982. This is the longest time-series of catch and effort statistics available. Unfortunately, these statistics do not provide good indices of the relative abundance of small tunas because in the case of the gillnet fishery different species and/or species groups were sought during this time interval and the different types of purse seines have been grouped together in the case of this fishery. The third source of information was the tuna sampling programme initiated in May 1987 by the Marine Fisheries Department with assistance from the Indo-Pacific Tuna Development and Management Programme (IPTP). Data collected from May 1987 to December 1989 have been published in "Tuna sampling programme in Thailand" (IPTP, 1990b). The fourth source was a collection of papers on small tunas published by the Marine Fisheries Department by Klinmuang (1978; 1981) and Supongpan and Saikliang (1987).

3.1.2 Malaysia

Catch statistics of small tunas from 1977 to 1983 were obtained from the Fishery Statistical Bulletin (SEAFDEC, 1979; 1980; 1981; 1982; 1983; 1984; 1985) and thereafter from the Statistics Division, Department of Fisheries (IPTP database). The second source of information for Malaysia was the tuna sampling programme initiated by the Marine Fisheries Resources Research Center, Terengganu in 1982. This programme was continued through 1986; unfortunately, some data collected during this programme have been lost. Catch and effort statistics are available from 1982 through 1985 and length-frequency distributions for some months of 1983, 1985, and 1986 in the IPTP database. A more intensive tuna sampling programme initiated by the Center in January

1987, with the assistance of IPTP, was the third source of information for this analysis. Data collected from January 1987 to December 1989 have been published in "Tuna sampling programme in Malaysia" (IPTP, 1990a). The fourth source was catch statistics collected at the government landing site in Kuala Besut.

3.2 Landings

Landings of small tunas from the South China Sea coast of Thailand and Malaysia increased exponentially from 19,000 mt in 1980 to 141,000 mt in 1989. Thailand accounted for 93% and Malaysia the remainder of the 1989 landings (Table 2 and Figure 1).

Table 2. Small tuna landings from the South China Sea coasts of Thailand and Malaysia.

Country Gear	Thailand[1]				Malaysia[2]					TOTAL
	PS	GILL	UNCL	TOTAL	PS	GILL	TROLL	UNCL	TOTAL	
1970	-	-	4315	4315	-	-	-	-	-	-
1971	3960	-	1464	5424	-	-	-	-	-	-
1972	2784	626	2098	5508	-	-	-	-	-	-
1973	2699	3125	695	6519	-	-	-	-	-	-
1974	4318	2405	1992	8715	-	-	-	-	-	-
1975	5108	2845	3210	11163	-	-	-	-	-	-
1976	4287	3089	1514	8890	-	-	-	-	-	-
1977	5936	5249	111	11296	3454	1886	4667	95	10102	21398
1978	3800	3294	1161	8258	2675	2470	3777	32	8952	17210
1979	7547	5769	1397	14713	1936	1682	3098	162	6878	21591
1980	6225	6290	380	12895	891	2465	2833	197	6386	19281
1981	8691	11236	271	20198	1339	5413	8320	21	15093	35291
1982	18123	21243	295	39661	490	4054	8346	-	12890	52551
1983	67323	14476	202	82001	1479	5850	8818	11	16158	98159
1984	48949	20406	114	69469	2045	6130	6472	1	14648	84117
1985	60825	20375	33	81233	3275	5077	6242	6	14600	95833
1986	70969	19260	155	90384	1330	2842	8348	123	12643	103027
1987	77720	13041	-	90761	6470	3615	8620	41	18746	109507
1988	98911[3]	17114[3]		116025[3]	8529	872	6258	36	15695	131720
1989	123371[4]	7696[4]		131067[4]	2737	1848	5351	23	9959	141026

1/ - Statistics Division, Department of Fisheries, Thailand IPTP, 1991
2/ - 1977-1983 statistics - SEAFDEC, 1979-1985
    - 1984-1989 statistics - IPTP, 1991
3/ - Logbook survey, Statistics Division, Department of Fisheries, Thailand
4/ - Preliminary estimates from the tuna sampling programme

Small tuna landings from the South China Sea coast of Thailand increased gradually from 4,000 mt in 13,000 mt in 1980, then markedly to 82,000 mt in 1983 (Table 2 and Figure 1). After a slight decrease in 1984, landings have since increased and totalled 131,000 mt in 1989. Landings by purse seiners and gillnetters were of equal magnitude up to 1982, but thereafter, landings by purse seiners greatly exceeded those of gillnets. Purse-seine landings have increased by 150% and gillnet landings have decreased by 62% since 1984.

Small tuna landings from the east coast of Peninsular Malaysia fluctuated between 13,000-19,000 mt from 1981 to 1988 and decreased to 10,000 mt in 1989. Troll and gillnet landings decreased from 1987, whereas purse-seine landings decreased from 1988.

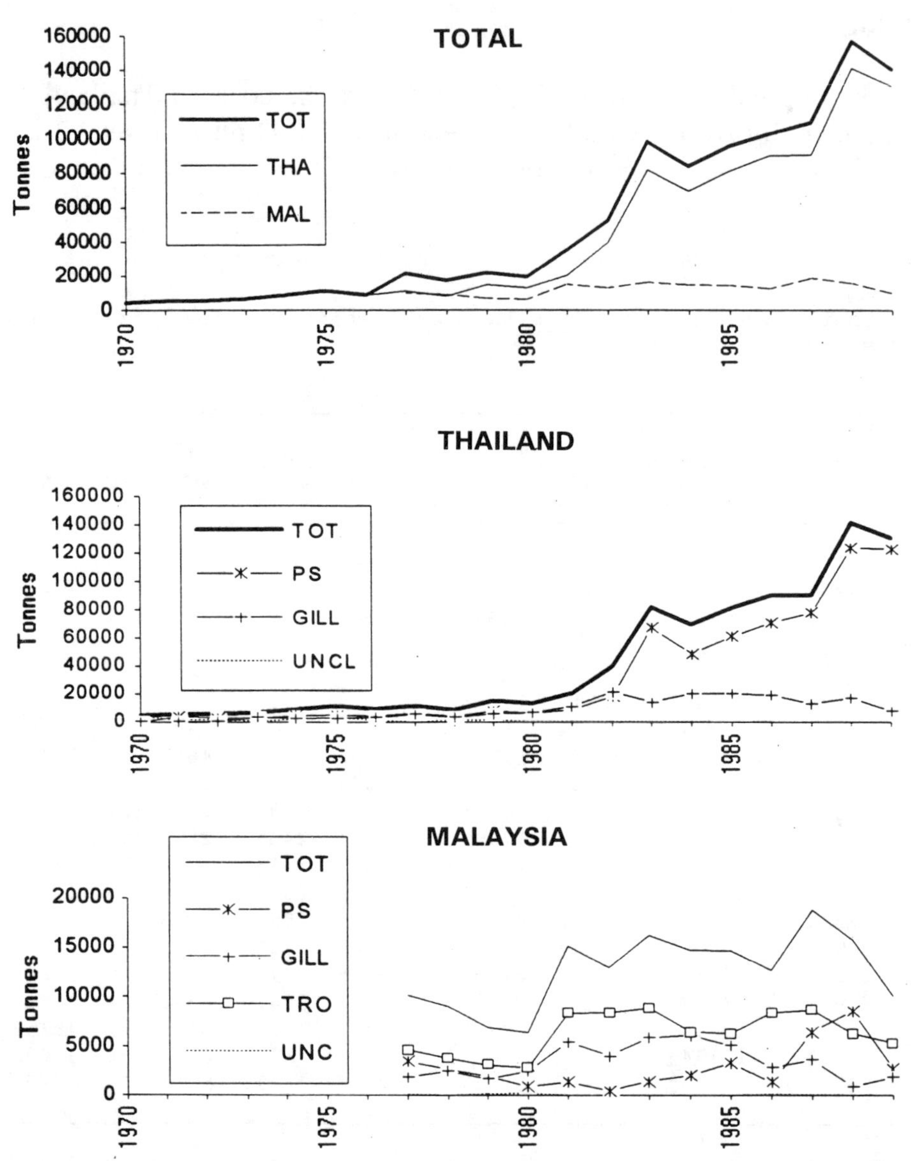

Fig. 1. Landings of small tunas from the South China Sea coasts of Thailand and Malaysia by fishing gears.

3.3   Catch and Effort

A long time-series of catch and effort statistics is available only for the Thai purse seine (all types combined) and Spanish mackerel gillnet fisheries and the Malaysian troll fishery. Annual effort of Thai purse seiners increased from 23,000 days in 1973 to 113,000 days in 1983 and catch from 4,000 mt to 67,000 mt for these respective years. Thai purse-seine effort and catch are still rising with respective increases of 28% and

100% since 1984. On the other hand, the annual catch of Thai gillnetters peaked in 1982 and has since stabilized at about 17,000 mt though effort increased 56% from 1984 to 1988. Effort of Terengganu trollers declined from 2,400 trips in the 1983-1985 interval to 1,700 trips in the 1987-1989 interval and annual catches decreased from an average of 905 mt to 356 mt during these respective intervals (Figure 2). Furthermore, the average number of trips per month of trollers at Kuala Besut decreased from 263 for the last semester of 1985 to 46 during 1990 and annual landings during these respective intervals declined from 99 mt to 30 mt (Table 3).

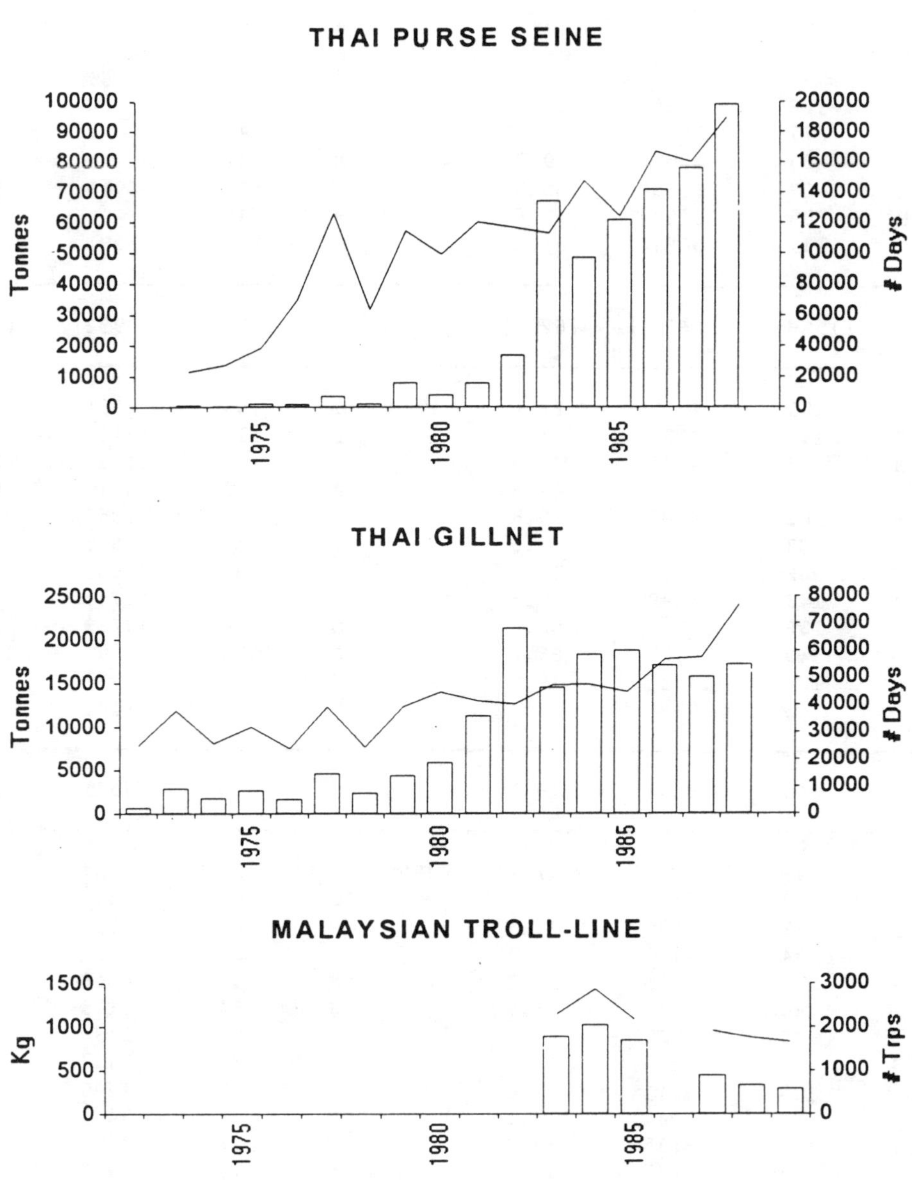

Fig. 2. Effort (lines) and catch (bars) of Thai purse seines and gillnets and Malaysian troll-lines.

Table 3. Monthly effort, catch and catch rates of trollers in Kuala Besut from July 1985 to December 1990.

Year	Mon	No.trips	Catch(kg)	Kg/trip	Year	Mon	No.trips	Catch(kg)	Kg/trip
1985	J	-	-	-	1986	J	181	7595	42.0
	F	-	-	-		F	90	2792	31.0
	M	-	-	-		M	16	255	15.9
	A	-	-	-		A	95	-	-
	M	-	-	-		M	163	2905	17.8
	J	-	-	-		J	103	65	0.6
	J	365	17569	48.1		J	164	-	-
	A	277	7914	28.6		A	86	-	-
	S	285	22702	79.7		S	181	-	-
	O	337	26955	80.0		O	190	17989	94.7
	N	205	13851	67.6		N	130	-	-
	D	111	9682	87.2		D	112	2391	21.3
		(1580)	(98673)	(62.5)			1511	(33992)	(39.8)
1987	J	101	2330	23.1	1988	J	163	3591	22.0
	F	147	1107	7.5		F	116	2102	18.1
	M	83	2096	25.3		M	158	4131	26.1
	A	152	7072	46.5		A	100	6770	67.7
	M	162	4993	30.8		M	127	3356	26.4
	J	103	5627	54.6		J	135	2561	19.0
	J	202	7080	35.0		J	87	2251	25.9
	A	143	5099	35.7		A	106	6757	63.7
	S	136	11815	86.9		S	84	19729	234.9
	O	165	14226	86.2		O	170	20101	118.2
	N	164	11351	69.2		N	34	1071	31.5
	D	85	2898	34.1		D	26	1181	45.4
		1643	75694	46.1			1306	73601	56.4
1989	J	49	866	17.7	1990	J	39	2525	64.7
	F	64	3142	49.1		F	162	5553	34.3
	M	72	2989	41.5		M	40	3119	78.0
	A	94	7416	78.9		A	15	1418	94.5
	M	79	2455	31.1		M	48	4046	84.3
	J	86	1512	17.6		J	64	2173	34.0
	J	78	1721	22.1		J	90	1193	13.3
	A	78	3416	43.8		A	34	2117	62.3
	S	108	8311	77.0		S	15	1285	85.7
	O	85	10041	118.1		O	11	3473	315.7
	N	38	8815	232.0		N	18	2207	122.6
	D	26	857	33.0		D	16	912	57.0
		857	51541	60.1			552	30021	54.4

The Gulf of Thailand and contiguous South China Sea have been demarcated into fishing areas for statistically purposes by the Department of Fisheries, Thailand (Figure 3). The areas fished by both Thai gillnetters and purse seiners have changed considerably since the early seventies. Thai gillnetters operated principally off the east coast (Area 1) and inner Gulf (Area 2) and in some years off the southwest coast (Area 4) during the seventies, with most of the small tuna catch made in Areas 1 and 2. During the eighties, Thai gillnetters operated primarily in Areas 4 and 3 (west coast) and to a lesser extent in Area 2. Most of the catch during this latter period was made in Areas 4 and 3 (Figure 4).

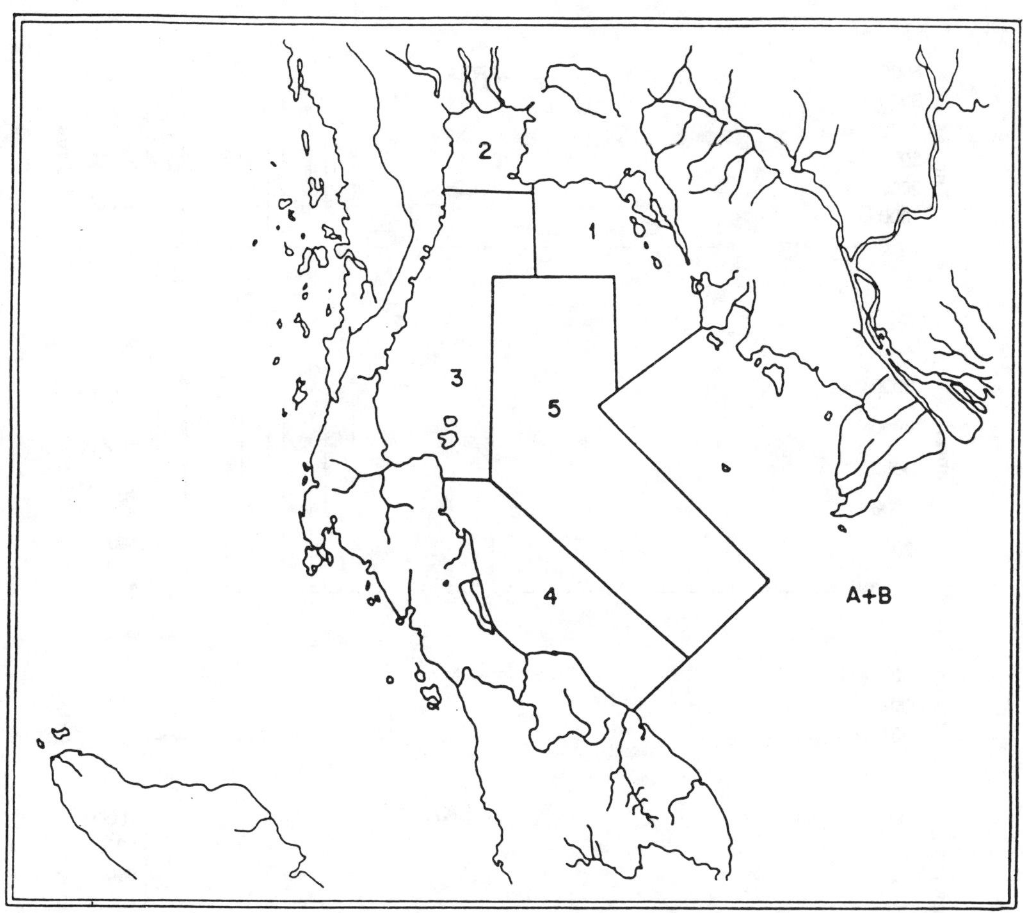

Fig. 3   Fishing grounds in the Gulf of Thailand.
          (1 - east coast, 2 - inner Gulf, 3 - west coast, 4 - southwest
          coast, 5 - middle Gulf, A + B - Vietnam and South China Sea).

Thai purse seiners fished essentially in the upper half of the Gulf of Thailand (Areas 3 and 1) for small pelagic species during the seventies (Figure 5). Small tuna catches were negligible in all areas. During the eighties, Thai purse seiners expanded operations to the middle (Area 5) and lower half (Area 4) of the Gulf of Thailand. The largest catches were reported from Area 5 in 1983, 1987, and 1988 and Area 4 in 1986.

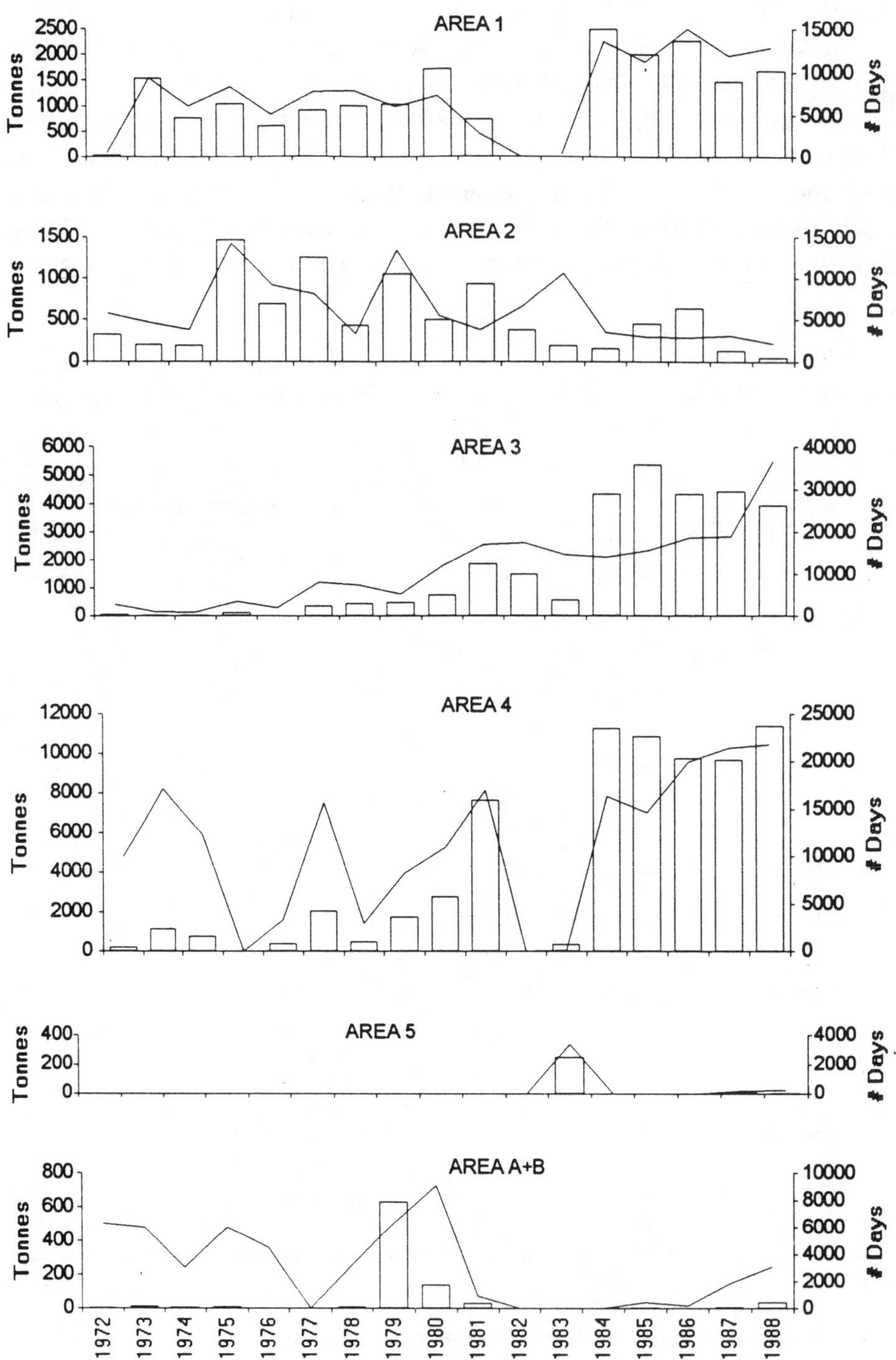

Fig. 4. Effort (lines) and catch (bars) of Thai gillnets by fishing areas from 1972 to 1988.

The above information was obtained by the Statistics Division, Department of Fisheries from logbooks completed for selected purse seiners at landings sites throughout the Gulf of Thailand. The selected vessels included one-boat, luring, and tuna purse seiners. The tuna sampling programme of the Marine Fisheries Department monitors the landings principally of tuna purse seiners at two major tuna landing sites off the south coast of Thailand. Information generated by this programme shows the fishing pattern of tuna purse seiners to be different from that of purse seiners depicted by the Statistics Division (Figure 6). The Statistics Division information shows that purse seiners operated almost entirely in the Gulf of Thailand (Areas 1, 2, 3, 4, and 5) and obtained most of the tuna catch from the middle of the Gulf (Area 5) during 1987 and 1988. Conversely, the tuna sampling programme shows tuna purse seiners operated principally in, and obtained most of the tuna catch from the South China Sea (Area B) during 1988 and 1989. During 1987, tuna purse seiners catch and effort was about equally distributed in the middle of the Gulf (Area 5) and the South China Sea (Area B).

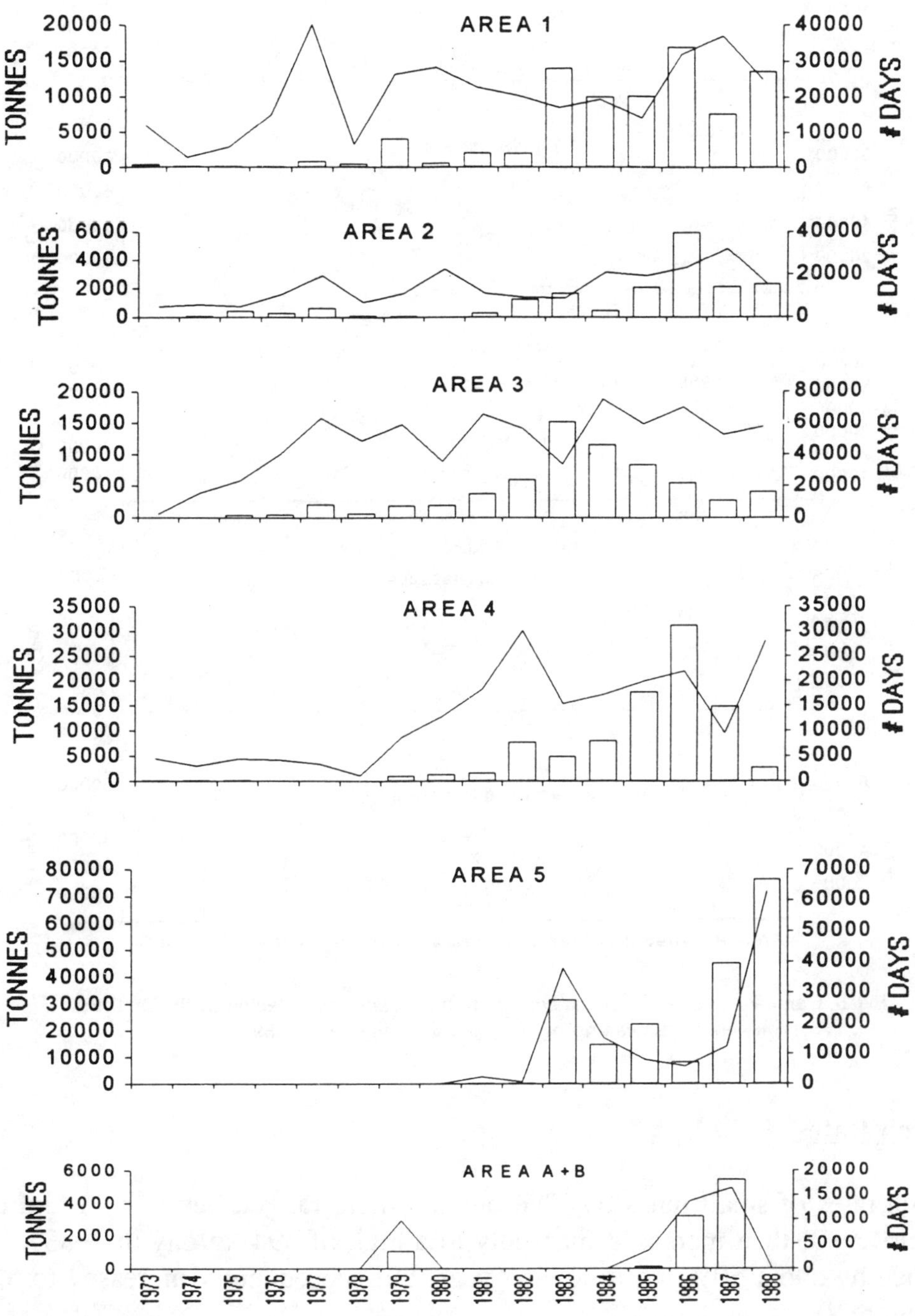

Fig. 5. Effort (lines) and catch (bars) of Thai purse seines by fishing areas from 1973 to 1988.

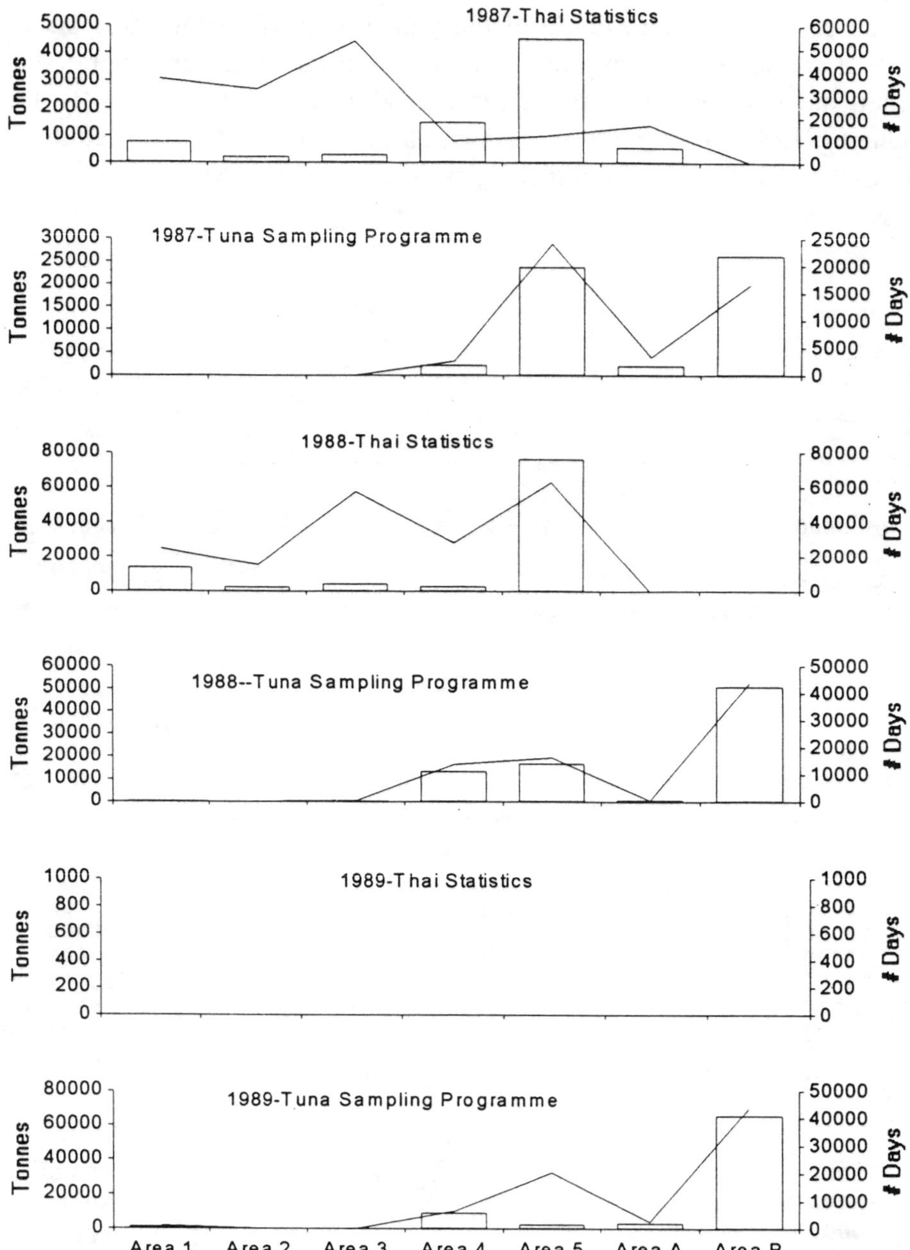

Fig. 6. Comparison of effort (lines) and catch (bars) statistics collected by the Statistics Division and the tuna sampling programme during 1987-1989.

3.4   Catch Rates

Catch rates of small tunas by Thai purse seiners ranged between 11 and 69 kg/day from 1973 to 1980, then increased markedly to a high of 594 kg/day in 1983. Catch rates declined by about 50% the following year, but subsequently increased to 522 kg/day in 1988 (Figure 7).

Small tuna catch rates by Thai gillnetters ranged between 25 and 131 kg/day from 1972 to 1980. Catch rates increased abruptly to 270 kg/day in 1981 and 527 kg/day in 1982, but subsequently declined to 223 kg/day in 1988 (Figure 7).

Trollers based in Terengganu, Malaysia averaged 372 kg/trip of small tunas during 1983-1985 and only 209 kg/trip during 1987-1989. Monthly catch rates by trollers fluctuated between 226 and 544 kg/trip from August 1982 to December 1984. During 1985, monthly catch rates were much more erratic and ranged from 27 to 967 kg/trip. If the exceptionally high monthly catch rate is discounted, the catch rate averaged 320 kg/trip for 1985.

3.5   Species Composition

Longtail tuna is the dominant tuna species on the continental shelf of the South China Sea, followed by kawakawa and frigate tuna. The percentage of longtail, kawakawa, and frigate tuna was highest in purse seine (55-65%), troll (22-45%), and gillnet (17-32%) catches, respectively. The proportion of longtail tuna in the gillnet catches of Thailand has been declining in recent years, with a corresponding increase in the proportion of frigate tuna (Table 4).

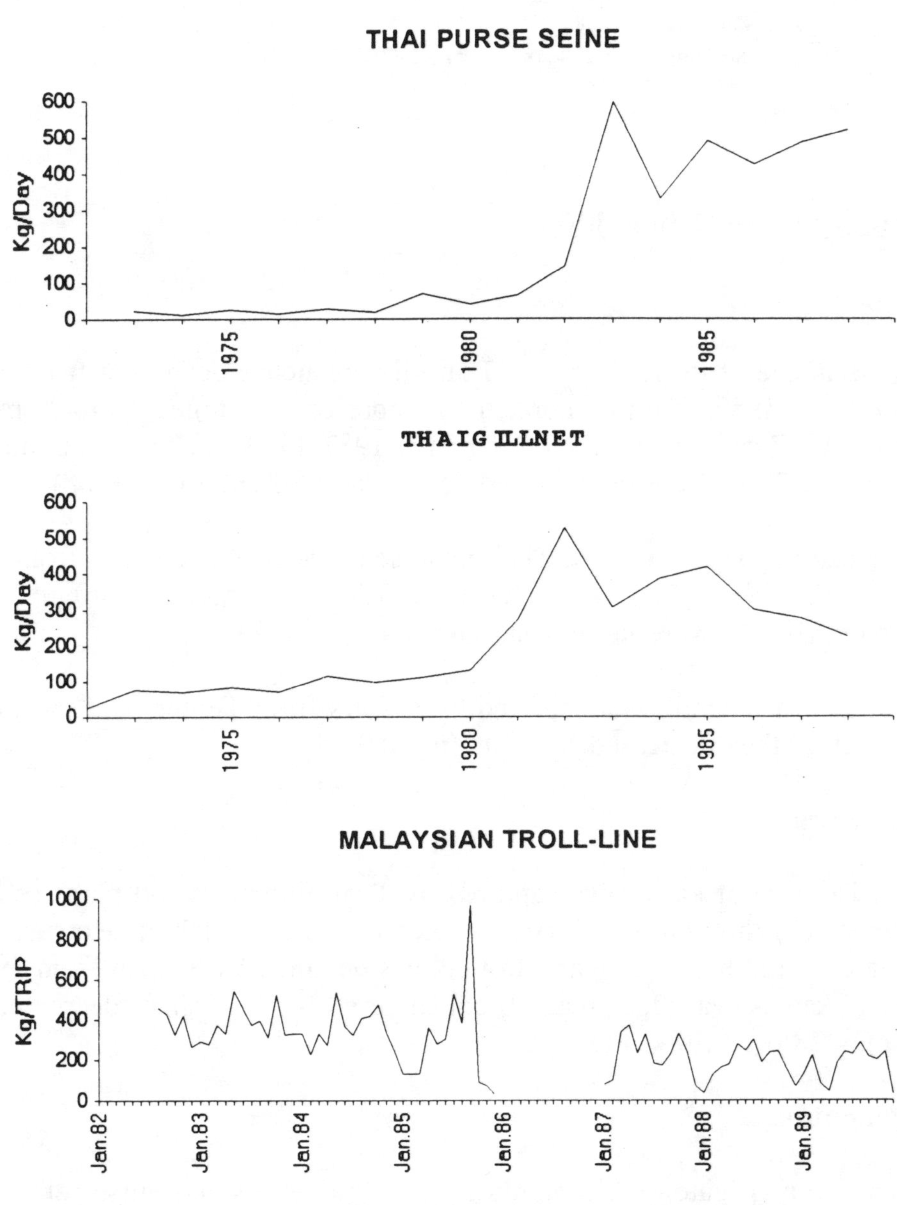

Fig. 7. Annual catch rates of small tunas by Thai purse seines and gillnets from 1972 to 1988 and monthly catch rates by Malaysian troll-lines from August 1982 to December 1989.

Table 4. Species composition (percent) of small tuna landings by various gears in the South China Sea.

Country	Thailand						Malaysia		
gear	Purse seine			Gillnet			Troll line		
Species	LOT	KAW	FRI	LOT	KAW	FRI	LOT	KAW	FRI
Year 1977	–	–	–	57[1]	26[1]	17[1]	–	–	–
1983	64[2]	26[2]	10[2]	–	–	–	61	36	3
1984	–	–	–	–	–	–	74	22	4
1985	–	–	–	–	–	–	60	30	10
1986	–	–	–	–	–	–	–	–	–
1987	56[3]	24[3]	20[3]	–	–	–	52	45	3
1988	65	16	19	43	28	29	56	42	2
1989	55	22	23	32	36	32	60	39	1

Source – [1] – Klinmuang, 1981
[2] – Supongpan and Saikliang, 1987
[3] – May – Nov 1987

3.6   Changes in Size Composition

3.6.1 Longtail tuna

The mean size of longtail tuna in Thai gillnet catches decreased from 43.3 cm in 1977 to 38.7 cm in 1989. Furthermore, the number of fish larger than 48 cm accounted for 31% of the 1977 catch and only 1-2% of the 1987-1989 catches. The mode decreased from 49 cm in 1977 to 43 cm in 1987 and 35 cm in 1988-1989 (Figure 8).

Mean size of longtail tuna in Thai purse seine catches decreased from 42.3 cm in 1983 to 38.8 cm in 1989. In 1983, 77% of the fish were larger than 40 cm, whereas in 1988-1989 only 51-53% were larger than this size.

Mean size of longtail tuna captured by trollers from Terengganu increased slightly from 1983 to 1987 then decreased by 4 cm in 1989.

3.6.2 Kawakawa

The mean size of kawakawa captured by Thai gillnetters and purse seiners has not changed appreciably through the years for which data are available. However, mean size of kawakawa captured by Terengganu trollers has declined by 8.8 cm from 1987 to 1989. The number of kawakawa larger than 30 cm captured by the fishery decreased from 90% in 1983 to only 28% in 1989 (Figure 9).

3.6.3 Frigate tuna

Mean size of frigate tuna captured by Thai gillnetters and purse seiners has not changed significantly through the years. However, the mean size of frigate tuna captured by Terengganu trollers increased 7.4 cm from 1983 to 1987 and thereafter decreased by 5.5 cm to 1989 (Figure 10).

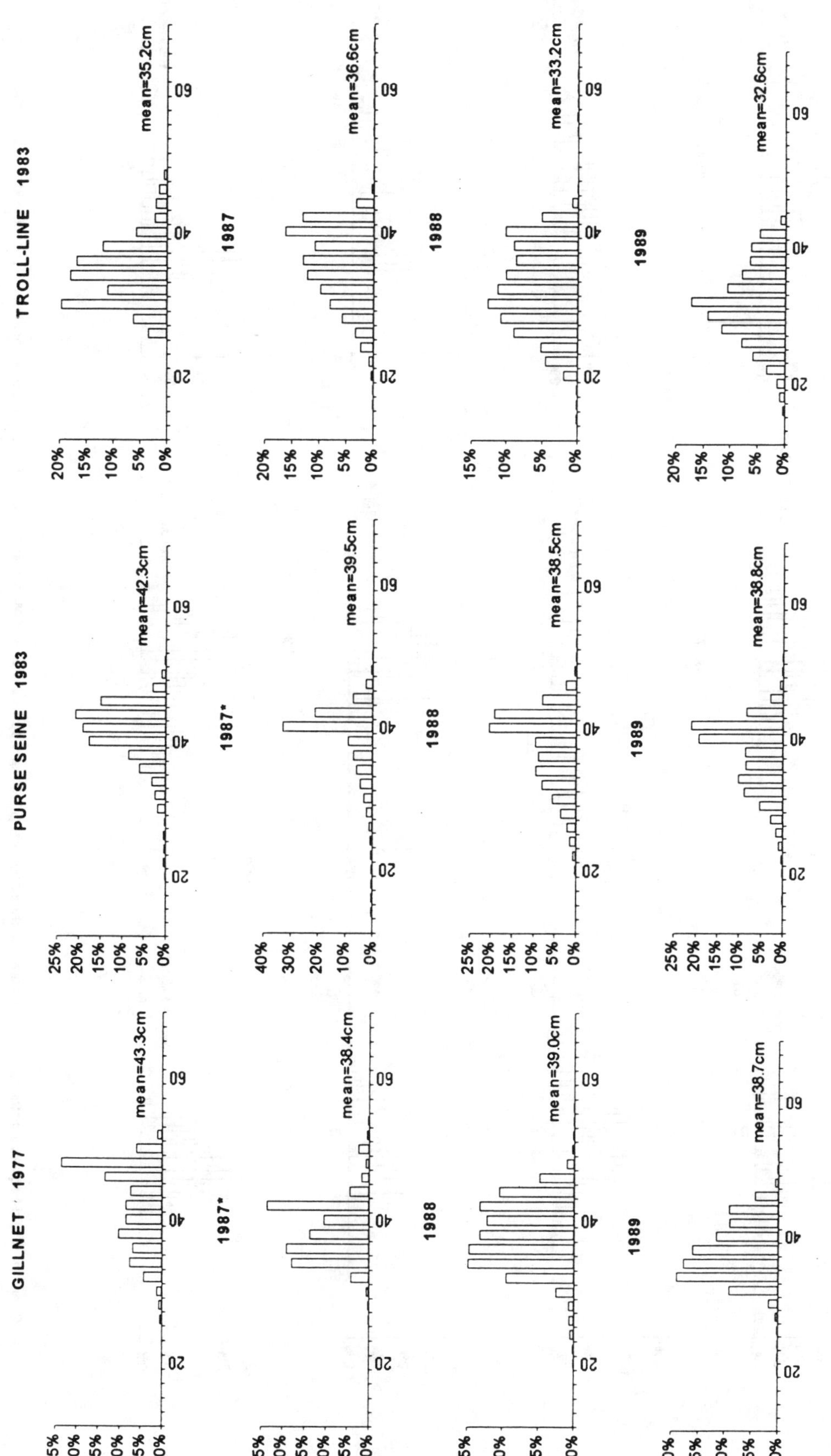

Fig. 8. Annual length-frequency distributions of longtil tuna captured by Thai gillnets and purse seines and Malaysian troll-lines (1987* - includes information only from May to December).

314

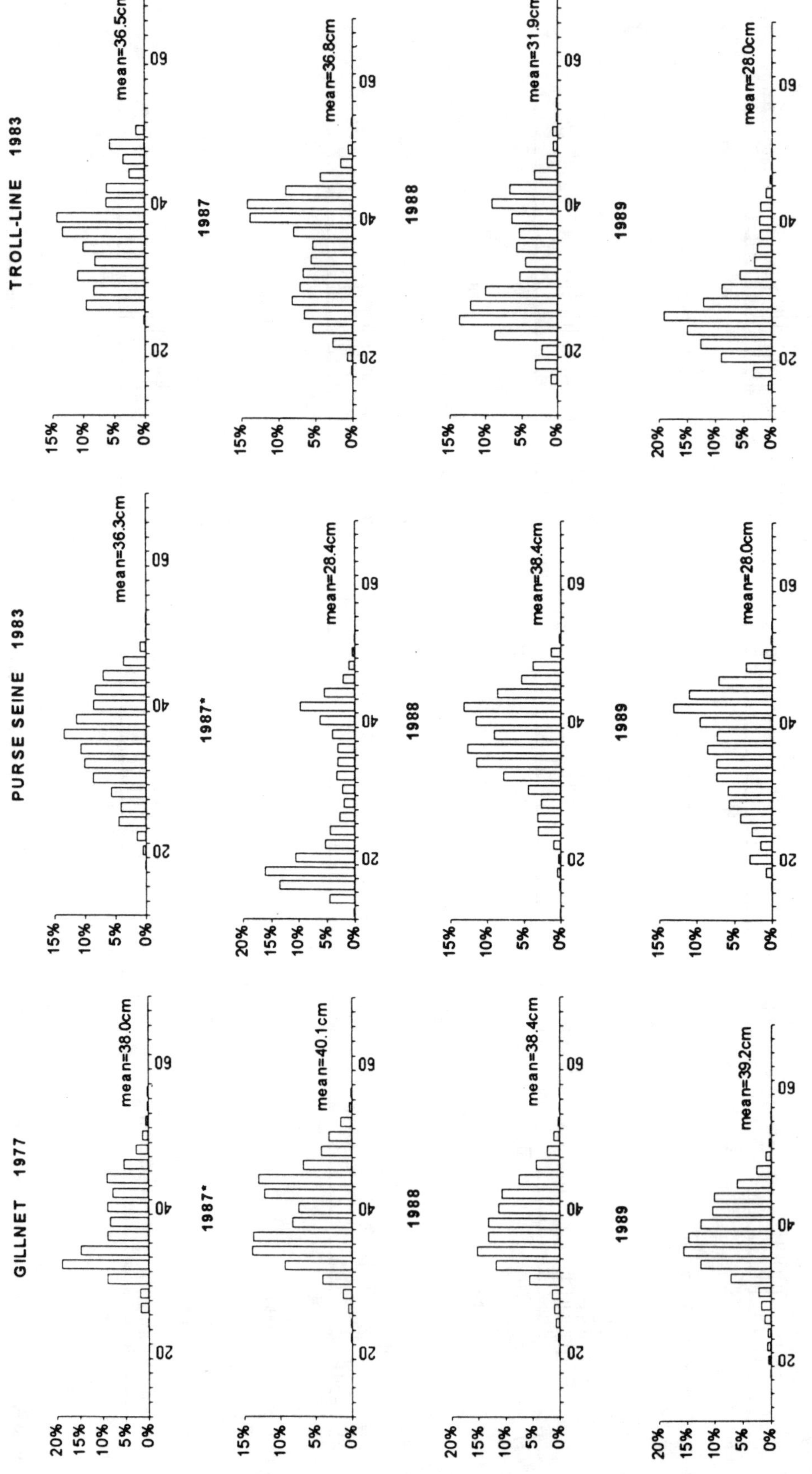

Fig. 9. Annual length-frequency distributions of kawakawa captured by Thai gillnets and purse seines and Malaysian troll-lines (1987* - includes information only from May to December).

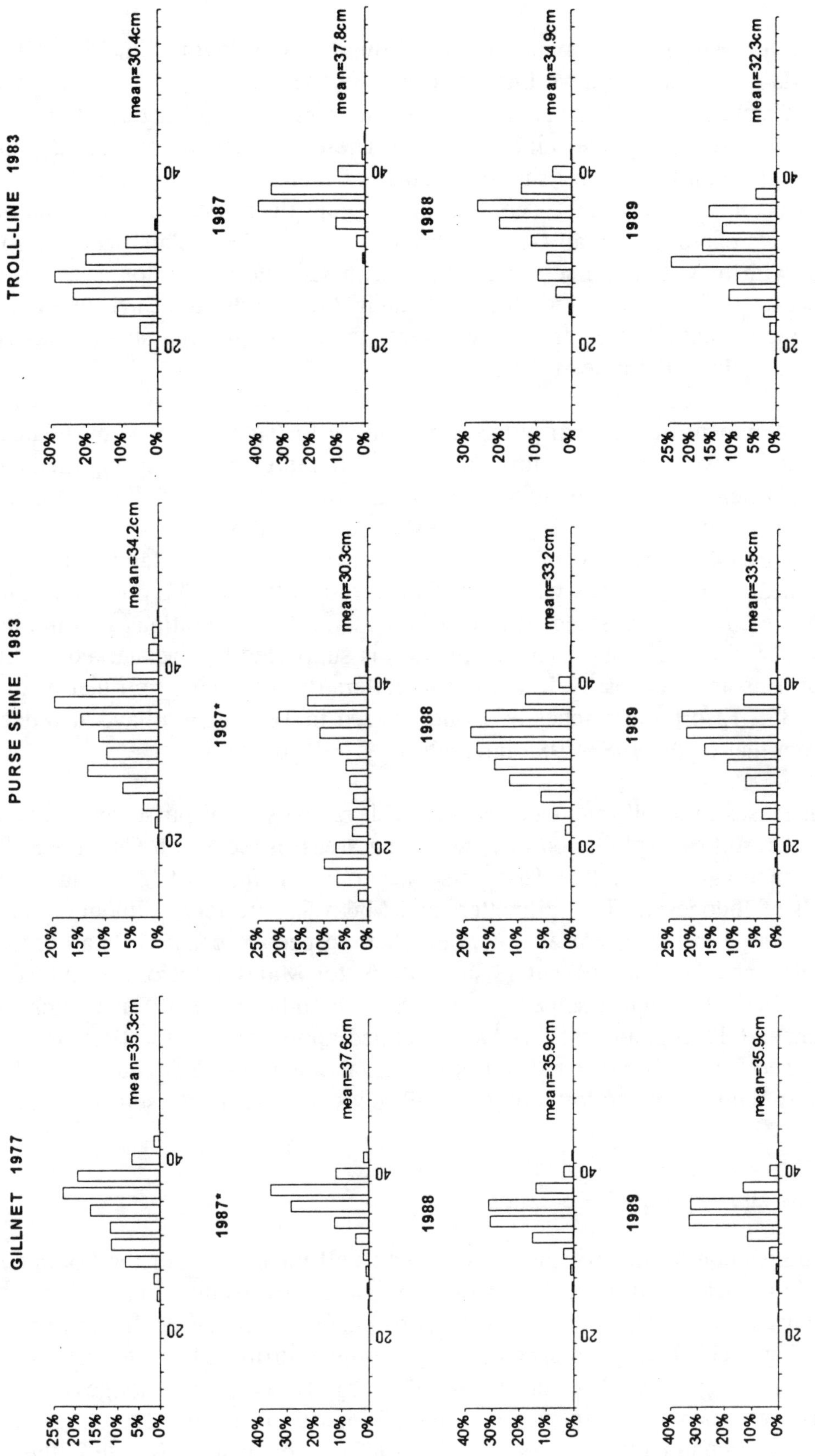

Fig. 10. Annual length-frequency distributions of frigate tuna captured by Thai gillnets and purse seines and Malaysian troll-lines (1987* - includes information only from May to December).

4. INTERACTIONS BETWEEN FISHERIES

4.1 Catches and Catch Rates

Of the five basic fisheries exploiting small tunas in the South China Sea off Thailand and Malaysia, catches have increased in recent years only in the Thai purse-seine fishery. Catches of small tunas have decreased in the Thai gillnet and the Malaysian troll, gillnet, and purse-seine fisheries. These declines are attributed to lower availability of small tunas to these fisheries because of increased exploitation by the Thai purse-seine fishery in recent years. The Thai purse-seine fleet is comprised of the largest vessels of all fleets exploiting small tunas in the South China Sea. This fleet is also the best equipped, with most tuna purse-seine vessels with satellite navigation systems, radar, and sonar. Because of their size and electronic navigational aids, the Thai tuna purse-seine fleet is able to range farther than other fleets, thereby exploiting small tunas before they become available to the other fleets.

The development of the Thai purse-seine fishery for small tunas is postulated to have evolved as follows, based on information of catch and effort by fishing areas. Thai purse-seine fishermen started to focus on small tunas in about 1982 off the east coast (Area 1) of the Gulf of Thailand. They soon expanded operations to the west (Area 3) and southwest (Area 4) coasts and middle (Area 5) of the gulf of Thailand and they probably confined their operations within the Gulf through 1984. The poor tuna catch of that year prompted Thai purse-seine fishermen to expand their operations into the South China Sea in the following years. This contention is supported by the marked decline in the number of trips and catches by Kuala Besut trollers during 1986 from that of the last semester of 1985. Effort and catches by trollers based in Terengganu have also decreased since 1987 from that of the 1983-1985 intervals.

The increases in small tuna catches and catch rates by Thai purse seiners since 1984 is a direct result of exploitation of new fishing areas in the South China Sea. Available information indicates that Thai purse seiners are fishing off Malaysia and the Natuna Islands of Indonesia. Thai gillnetters and Malaysian trollers, gillnetters, and purse seiners operate in more coastal waters off their respective coasts. Small tuna catches have declined for Thai gillnetters since 1986, for Malaysian trollers and gillnetters since 1987 and Malaysian purse seiners since 1988. In addition, small tuna catch rates of Thai gillnetters and Terengganu trollers have declined appreciably since 1985 and 1987, respectively. The availability of small tunas in coastal waters of Thailand and Malaysia has been greatly reduced by exploitation of small tunas by Thai purse seiners in offshore waters.

4.2 Changes in Species Composition

The species composition of the unexploited small tuna stocks of the South China Sea is probably 65% longtail tuna, 25% kawakawa, and 10% frigate tuna. This composition appears to be changing, with the percentage of longtail declining and that of frigate increasing. This change in species composition is attributed to life-history characteristics of the species. Longtail tuna is the largest species with longevity of approximately five years. This species is essentially neritic in habitat and abundant only in areas with broad continental shelves (Yesaki, 1987). Conversely, frigate is the smallest species with longevity of about two years. This species is distributed near the outer-continental shelf and contiguous oceanic zone. The longer-lived species would be most affected by exploitation and would be replaced by a shorter-lived species under continued exploitation. Frigate tuna would especially be unaffected by exploitation, in this instance, because a significant portion of the stock would be in the oceanic zone and not available to the various fisheries of Thailand and Malaysia. The Thai tuna purse-seine fishery extends the farthest offshore of all these fisheries, but is confined to the continental shelf because of the light construction of its nets.

4.3 Changes in Size Composition

Size of longtail tuna has decreased in the three fisheries for which there is information. On the other hand, sizes of kawakawa and frigate tuna have decreased only in the Malaysian troll fishery and not in the Thai purse-seine and gillnet fisheries. The diminution in size of longtail is again attributed to its life-history characteristics. This species is the longest-lived of the three tunas considered in this analysis, so would be the first to show the effects of exploitation.

The most pronounced changes in size composition have occurred in the Malaysian troll fishery. Mean size of longtail and frigate increased from 1983 to 1987, whereas that of kawakawa did not change appreciably. These size increments may have resulted from trollers expanding operations to more offshore areas frequented by larger fish. Mean sizes of all species have decreased since 1987. A possible explanation for these diminution may be the use of hooks designed for smaller fish and/or greater effort expended around FADs to compensate for declining catch rates which started in 1986. Smaller fish were captured around FADs than from free-swimming schools during tagging experiments conducted during 1990 aboard Terengganu trollers. Annual catch rates for this fishery decreased from 394 kg/trip in 1983 to 239 kg/trip in 1987 and 182 kg/trip in 1989.

5. FUTURE STUDIES

5.1 Work Contemplated

An analysis of the data collected by the tuna sampling programme in Thailand and Malaysia is planned to assess the status of the small tuna stocks of the South China Sea. A detailed examination of these data should provide better information on interactions of the various fisheries between and within countries.

5.2 Data Needed

There is an urgent need for the tuna sampling programme for small tunas in Thailand and Malaysia to be continue, as four years of data are insufficient to assess adequately the status of stocks. The total landings of small tunas from the South China Sea is still increasing, but evidence indicates that some fisheries and some stocks are being affected by exploitation. Continued monitoring of the small tuna fisheries of both countries is necessary to complete an adequate assessment of the small tuna stocks on which to base meaningful management advice.

5.3   Institutions Involved

The institutions involved in continuing the tuna sampling programmes would be the Marine Fisheries Department of Thailand, the Department of Fisheries, Malaysia and IPTP.

5.4   Assistance Required

The IPTP should continue to provide funds and technical assistance to the Marine Fisheries Department and the Marine Fisheries Research Center, Department of Fisheries, Malaysia in order to maintain the tuna sampling programme.

6.   REFERENCES CITED

Cheunpan, A. 1986. Review of tuna fisheries/resources in the Gulf of Thailand. Paper presented at the Meeting of Tuna Research Groups in the Southeast Asian Region, Phuket, Thailand, 27-29 August 1986.

IPTP. 1990a. Tuna Sampling Programme in Malaysia. *Indo-Pac.Tuna Dev.Mgt. Programme,* IPTP/90/MAL/SP:55 p.

IPTP. 1990b. Tuna Sampling Programme in Thailand. *Indo-Pac.Tuna Dev.Mgt. Programme,* IPTP/90/THA/SP:67 p.

IPTP. 1991. Indian Ocean and Southeast Asian Tuna Fisheries Data Summary for 1989. *Data Summ.Indo-Pac.Tuna Dev.Mgt.Programme,* II:96 p.

Klinmuang, H. 1978. Preliminary studies on the biology of tunas in the west of the Gulf of Thailand and off the east coast of Peninsular Malaysia. *Pelagic Fish.Rep.Mar.Fish. Div.Dep.Fish., Bangkok,* 5:27 p.

Klinmuang, H. 1981. Studies on the size frequency distribution and length-weight relationship of the tunas in the Gulf of Thailand. *Pelagic Fish.Rep.Mar.Fish. Div.Dep.Fish., Bangkok,* 24:39 p.

Munprasit, A., Y. Theparoonrat, S. Sae-Ung, S. Soodhom, Y. Matsunaga, B. Chokesanguan, and S. Siriraksophon. 1989. Fishing gear and methods in Southeast Asia: II. Malaysia Training Dept., *Southeast Asian Fish.Dev.Cent.,* TD/RES/24:338 p.

Okawara, M., A. Munprasit, Y. Theparoonrat, P. Masthawee, and B. Chokesanguan. 1986. Fishing gear and methods in Southeast Asia: I. Thailand Training Dept., *Southeast Asian Fish.Dev.Cent.,* TD/RES/9:329 p.

SEAFDEC. 1979. Fishery Statistical Bulletin for South China Sea Area 1977. *Southeast Asian Fish.Dev.Cent.,* 162 p.

SEAFDEC. 1980. Fishery Statistical Bulletin for South China Sea Area 1978. *Southeast Asian Fish.Dev.Cent.,* 200 p.

SEAFDEC. 1981. Fishery Statistical Bulletin for South China Sea Area 1979. *Southeast Asian Fish.Dev.Cent.*, 227 p.

SEAFDEC. 1982. Fishery Statistical Bulletin for South China Sea Area 1980. *Southeast Asian Fish.Dev.Cent.*, 268 p.

SEAFDEC. 1983. Fishery Statistical Bulletin for South China Sea Area 1981. *Southeast Asian Fish.Dev.Cent.*, 275 p.

SEAFDEC. 1984. Fishery Statistical Bulletin for South China Sea Area 1982. *Southeast Asian Fish.Dev.Cent.*, 231 p.

SEAFDEC. 1985. Fishery Statistical Bulletin for South China Sea Area 1983. *Southeast Asian Fish.Dev.Cent.*, 178 p.

Supongpan, S. and P. Saikliang, 1987. Fisheries status of tuna purse seiners (using sonar) in the Gulf of Thailand. *Rep.Mar.Fish.Div.Dep.Fish., Bangkok,* 3:78 p.

Yesaki, M. 1987. Synopsis of biological data on longtail tuna, *Thunnus tonggol. Indo-Pac.Tuna Dev.Mgt.Programme,* IPTP/87/WP/16: 56 p.

CURRENT STATE OF THE MEXICAN TUNA FISHERY AND ECOLOGICAL INTERACTIONS BETWEEN LARGE AND SMALL TUNAS IN THE PELAGIC PACIFIC ENVIRONMENT

Arturo Muhlia-Melo
Centro de Investigaciones Biológicas
de Baja California Sur, A.C.

ABSTRACT

Catches of yellowfin tuna and other tuna species by the Mexican tuna fishery are briefly described. The Mexican fleet shifts operations seasonally along coastal waters and offshore as far west as 150°W. A suggestion of species interaction between *Auxis* spp. and yellowfin is advanced based on a prey-predator association.

1.   INTRODUCTION

A historical analysis of the development of the Mexican tuna fishery in the eastern Pacific Ocean from 1937 to 1985 was presented in Muhlia-Melo (1987). During the period of 1987-1991 the number of active purse-seine vessels decreased from 55 in 1987 to 50 in 1991; the carrying capacity also decreased from 47,926 mt to 42,126 mt (Table 1). The baitboat fleet also declined from 16 vessels in 1988 to 7 in 1991. The carrying capacity of this fleet decreased from 1,840 mt in 1989 to 691 mt in 1991 (Table 1).

According to an analysis of relative fishing power of the Mexican purse-seine fleet during 1984-1986, the most efficient purse-seine vessels operating in the eastern Pacific were those of 1,090 mt of carrying capacity (Ortega-G and Muhlia-M, 1991).

Table 1.   Carrying capacity of Mexican tuna fleet in the eastern Pacific Ocean, 1985-1991.

unit: mt

Year	Purse Seiners	Baitboats	Total
1985	45,935	1,458	47,393
1986	38,980	1,127	40,107
1987	47,926	1,269	49,195
1988	47,539	1,738	49,277
1989	46,087	1,840	47,927
1990	46,033	1,840	47,873
1991	42,126	691	42,817

Data source: Bayliff, 1986; Bayliff, 1987; Bayliff, 1988; Bayliff, 1989; Bayliff, 1991; Bayliff, 1992a; and Bayliff, 1992b.

2. DISTRIBUTION OF FISHING EFFORT AND CATCH BY SEASON

From 1985 to 1991 the Mexican purse-seine fleet operated in the fishing areas in a typical pattern as defined by its operation in the 1985-1986. The fishing pattern was as follows:

Spring: Fishing concentrated around the southern end of the Baja California peninsula and offshore of central Mexico as far south as 5°N, and to the west as far west as 133°W (Figure 1A).

Summer: Fishing concentrated along the west coast of Baja California, off central Mexico between 5°N and 20°N and westward to 115°W, and west of 125°W (Figure 1B).

Autumn: Fishing concentrated off central Mexico and in the Gulf of California (Figure 1C).

Winter: Fishing concentrated along the Mexican coast (Figure 1D) and off northern South America.

3. CATCH BY SPECIES

From 1985 to 1991 the main species in the catch (Table 2) were yellowfin (*Thunnus albacares*), skipjack (*Katsuwonus pelamis*), bonito (primarily *Sarda chiliensis*), and northern bluefin (*Thunnus thynnus*). Figure 2 presents the variation in the catch and species composition of this fishery. Mexico's total production of tunas during this period shows a progressive increase from 87,942 mt in 1985 to 147,530 mt in 1989. However, after 1989 a slight decline is observed.

Table 2. Catches of tunas in the eastern Pacific Ocean by the Mexican fishery, 1985-1991.

Unit: mt

Year	Yellowfin	Skipjack	Bonito	Bluefin	Others	Total
1985	80,503	6,141	139	676	483	87,942
1986	105,188	7,994	317	189	46	113,734
1987	99,246	6,617	137	119	369	106,488
1988	105,542	18,685	6,004	447	130	130,808
1989	118,007	18,109	11,294	57	63	147,530
1990	117,021	7,209	9,979	50	349	134,608
1991	116,172	12,521	1,059	9	230	129,991

Data source: Bayliff, 1986; Bayliff, 1987; Bayliff, 1988; Bayliff, 1989; Bayliff, 1991; Bayliff, 1992a; and Bayliff, 1992b.

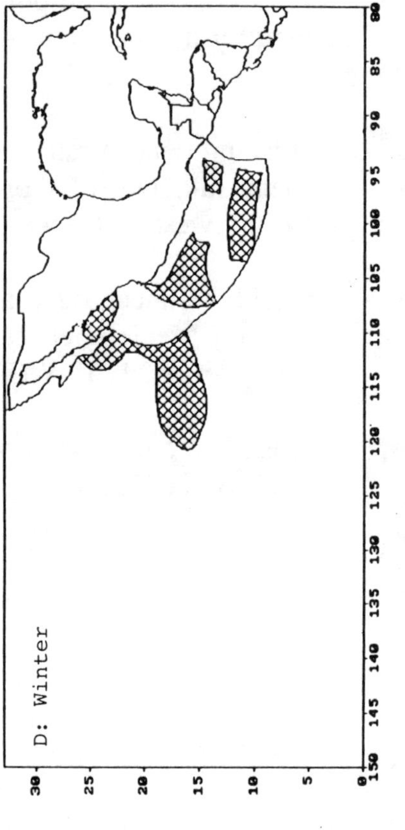

Figure 1C. General geographical distribution of fishing operation of the Mexican purse-seine in autumn.

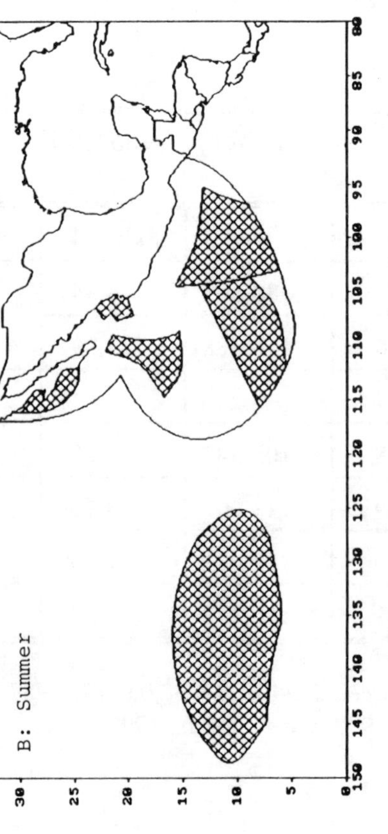

Figure 1D. General geographical distribution of fishing operation of the Mexican purse-seine in winter.

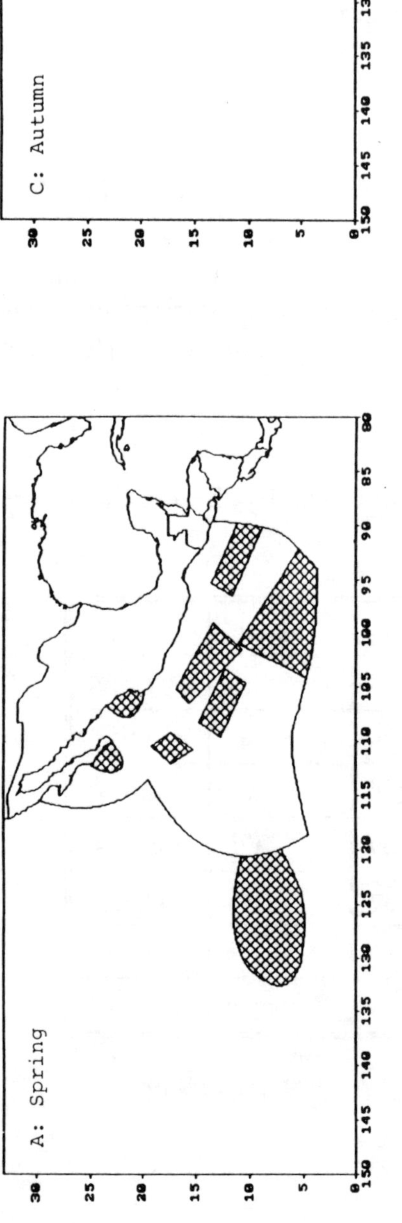

Figure 1A. General geographical distribution of fishing operation of the Mexican purse-seine in spring

Figure 1B. General geographical distribution of fishing operation of the Mexican purse-seine in summer.

Figure 3 presents the secondary species composition of the Mexican tuna fishery in the eastern Pacific Ocean (without yellowfin). Skipjack catches show an increase in 1988 and 1989 and a decline in 1990 and 1991. As noted in Figure 3 and Table 1, catches of bonito increased from 317 mt in 1986 to 11,294 and 9,979 mt in 1989 and 1990, respectively. Bonito is considered an incidental catch since the target species of the purse-seine and baitboat fisheries are yellowfin and skipjack.

New regulations on the purse-seine fishery (Bayliff, 1991) and political pressures to decrease the incidental mortality of dolphins will force fisherman to make more sets on free-swimming schools of tuna and schools of tuna associated with floating debris; fewer sets will be made on schools of tuna associated with dolphins. Due to these changes, the catch of skipjack and secondary species such as bonitos is expected to increase.

Larvae of *Auxis* spp. have been found in the vicinity of the mouth of the Gulf of California. Spawning is coastal in the northern part of this region; further south spawning is more oceanic (Klawe *et al.*, 1970).

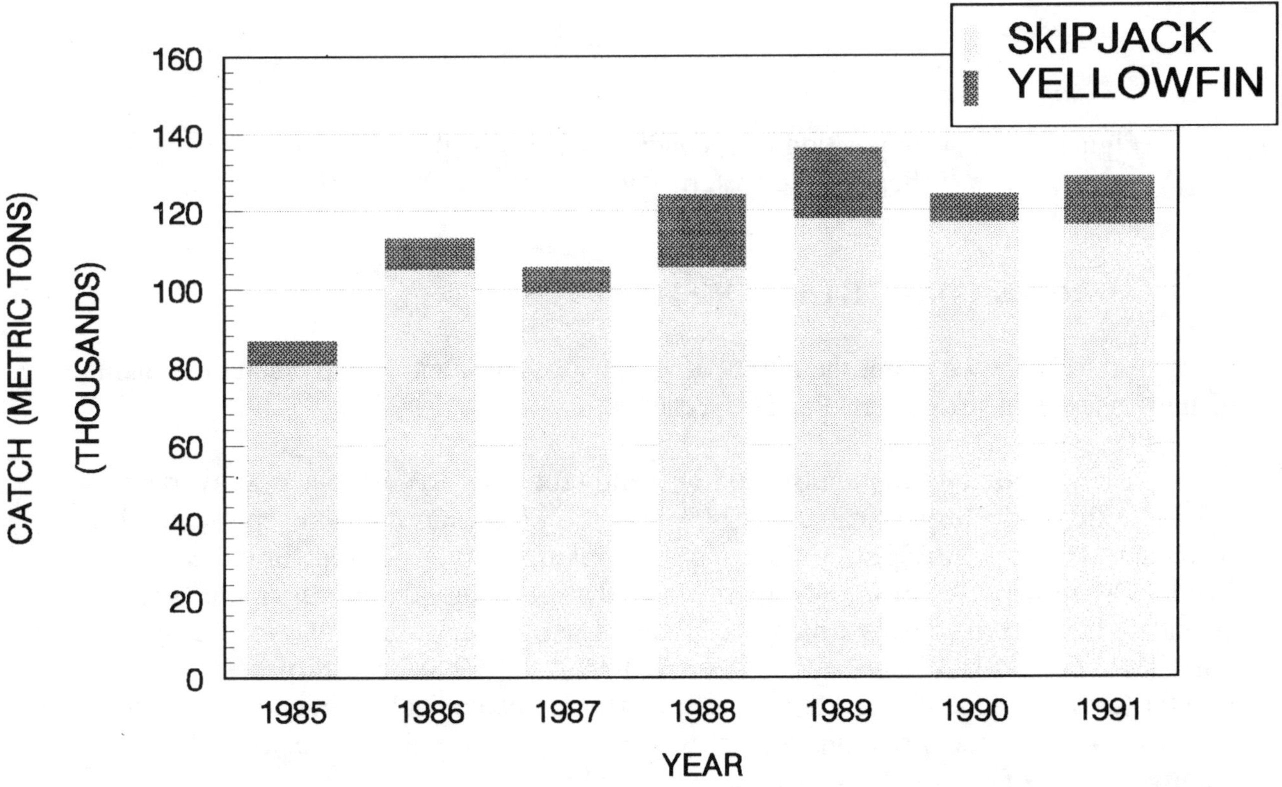

Figure 2.   Composition of principal species of the Mexican tuna fishery in the eastern Pacific Ocean, 1985-1991.

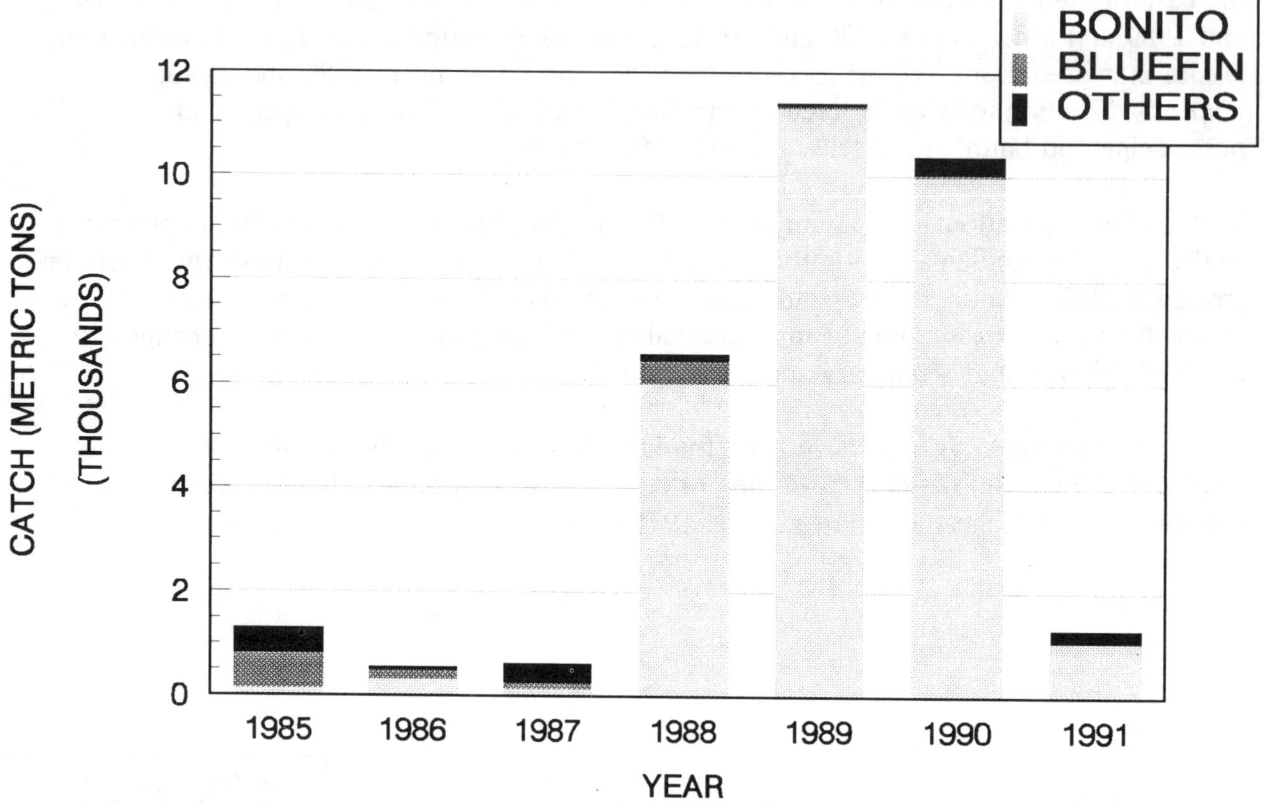

Figure 3.   Composition of secondary species of the Mexican tuna fishery in the eastern Pacific Ocean, 1985-1991.

4.   AUXIS AS FORAGE ORGANISM

Little is known about the adults of *Auxis* spp. and about the population dynamics of these species in the eastern Pacific Ocean.

The annual consumption of food by yellowfin tuna was estimated to average 4.3-6.4 million mt; of the total about 34% (1.5 to 2.2 million mt) was represented by *Auxis* spp. (Olson and Boggs, 1986). The total biomass of *Auxis* spp. in the eastern Pacific Ocean was estimated at 11 to 17 times that of tunas. Thus, *Auxis* spp. are considered forage organisms which occupy an important position in the food chain of tunas (Uchida, 1981; Watanabe, 1964; Galvan-Magaña, 1990) and billfishes (Abitia-Cárdenas and Galvan-Magaña, in press). An ecological interaction may occur in this fishery due to the direct predator-prey relationship since *Auxis* spp. constitutes a significant part of the food chain of adult yellowfin tuna.

Since *Auxis* represents one of the most important food items of larger yellowfin tuna and other large predatory fish, fluctuations in the abundance of *Auxis* may be one of the causes of fluctuations in the abundance of the predator species. Accordingly, it would be useful to monitor the abundance of *Auxis* and if possible, determine the causes of fluctuations in abundance.

5. REFERENCES CITED

Abitia-Cárdenas L. A. and Galván-Magaña F. (in press). Trophic energetic spectrum of striped marlin *Tetrapturus audax* Philippi, 1887) from the area of Cabo San Lucas, Baja California Sur, Mexico. *J.Fish Biol.*

Bayliff, W. H. (editor). 1986. Annual report of the Inter-American Tropical Tuna Commission, 1985. *Annu.Rep.I-ATTC*, (1985):248 p. (In English and Spanish).

Bayliff W. H. (editor). 1987. Annual report of the Inter-American Tropical Tuna Commission, 1986. *Annu.Rep.I-ATTC*, (1986):264 p. (In English and Spanish).

Bayliff, W. H. (editor). 1988. Annual report of the Inter-American Tropical Tuna Commission, 1987. *Annu.Rep.I-ATTC*, (1987):222 p. (In English and Spanish).

Bayliff, W. H. (editor). 1989. Annual report of the Inter-American Tropical Tuna Commission, 1988. *Annu.Rep.I-ATTC*, (1988):288 p. (In English and Spanish).

Bayliff, W. H. (editor). 1991. Annual report of the Inter-American Tropical Tuna Commission, 1989. *Annu.Rep.I-ATTC*, (1989):270 p. (In English and Spanish).

Bayliff, W. H. (editor). 1992a. Annual report of the Inter-American Tropical Tuna Commission, 1990. *Annu.Rep.I-ATTC*, (1990):261 p. (In English and Spanish).

Bayliff, W. H. (editor). 1992b. Annual report of the Inter-American ropical Tuna Commission, 1991. *Annu.Rep.I-ATTC*, (1991):271 p. (In English and Spanish).

Galvan-Magaña F. 1990. Composición y análisis de la dieta del atún aleta amarilla (*Thunnus albacares*) en el Océano Pacífico mexicano durante el periodo 1984-1985. M.S. Thesis. Instituto Politécnico Nacional de México (CICIMAR).

Klawe, W.L., J.J. Pella, and W.S. Leet. 1970. The distribution, abundance and ecology of larval tunas from the entrance of the Gulf of California. *Bull.I-ATTC*, 14(4):505-44.

Muhlia-Melo, A. 1987. The Mexican tuna fishery. *Rep.CCOFI*, (28):37-42.

Olson, R.J., and C.H. Boggs. 1986. Apex predation by yellowfin tuna *Thunnus albacares*: independent estimates from gastric evacuation and stomach contents, bionergetics, and cesium concentrations. *Can.J.Fish.Aquat.Sci.*, 43:1760-75.

Ortega-G., and Muhlia-M., A. 1991. Analysis of the relative fishing power of the Mexican purse-seine fleet operating in the eastern Pacific. *Ciencias Marinas*, 18(1):55-78.

Uchida, R.N. 1981. Synopsis of biological data on frigate tuna, *Auxis thazard*, and bullet tuna, *A. rochei*. *NOAA Tech.Rep.NMFS Circ.*, (436): 63 p.

Watanabe, H. 1964. Frigate mackerels (genus *Auxis*) from the stomach contents of tunas and marlins. *In* Proceedings of the Symposium on Scombroid Fishes, Part II. *Symp.Ser.Mar.Biol.Assoc.India*, 1:631-41.